Robert Winston is one of the country's best-known scientists. As Professor of Fertility Studies at Imperial College, University of London, and Director of NHS Research and Development and Consultant Obstetrician and Gynaecologist at Hammersmith Hospital, he has made advances in fertility medicine and been a leading voice in the debate on genetic engineering. His television series include *Making Babies*, *The Human Body*, *Human Instinct*, *Walking with Cavemen*, *The Human Mind* and *Child of Our Time* and have made him a household name across Britain. He became a life peer in 1995. His most recent books *Human Instinct* and *The Human Mind* are also published by Bantam Books.

The Story of God

ROBERT WINSTON

BANTAM BOOKS

LONDON • TORONTO • SYDNEY • AUCKLAND • JOHANNESBURG

THE STORY OF GOD
A BANTAM BOOK: 0553817434
9780553817430

Originally published in Great Britain by Bantam Press,
a division of Transworld Publishers

PRINTING HISTORY
Bantam Press edition published 2005
Bantam edition published 2006

1 3 5 7 9 10 8 6 4 2

Copyright © Professor Robert Winston 2005

By arrangement with the BBC.
BBC logo © BBC 1996
The BBC logo is a trademark of the British Broadcasting Corporation
and is under licence.

Set in 11/13pt Sabon by
Falcon Oast Graphic Art Ltd.

Bantam Books are published by Transworld Publishers,
61–63 Uxbridge Road, London W5 5SA,
a division of The Random House Group Ltd,
in Australia by Random House Australia (Pty) Ltd,
20 Alfred Street, Milsons Point, Sydney, NSW 2061, Australia,
in New Zealand by Random House New Zealand Ltd,
18 Poland Road, Glenfield, Auckland 10, New Zealand
and in South Africa by Random House (Pty) Ltd,
Isle of Houghton, Corner of Boundary Road & Carse O'Gowrie,
Houghton 2198, South Africa.

Printed and bound in Great Britain by
Cox & Wyman Ltd, Reading, Berkshire.

Papers used by Transworld Publishers are natural, recyclable products
made from wood grown in sustainable forests. The manufacturing
processes conform to the environmental regulations of the
country of origin.

To a most dear friend, Richard Dale

Contents

Contents

Acknowledgements

I would not have dreamed of writing this book had it not been for a lunch with Lorraine Heggessey when she was Controller of BBC1. I tried to persuade her to commission a totally different project, and was sure she was joking when she suggested instead a TV series on the subject of God. A less vibrant and engaging friend could not have persuaded me; so this book is in part the result of her charm and enthusiasm. I am much in her debt for her many years of encouragement and support.

Writing a book to coincide with a major television series is invariably demanding – not least because of the need to conduct research and write extremely rapidly while both mentally and physically occupied with filming. So I owe gratitude to a number of people who have made this job much easier and more pleasurable. First – and not for the first time – I am immensely grateful to Matt Baylis, who did so much to ease my way into this subject. His huge knowledge and interest have been a constant intellectual stimulus. His ability to find interesting themes, his extraordinary intelligence, his sensitivity to research topics which aroused my interest and enthusiasm, and his extensive understanding of world religions have all been invaluable. He has always been available at short notice to dig out useful material and to check references, and has been immensely reassuring when my confidence was flagging.

When I started on this project, Leo Singer most ably helped with a good deal of the early research material. Again, this is not the first time we have worked productively together. I am immensely grateful to him for suggesting certain themes in this book and for appraising three of the central chapters.

During many convoluted conversations, a number of friends have made highly useful suggestions about different areas covered in this book. These include Professor Mark Geller, whose extensive knowledge of ancient Sumer was a great help, and Michael Pollak, who suggested important spiritual themes for the last chapter. Some of the contributors to the television programme I regard as friends, and their thinking has been highly influential. I am deeply grateful in particular to the Chief Rabbi, Dr Sir Jonathan Sacks, Rabbi Ken Spiro, The Ven. Bellanwilla Wilimaratne Thero, Professor Richard Dawkins, Dr Jim Virdee, Professor Aviad Kleinberg, Dr Stephen Unwin, Dr Irving Finkel, Dr Dean Hamer and Dr Jean Clotte, for exciting conversations that helped to crystallize my thoughts.

As with *The Human Mind*, Dr Joel Winston read the entire manuscript, highlighting and correcting a number of inconsistencies with his usual intelligence. Lira, my wife, supportive as ever, lovingly and assiduously read through nearly all the material and made the most valuable criticisms. Several other friends read parts of various chapters and made a number of really useful suggestions – I am grateful to them all.

I have been truly lucky with a wonderful copy-editor, Gillian Somerscales. Her sensitivity, intelligence and wise suggestions have all made a big difference to this book. I am amazed how quickly she has worked. It goes without saying how grateful I am to Sally Gaminara and her team at my publishers, Transworld. Her continued trust in me is deeply appreciated. Thanks also to Katrina Whone,

Managing Editor at Bantam Press, and Simon Thorogood.

Although I have used relatively little of the material covered in the TV series, the book has certainly benefited from the stimulus of the Dangerous Films team with whom I worked. Tim Kirby, the series producer, has been extraordinary – his intelligence, his grasp of detail, his focus, his exceptionally wide knowledge about so many different things have been an inspiration. Many ideas in this book grew out of conversations with him. I am also very grateful to Ros Homan, producer, and to the production team, Katie Churcher, Zahra Mackaoui and Patricia Thompson. Massive thanks to the delightful Hattie Bowering, who has put up with my evil temper and been highly supportive when things were not going smoothly, and who was incredibly efficient with so many difficult arrangements and rearrangements. It has been a real pleasure to work with a phenomenally skilled film crew, and my deepest thanks go to Douglas Hartington and Andy Thompson in particular – wonderful cameramen, and entertaining company on difficult locations.

Finally, as ever, I am very grateful to my old friend Maggie Pearlstine and to her admirable colleague, Jamie Crawford. They have excelled as my literary agents, as they always do. They are a constant source of encouragement and support; I am ever in their debt.

Time Line

27,000 BCE	Hands painted in the Grotte de Gargas, France
25,000 BCE	'Venus of Willendorf' carving
15,000 BCE	Cave paintings at Lascaux, France
9500 BCE	Farming at Abu Hureyra
3000 BCE	Beginnings of cuneiform and hieroglyphic writing
3000 BCE	Indus valley cities, Mohenjodaro and Harappa, flourished
2750–2500 BCE	The *Epic of Gilgamesh* written
2575–2150 BCE	Great Pyramids in Egypt built
2000–1900 BCE	Time of Abraham
2000 BCE	Hinduism founded
1600 BCE	Beginnings of alphabetic writing
1600–1500 BCE	Zoroastrianism probably founded
1369–1332 BCE	Akhenaten's monotheistic cult in Egypt
1300 BCE	Time of Moses
1200–1070 BCE	Period of biblical Judges and Jephthah
1000–900 BCE	Solomon builds Jerusalem Temple
701 BCE	Sennacherib destroys Lachish
640–609 BCE	King Josiah discovers the 'lost book'
587 BCE	Nebuchadnezzar destroys Jerusalem; Babylonian exile
537 BCE	Babylonian exile ends; Temple rebuilding commenced

563–483 BCE	Buddha lived
384–322 BCE	Aristotle lived
73–4 BCE	Herod the Great lived
1–33 CE	Christ lived
62 CE	St Paul executed
70 CE	Romans burned Second Temple; Jerusalem destroyed
1–100 CE	Teotihuacan built
200 CE	Mishnah redacted by Rabbi Judah the Prince
325 CE	Council of Nicaea
337 CE	Emperor Constantine converted to Christianity on his deathbed
570–632 CE	Muhammad lived
680 CE	Imam Hussain killed at Karbala
1172 CE	Almohads captured Seville
1204 CE	Maimonides died
1209 CE	Albigensian Crusade
1225–1274 CE	St Thomas Aquinas lived
1455 CE	Gutenberg printed the Bible
1517 CE	Martin Luther nailed his manifesto to the cathedral door
1519 CE	Cortés invaded South America
1613 CE	Galileo stated that the earth revolves around the sun
1648 CE	Bogdan Chmielnicki leads uprising in Ukraine
1687 CE	Newton published his theory of gravity
1822 CE	Church ban on Galileo's *Dialogue* lifted
1859 CE	Darwin's *On the Origin of Species* published
1905 CE	Einstein published the Special Theory of Relativity
1917 CE	Rutherford split the atom

Prologue
Wrestling with God

If you look carefully, you might just see one of those slim, elongated boxes attached to the upper part of the front doorpost of one of the houses near where you live. They appear on houses in many parts of the world – wherever Jews have lived, for that insignificant-looking little box will have been nailed there by a Jewish inhabitant. Most Jews, no matter however religiously unobservant, tend to attach such a box to the right-hand side of their front door. Sometimes one of these boxes may be found on the doorpost of a house where a Jewish family once lived. Although the family has moved on, the box has been neglectfully left in place, perhaps repeatedly painted over by builders or decorators but still just visible.

It is not a good-luck charm – although, in certain minds, it may have a somewhat similar symbolism. The box contains a tightly rolled piece of parchment on which a qualified scribe will have written Hebrew letters in special ink. The text contains the commandment from Deuteronomy to attach a sign to all the doorposts of your house. It starts with the affirmation, central to the Jewish faith, of the existence of a single God; an exhortation repeated every morning and evening, learned by every Jewish child as soon as he or she is able to pronounce Hebrew, and the last words muttered on the deathbed of an Orthodox Jew: 'Hear O Israel, the Lord our God, the Lord is One.' Every letter of this phrase written on the parchment in the box must be exactly the same as the corresponding letters in the Hebrew Bible; there must be no mistakes, and the script should be clear enough to be completely readable by an average eight-year-old. And on the reverse side of the parchment will be written 'Shaddai' – 'Almighty': just three Hebrew letters, because biblical Hebrew is written without vowels.

The oldest of these parchments to have been clearly identified comes from one of the caves in Qumran, where the various Dead Sea Scrolls lay for some two thousand years until discovered in 1947. The little box is called a *mezuzah* in Hebrew, which probably means 'doorpost', though the precise origin of this word is quite obscure. Certainly the word *mezuzah* is very ancient and originally may have come from a similar-meaning Assyrian word, *manzazu*.

My house bears such a little box fixed by the front door, and when I leave home on a workday morning, my head crammed with the usual worldly thoughts and worries, my hand may occasionally stretch out for a moment to touch it. Many strictly Orthodox Jews, doing the same thing, then often also touch the box with their lips. With this tiny symbolic gesture we link ourselves to the beliefs and practices of our ancestors, and express our love and respect for the laws that we consider were given to those ancestors by God, when they lived as nomads in the deserts of Sinai and Canaan. Then, having passed through our doorways, my fellow co-religionists and I head into the morning mayhem of a working day, each of us swiftly becoming just another person struggling to get to work on time.

Simultaneously, as I am closing my front door, an elderly vegetable merchant in Tashkent, some 4,500 miles from where I live in north-west London, escapes from the intense noon heat. Wearing a white lace cap, he stretches out an old, rather faded, small woollen carpet on the stone floor in a quiet area in his cool warehouse. Having smoothed the carpet with his hands, he slips off his sandals and prepares to recite Zuhr, the midday prayers. Facing towards the holy city of Mecca, he stands upright. Hands out level with his shoulders, eyes closed, he acknowledges Allah's greatness: 'Allahu Akbar.' Having placed his hands across his chest, right over left, he calls on God's glory. Then he bows and stands upright before prostrating himself with his hands and his forehead touching the carpeted area in front of him. Then he kneels, palms on knees, for a moment in silent prayer. Having repeated these movements, he stands, puts on his sandals and rolls up the little carpet. Then he walks from the storeroom towards the front of his little shop and sits on a battered old stool sipping a cup of coffee his son has just brewed to await the first customer of the afternoon.

Just as that elderly gentleman takes his first sip of the thick, sweet coffee, some four thousand miles away in Phnom Penh, Cambodia, a young woman carefully lays out some tropical fruits and flowers on a burnished metal platter. Then she lights some incense sticks at a small, brightly coloured box – a box that to our western eyes resembles nothing so much as the top of a very clean bird table. She bows briefly, then stands up for a second, her face preoccupied yet at peace. She bows once more, and then turns to cook rice for the family evening meal, her head, like mine and the greengrocer's, filled for the most part with the usual banal human concerns – children, husband, money. But, like me in London and like the gentle merchant in Tashkent, she has, for a few brief moments, given her mind and her body over to something that was not physically present. Whatever she was contemplating was not obviously related to the usual twin goals of survival and reproduction that seem to govern so much of our existence. She was paying her respects to something that she feels is bigger than herself – and also to those who had gone before, venerating her ancestors who, she believes, live on in spirit form.

These examples of ritualized behaviour or prayer are repeated many times over in many places and nearly all human cultures. A description of a few more of them might just make mildly interesting reading for some of us, but it would not get us very far towards drawing any conclusions about them. I am a medical scientist, who has spent his career fascinated with one of our most basic human instincts: the compulsion to reproduce. For some of my scientific colleagues, humans are mere expressions of an intricate genetic programme which creates that drive to have babies. For some scientists, this belief – and I deliberately call it that – has some of the consolations of religion. It attempts to make sense of every corner of existence, and of our place within all that exists. It unifies a whole array of disparate, discordant phenomena – wars, internet dating sites, luxury cars – into a single coherent system. Wars are nothing more than competition for resources, related to our bid to gain the optimum chance of survival. Internet dating is just a human way of co-opting a technological tool to assist in mating procedures. Luxury cars are the means by which we display our 'good' genes and attract sufficiently choosy partners with whom to make babies.

Scientists tend to build a reputation on refuting the theories of those who have gone before. There is a perception that, as humans gain increasingly sophisticated knowledge, science rewrites the facts of what we thought we knew in past times. Science seems to provide answers to the mysteries of our existence. Yet, whatever we hypothesize, observe, measure or record about the natural world leaves more, and increasingly more, unanswered questions. Though a very superficial view might suggest the contrary, science does not give us certainty about ourselves or our origin. In spite of the ideas of some nineteenth-century theorists heavily influenced by Darwin, human history is not a process of ever better, clearer ideas replacing the nonsense and superstitions of our forefathers. Some ideas endure, whatever revolutions are taking place in the realms of culture, politics, economics or medicine; and the most enduring idea of all is the idea of a supernatural dimension to our existence. Throughout this book I call it the 'Divine Idea' – but I am not referring solely to the Judaeo-Christian belief in a single God, who created the world and cares what we do in it. Within the Divine Idea I also include a belief in many gods, as in Hinduism or in the religion of the ancient Greeks, and a belief in 'spirit-beings', whether they be departed ancestors or parts of the natural world. All are different expressions of 'something beyond' this current human life.

For some people, the very fact that the 'idea of God' has survived is proof enough of God's reality. Some Jewish sources argue in a similar way that the extraordinary history and survival against all odds of the Jewish people is all the proof we need that God signed a meaningful contract with us in the desert. But these are simplistic theories; things survive for all sorts of random reasons, and not everything that survives is necessary for human continuity. The small, rather dangerous, blind-ending piece of bowel that we are all born with – the appendix – is quite a good example. In any case, this book is not an enquiry into the existence of God: that is a territory best left to the theologians and the theoretical physicists, although I shall certainly touch on what they have to say. Indeed, at many points along the journey traced in this book it might seem to readers that this is exactly what I am doing – discussing and evaluating the various arguments for supernatural beliefs in a bid to find out which, if any, is the most persuasive. My central purpose, however, is to tell the story of an idea, the story of how humans have approached that idea, and the story of how that idea has shaped human life. It does not matter whether you believe there lurks a real God or gods behind the idea. The *idea* is real; and, as a scientist who studies 'real things', I believe it deserves to be examined.

In the book of Genesis there is an extraordinary, puzzling episode involving Jacob, who in many ways is a deeply flawed character. He has deceived his blind and helpless father, when the old man was virtually on his deathbed. He has also schemed against his brother and deprived him of his father's blessing and of the greater part of his inheritance. Now, years later, fearful for his life – close to the end of his journey to be reunited with his cheated and aggressive brother awaiting him the other side of a river with four hundred men – he has one of the most enigmatic encounters in the Bible. Having temporarily stripped himself of all possessions, having divided his group of followers into two separate camps because he expects to be attacked, having sent his family and his closest supporters to a safe location, he waits, completely alone, near the river bank. The growing dark adds to his fears and as night falls, he finds himself grappling, using all the strength he can possibly muster, with a mysterious man. The implacable, silent wrestler grasps him in a hostile grip – but, though hostile, he is *face to face* in the dark with his opponent. Jacob's body is knotted with that of the man. Perhaps the greatest biblical commentator of all time, Rashi,[1] implies that the Hebrew word 'to wrestle' used here means something very personal – being bound to each other, their bodies intimately entwined. This is a deeply passionate experience, perhaps not so dissimilar from an act of love. Eventually the man, having wrestled all night with Jacob, and injured him in the hip, leaves just before dawn so that Jacob never actually sees his face. Do they recognize each other at all?

Who is this wrestling visitor, who evaporates into the pre-dawn gloom? The wrestling match has had no defined outcome – Jacob is wounded, but not beaten. The man with whom he wrestles begs for release – 'Let me go, for dawn breaks.' Replies Jacob: 'I shall not let you go, till you bless me.' In the dark, his opponent asks Jacob his name. On hearing the answer he says, 'From now on, you will be called Israel' – and yet, curiously, Jacob's name remains unchanged as the rest of the story unfolds and he meets his wronged brother.

Many English translations of the Bible call Jacob's antagonist an angel. Some Jewish sources argue he is Jacob's guardian angel, or saviour. Other commentators suggest that the man Jacob wrestles with is Jacob himself; that his opponent is his own conscience, working feverishly all night as he is about to meet one of the greatest dangers he has ever faced. But perhaps this wrestler is God. God's face is nearly always hidden in the Bible. And surely only God can have the authority to change Jacob's name.

Whatever the meaning of Jacob's story, it gives a powerful image for the impulse underlying this book. Virtually all of us have, at one time or another, irrespective of our background, education, training, profession or family, wrestled with God. Often this wrestling match starts when we are children, sometimes it is profound when we are adolescents, and for some the wrestling continues for most of our lives. For some, too, the wrestling is most violent when we are frightened, dismayed or distressed – or face death. Many people, in other situations, times and places, have found their own routes towards an understanding of the bewildering, fascinating mystery of human existence, their own ways of perceiving the unity behind the many and various. And they have done so most frequently by referring to a powerful, supernatural being or beings, a God or gods. Their ideas of what such an entity is like, its involvement in the world, its relationship to humans, are almost as many as there are humans walking on the planet. As we look at these ideas in detail, we will see many similarities – reflecting both the way our brains work and the concerns and needs we all have as humans. But we will also see, I hope, something else.

I think humans have always wrestled with the Divine Idea – an idea that unites and separates, creates and destroys, consoles and terrifies. It is virtually certain that religious belief is as old as our species. And it is equally possible that uncertainty, doubt and scepticism about God have existed since prehistoric times. We find the earliest traces of the Divine Idea in the 30,000-year-old caves and graves of prehistoric humankind – but it is probably older than the Cro-Magnon people who lived in those caves. Throughout human history, it is an idea that seems sometimes to have caused whole populations to rise up and slaughter one another. Subjects have rebelled against their rulers and individuals have rejected the families that raised them and the societies in which they lived because of the Divine Idea. It is also a kind of bond, a mode of human expression that links me, a Jew, the merchant in Tashkent and the woman in Cambodia, although we come from quite different cultures. Like so many humans around the globe, we take some precious time to pay recognition to something beyond the life we can explain.

All paths to the divine involve a wrestling match. Wherever God is considered, whether in polytheistic terms or in a monotheistic framework like Judaism, Christianity or Islam, there are radically conflicting elements. Spirituality on its own could not have been sufficient for human consciousness: we needed to formalize our beliefs, to give them some structure, to arrive at a framework for the rules of living. Religion has endured since the dawn of human consciousness precisely because it encompasses so much of being human: division and unity, hate and love, anger and pity, precise law and simple piety – and certainty and uncertainty. No idea has endured so long, gathered up so many disparate needs, wants and feelings, and inspired so many different paths towards understanding it.

In some ways, the wrestling match is typified by the apparent conflict between God and science. Now that we have science, it is said, do we need God? As I hope to explain in this book, this dispute, and therefore this question, are largely vacuous. God and science are essentially two totally different ways of looking at the natural world, though each gives important insights into the nature of the other. Many people, including a considerable number of scientists, claim that a belief in God is harmful, citing in evidence the bitterness, damage, death and destruction that religion has caused throughout the ages. But this is rather like denigrating the pursuit of scientific investigation by reference to the amount of harm that has been done in developing terrible weapons and in damaging the earth's natural environment. We must not confuse religion with God, or technology with science. Religion stands in relationship to God as technology does in relation to science. Both the conduct of religion and the pursuit of technology are capable of leading humankind into evil; but both can promote great good.

This book is not an exhaustive history of the struggle between science and the divine. I have compressed the story, preferring to focus on some influential religious movements and some of the more interesting examples from science. I have also examined some of the applications of technology used to investigate the history of humankind's spirituality. In presenting an overview, I deal hardly at all with many interesting issues – for example, the presumed conflict between religions and the study of anatomy, and the fascinating history of religious attitudes towards the practice of medicine through the ages. Instead, I have chosen to amplify some of the ideas that are considered in the accompanying BBC television series, *The Story of God*. This book is not, though, simply an account of what we have broadcast. It is intended to cover a much larger canvas, using only a very small amount of the material seen on screen. It is intended to stand on its own, to be sufficiently interesting in its own right. I hope also that a personal account of some of my own struggles with God, and an impression of how I continue to attempt to resolve that conflict, as an averagely rational scientist and as a Jew, will be of some interest.

1

Religion's Roots: What Did Prehistoric Humans Believe?

As I write this, controversy rages over a tiny skeleton found on a remote Indonesian island. Back in October 2004, a team of Australian and Indonesian palaeoanthropologists found the remains of a miniature human being, around 3 feet tall, which apparently lived on Flores around 18,000 years ago. With a brain just one-quarter the size of ours, this creature made its own miniature tools to hunt giant rats and small elephants the size of ponies. Further remains were found, dating from only 12,000 years ago – a fact that astonished the scientific community. Since modern, full-sized humans were known to have come to the area 45,000 years ago, the finding of the pygmy skeleton, affectionately dubbed 'Flo', suggested that there might have been contact between the two groups. Local myths on the island refer to a human-like creature called Ebu Gogo, which stood just 1 metre tall and interacted with the islanders. Across the centuries, people in the area have claimed to have seen the tiny humans, the last sighting being around a hundred years ago.

The world's press immediately leaped upon the story,

informing us that 'hobbits' were real, and speculating that they might still be around in some isolated parts of the globe. Could it be possible that the 'little people' – the assorted fairies, elves, pixies, dwarves, goblins – of European folk tales had their origins in the same circumstances, of early humans rubbing shoulders with pre-human antecedents? Some linguists joined the fray, noting that the Keo and Ngada languages spoken on Flores were unnaturally simplified and suggesting that they could have been learned by these tiny people as they were integrated into and absorbed by modern groups.

As speculation bubbled away, other scientists were quick to quash it. A top Indonesian palaeoanthropologist, Professor Teuku Jacob, stated that 'Flo' was simply a modern human who had suffered from a congenital disease. He debunked the claims of the original team in a rival scientific journal, and then had the skull and jaw removed to his own laboratory, apparently without official permission, although he agreed to return them at the beginning of 2005.

The bitter feuding surrounding this single small skeleton should remind us that certain subjects are likely to raise strong emotions – and that human origins are perhaps the hottest topic of all: the central arena where scientific inquiry has clashed frequently with other theories and belief systems.

In pursuit of the Divine Idea

Humans evolved from hominids; apes that naturally stand upright on two legs. As I sit writing that statement, it hits me with unexpected force that I cannot avoid controversy with this book. My view of the world is mostly scientific and my scientific position, baldly presented, will be anathema to many. A recent Gallup poll of 1,028

American teenagers aged from thirteen to seventeen years found that 38 per cent of them rejected the idea of evolution, believing that 'God created human beings pretty much in their present form at one time within the last 10,000 years or so'. And George Bishop, Professor of Political Science at the University of Cincinnati, writing of university graduates, notes that 'the percentage of Americans who identify themselves with the biblical creationist world view is about 44 per cent; about four in 10 subscribe to the theistic evolutionist view; and only one out of 10 endorse the Darwinist position of natural science'. The patterns of belief may be different in the UK, but it seems that, as a scientist writing about religion, I will inevitably be venturing onto contested ground.

But my personal opinions will be equally controversial for many scientists. I am not an atheist. I do not pretend to understand the nature of God; I do not know whether our moral code is a human construct, a piece of genetic programming or a God-given gift; I do not fully understand the concept of a soul and I have no idea whether there is an afterlife – but I am prepared to accept that God may exist. Paradoxically, I am a practising Jew. Many of my closest scientific colleagues and friends find all this laughable. Some even think I am a little weak in the head.

This book will also be controversial among scientists for quite another reason. Much of what I write is, by its very nature, bound to be speculative. Take the title of this opening chapter: 'Religion's roots: what did prehistoric humans believe?' It is a futile question; how can we possibly know the answer? We have a few clues, of course. We know that prehistoric humans must have had the same genes as ourselves and lived in an environment not hugely dissimilar from that which still persists in parts of this planet. It is likely that prehistoric people had many emotions identical to our own – but we cannot know for certain what they thought, or what they dreamed.

Nevertheless, the fact that we stood upright on two legs would have had all sorts of consequences for our behaviour, our survival techniques and the kinds of societies we formed. Inevitably, it would also have affected our perspective on life. The ability to stand up, and therefore see across longer distances and in all directions, gave us the opportunity to see 'beyond'. Perhaps, just as our sensory organs were no longer restricted to what we could discern from the patch of land immediately around us, so our minds were freed from their most immediate concerns and allowed to roam more widely. It is likely that, from this evolving upright perspective, a range of mental processes came into action – including the ability to imagine, to plan and to solve problems.

Of course, the greatest legacy of human evolution is our relatively huge brain. One early hominid ancestor, *Australopithecus afarensis*, had a cranial capacity of around 450 millilitres, and *Homo erectus*, one or two million years later, a brain of about twice that volume. But *Homo sapiens* – the modern human – trumped all earlier versions. Humans have a brain size of around 1,450 millilitres, and with this increase came an unsurpassed cerebral cortex – the seat of the imaginative and thinking powers.

The evolutionary causes of this new improved brain are not certain. Group interaction, tool-making, walking upright, patterns of hunting and changes in diet may all have played their parts. Doubtless, people of a particular religious persuasion will argue that evolution was not the driving force, that this unique feature of our humanity is a gift from God. But for scientists, of course, the similarity between our brains and those of apes makes the idea of evolution irrefutable.

All the same, a big brain does not necessarily go hand in hand with the ability to imagine, or to develop a religious outlook. Another apelike species seems to have co-existed with us for over 100,000 years. *Homo nean-derthalensis* had a bigger brain than that of the modern human – around 1,500 millilitres. But what evidence we have suggests that Neanderthals had far less imagination than humans, and consequently were less adaptable and less able to cope with a changing environment.

Notwithstanding the recent mysterious (and possibly mischievous) reported sightings of *Homo floriensis*, the pygmy of Indonesia, it is safe to say that we humans are now the only members of the ape family still walking upright on this planet. Yet of all apes, body weight for body weight, we have been the weakest, the least agile, the least fleet of foot and so – with few natural weapons – the most defenceless. With our soft, fleshy young, remaining dependent for so long upon their parents, we were the perfect snack for the many predators around us. *Homo sapiens* was a deeply endangered species, and came into being in a period of climate change. Ingenuity and the associated ability to master this changing environment was a key factor in human survival. And this very human faculty would not only have helped our ancestors to survive; it would also have helped them to live a rich imaginative life, to consider the mysteries of their existence and develop a set of answers that we call religion. Throughout my own account, I shall call this the 'Divine Idea', referring to the many and various human notions about God, gods, spirits and supernatural forces.

Dead and buried

By the ancient banks of the Euphrates river, in what is now Syria, there is a remarkable record of human existence at the time of the last great climate change. In July 1973, in an effort to harness the Euphrates for irrigation and a hydroelectric power scheme, the Syrian government inaugurated the building of the Tabaqah Dam. Now completed, it is the largest earth dam in the world. Around 1 billion cubic feet of sand and gravel went into its construction; it stretches 2.7 miles across the north Euphrates river valley, standing 196 feet high. The artificial lake it forms, Lake Assad, is almost 50 miles in length. Diverted into Syria's agricultural areas, it will irrigate around 1.5 million acres of countryside. Through the ingenuity of modern humans, this water, supplying giant turbines, will eventually produce some 1,100 megawatts of electricity – enough power for a city of two million people.

The project involved flooding many archaeological sites, among them a slight mound called Abu Hureyra. In 1973 Andrew Moore, now a senior academic at Yale University, was a 27-year-old archaeologist studying for his doctorate at Oxford. He was exploring the large hill near the bank of the Euphrates, attracted by the huge number of shaped flint stones that covered the ground over an area of about 25 acres. Time was limited because, irrespective of the riches that might be found, Abu Hureyra would soon disappear for ever under the waters of Lake Assad. With the full agreement of the Syrian government, Moore commenced his meticulous excavations.

Abu Hureyra turned out to be one of the earliest sites where prehistoric humans are known to have settled and built villages. Excavations showed that 13,000 years ago there were forests stretching right down to the banks of the river, home to around 150 different edible plants. Now, the nearest forest is almost a hundred miles away. For at least five hundred years or more, the people of Abu Hureyra had plentiful plant food at their fingertips, without much need for farming. Moreover, the dense vegetation attracted many gazelles and other mammals, and the villagers were often able to slay complete herds of desert deer. But around 10,000 BCE disaster struck. Probably because of massive changes to ice formations in a very distant part of the globe – what is now Canada – there was a severe drought lasting well over five centuries, which turned much of this part of Asia into a desert.

The people of Abu Hureyra abandoned their villages because there was no longer enough food. What happened to these families is not known, but it is likely they formed small settlements around oases, where sufficient water allowed them to plant edible grasses. What we do know is that by the time the climate had changed once more, around 9500 BCE, a new and more sophisticated human settlement had been established at Abu Hureyra. Now people there were almost entirely dependent on farming. We know this because of the human remains found buried at this remarkable site.

The bones of 162 individuals, including 87 adults, were excavated. Of these, at least 44 were women. Careful examination of these bones and teeth revealed much about the daily life of these people, what they ate and their social behaviour. During this second period of human habitation, after the intense drought, the villagers cultivated a variety of cereals, such as oats and barley, and emmer seeds which could be ground to give a form of flour. All these needed vigorous grinding to make them palatable, and the excavated skeletons showed signs of the excessive physical strain produced by this work. The bones of the women – but not the men – had the big toes of their right feet bent upward. They had spent long periods kneeling. So the heavy grinding was almost entirely done by the womenfolk.

Grinding cereals was a continuous, relentless job, taking many hours. It had to be done daily, as these seeds do not keep well after removal of their husks. Anatomical examination of the parts of the bones to which muscles are normally attached was highly revealing. Grain would have been placed on a quern stone and the pestle, a stone held with both hands, pushed across the quern. At the end of this movement, the woman's upper body would have been almost parallel to the ground. Then she would drag the pestle stone back to the starting position. This laborious movement would have needed strong deltoid muscles in the shoulders; and, because their arms would also turn inward, the bones of all these women show they had prominent biceps. Perhaps not surprisingly, the hypertrophy of these two muscle groups was the same on both sides – showing that they used both arms for the pestle. These may not have been women you would necessarily want to meet on a dark night.

Working like this for many hours not only strains the toes and knees, it also puts considerable extra pressure on the hips and, especially, on the lower back. We can guess that a slipped disc or crushed vertebra must have been particularly likely if the person grinding repeatedly overshot the rim of the quern and had suddenly to support the weight of the massive pestle. As with the injuries to feet, knees and hips, Moore's team found that only the women had the characteristic damage to the lower vertebrae.

The discoveries at Abu Hureyra show important first steps in human civilization – constant attempts at not just survival, but perhaps a better life. Abu Hureyra is highly instructive for many reasons. The science employed at archaeological digs like this is testimony to the amount we can infer from such apparently limited evidence. And presumably, having the same genes as modern humans, these people could not have been so dissimilar from ourselves. While we cannot say what these people actually thought, it is highly probable that they would have had many of the feelings and anxieties that modern humans experience; and presumably they reflected on the constant drudgery of their existence. Though we have no evidence that they thought about a God, it seems unthinkable that they didn't. And there is a powerful clue: they buried all the bones of their people in just seven trenches.

Possibly the earliest archaeological evidence for 'religious' attitudes comes from the fact that our ancestors buried their dead, suggesting that they had ideas about an afterlife, or a 'soul' that survives beyond the expiry of the physical body. This sort of imagination depends on so-called 'decoupled thinking': what the French philosopher Jean-Paul Sartre once referred to as 'the ability to think of what is not'. This capacity to think about situations that aren't actually happening in front of us, but might happen, is undoubtedly a driving force in the history of human civilization. It is not hard to see how our ancestors might have applied this form of thought to many of the knotty problems of their daily existence: How do we make sure there is enough corn for everyone? Why do people get sick? What happens to people when they die? The Divine Idea marks an attempt to find answers to questions like these.

Some anthropologists and sociologists have argued that this is indeed the *only* real function of religion. Karl Marx called religion the 'opium of the people', expressing the notion that humans use ideas about the supernatural to console themselves, and to answer questions they cannot otherwise answer. For the early twentieth-century anthropologist Bronislaw Malinowski,[1] the idea of the soul arose as a response to the problem of death, and all religious beliefs and rituals were ways of handling 'crises' within individual and community life, like death and puberty.

Whenever we look at the evidence for religious belief in our ancestors and antecedents, we see that it centres on the business of death and the dead. This is not surprising. Any of us who has ever glimpsed a dead body will know the strange mix of feelings it incites: fear, pity, curiosity. But why should this be? If death is a natural part of our existence, why don't we feel the same responses to other natural phenomena, like pregnancy, love-making or childbirth?

For a number of reasons, dead bodies create an 'Error' message in our brains. The evolutionary demands of survival result in our having a well-developed detection system that tells us what, in our surroundings, is living and what is inanimate; what is a hyena and what is merely a hyena-shaped rock. A dead body, in its resemblance to a human being, is clearly animate. Yet it is not moving, not alive. Some of us might have similar feelings when we come up close to statues or shop dummies or even suits of armour in a museum – there is a spooky sense that this is a real person, but crucially, a real person *doing the wrong thing*: not breathing, not moving. Our brains do not relish this kind of anomaly, and when we come across it the confusion it generates tends to issue in strong emotions, like fear and disgust.

Anthropologists have suggested that there is a similar tendency in the religious outlooks of various peoples. Mary Douglas, author of *Purity and Danger*, holds a view that is interesting, though I disagree with it. Many readers will know that modern Jews observe a range of strict dietary rules – for instance, abstaining from pork and shellfish – which are set out in the Old Testament and reflect the practices of our ancestors when they lived in ancient Israel. Mary Douglas argues that the whole array of food prohibitions in Judaism is an expression of this 'anomaly response' in the human brain. Ancient Israelite thought, she argues, assigned a certain number of animals to each zone of the environment – water, land, air – on the basis of how they looked, moved and ate. Animals that were *traif* – that is, forbidden under the Law – were simply odd ones out. For instance, the Israelites were not allowed to eat shellfish because they lie motionless in their tough casings on rocks or on a river bed, whereas the permitted 'format' for water creatures is that they have scales and swim about. But there are faults with this theory – it doesn't hold true for every permitted and forbidden thing in Israelite law. It looks particularly shaky, for example, when you consider that camels were forbidden food – after all, what could be more appropriate to its environment than a camel in the desert?

But – unlike camel meat – dead bodies *are* dangerous. Organic material decomposes, and as a body decays it becomes a rich source of unhealthy bacteria which can make us very ill. In addition to that, they attract all sorts of other things which might not help our own survival – flies, hyenas, wolves, bears. It is not surprising that both Muslims and Jews are required to bury the dead as soon as possible after death (it is rare to wait more than twenty-four hours unless the Sabbath intervenes); for both religions originated in cultures living in hot climates, where it was necessary to dispose of corpses rapidly. Just as our capacity for discriminating between the animate and the inanimate lies behind the feelings of threat and unease we feel near dead bodies, so our brain is skilled at detecting sources of pollution and poison in our environment. We experience strong feelings of disgust at the smell of rotting things, or at the sight of inedible foodstuffs. And we convey these feelings very swiftly and forcefully to our fellow human beings in our facial expressions.

The fact that we carry very sophisticated communication equipment in our skulls tells us something else. We evolved as social animals and depend on a close network of communication and co-operation with others, in order to increase our chances of survival and reproduction. This adds another layer of complexity to our feelings about dead bodies. In modern societies, it is regrettably possible to live life in complete isolation from and ignorance of our neighbours. Newspapers often deplore, quite reasonably, the fact that some old people lie decomposing in their houses or flats for weeks before anyone even notices their absence – a sad consequence of modern, especially urban, life. For our ancestors, as for people living in other more traditional forms of society today, this would never have been the case. If someone died in the pack, or the camp or the village, then you knew who they were. If they were not directly or indirectly related to you, then they were bound to be someone with whom you co-operated and communicated on a daily basis. So here, then, is another crisis for the brain to handle. This is the body of Ungu, your old mate and hunting partner. You feel sad, because you had a bond with this person. But this weird person-that's-not-a-person is also a source of pollution and disease – it disgusts you to see it and smell it, and your instincts tell you to get it out of the way as soon as possible.

Death sets off further crises in the human brain because, quite simply, we are wired to survive. We therefore experience strong emotions at the death of another human which motivate us to avoid death however we can. Human death is also threatening because it is irreversible. Trees and bushes 'die' every year but are reborn in the spring. Bison and antelope can die and be eaten, but there will be more. When a person dies, on the other hand, something is lost irrevocably. Ungu's sons and grandsons may endure, but Ungu is gone for ever.

So death poses a crisis for the group, not just through the loss of one of its members, but also because of the presence of the body and the issue of what is to be done with it. A crisis requires a response.

Humans are not, in fact, the only species to respond to death. In 1989 a pair of researchers, Christoph Boesch and Hedwig Boesch-Ammerman, working in the Tai Forest in the Ivory Coast noticed some interesting behaviour among the pygmy chimpanzees they were studying.

These two biologists observed what happened after an influenza epidemic among the chimps, killing Pulin and

36

leaving Pippi, her daughter of three years and five months, orphaned. Pippi passed at least two sad nights in a fig tree watching helplessly just above her mother's corpse. None of the adult chimpanzees seemed particularly concerned. When the scientists came across the corpse and finished putting Pulin's body into a flour sack to move it in order to reduce the risk of further infection, little Pippi, who had been coughing rather weakly while eating figs, started to whimper, quickly climbed down from her tree and ran into the bush, crying loudly. Nine days later she was seen walking with a large group of chimps, fending for herself but not apparently adopted by any of the females. Nevertheless this baby survived and was later reported to look healthy. As a human might do, she grieved for a certain time and then got on with her life.

The behaviour surrounding the death of another, older chimp, Tina, was even more revealing. Tina was ten years old when she died after being attacked by a leopard. Several of her fellow chimps gathered round the body, making loud noises. At the end of this episode, some dozen adults sat around the body in silence. The silence was broken by some males displaying aggressive behaviour and dragging the corpse around.[2] High-ranking females were allowed to inspect the body by the males, who seemed almost to be guarding Tina's body, and were not slow to chase off interlopers from the lower ranks of chimp society. Two high-ranking males began to groom the corpse, and continued to do so for over an hour. Meanwhile, other chimps inspected the place where Tina was killed, occasionally laughing, as if to dispel the tension.

Tempting as it is to anthropomorphize, it is risky to attach human values to the behaviour of these chimps. This is not 'chimp religion'. It may, however, be symbolic activity. As we evolved into *Homo habilis*, *Homo erectus* and finally *Homo sapiens*, we probably engaged in similar forms of behaviour when one of our group died, before removing the corpse and placing it somewhere where it would not pollute the camp or attract predators. The problem is that we cannot *know*: behaviour like waving your arms around for an hour doesn't leave any obvious traces. A corpse placed in a tree or up a mountain just rots away or gets picked to bits by carrion, leaving nothing for later archaeologists to muse over. The very term 'prehistoric' sums up the task we face: history begins with writing, a system whereby our ancestors could render their ideas in marks and signs that later generations could find and 'read'. The alphabet of prehistoric behaviour is much more opaque – we can only look for clues, and those we find may point in several directions at once.

But deliberate burial gives us something to work with. A body placed somewhere beyond the reach of predators or carrion can provide us with evidence to examine. It may also tell us something about the beliefs of the people who buried it. Going out of one's way to place a body somewhere tells us that a certain level of respect was accorded to human remains. Perhaps their careful attention to the body of a dead person indicates that our ancestors imagined that person was still 'around' in some way, at least around enough to care about how we treated their body.

The burials at Abu Hureyra are, in archaeological terms, very recent, but we have evidence for this kind of deliberate burial from much earlier periods. In Skhul and Qafzeh caves at Mount Carmel in Israel, archaeologists have found the remains of men, women and children from around 100,000 years ago. Some skeletons were buried along with very simple possessions, such as stone tools, deer antlers and a boar's jaw. The presence of these artefacts in graves suggests a further dimension to prehistoric belief. There may have been the idea that the dead might 'need' these items, and so were somehow still alive. Anthropologists point out that in many societies the dead are believed to be 'angry' with the living, and this anger is viewed as a possible cause of misfortune within the group. Burying valuable items with the dead is one way of winning their favour and avoiding their anger. So we begin to see how 'religion' provides answers for the problems of existence.

In Upper Palaeolithic times (from about 40 million years ago to about 10,000 BCE) humans began more seasonal patterns of migration, living in established sites for months at a time, and ingesting a more widely varied diet. This era was characterized by a flowering of symbolic action. We find buried individuals placed in a fetal position – as they were at Abu Hureyra. This possibly indicates notions of a cycle of death and rebirth, though it is unlikely that prehistoric humans would have had much knowledge of embryology and the position a baby adopts inside the uterus. Perhaps it just represents an attempt to place the dead in a sleeping position. Very often, dead bodies are found having been oriented towards the east, in other words, the rising sun. In other sites, they may all face the west.

Once again, the archaeological record can only point to belief; it can do no more. The anthropologist Gerardo Reichel-Dolmatoff studied modern burials among the Kogi Indians of Colombia. He once witnessed the interment of a young girl. During the ceremony, the shaman, or ritual specialist of the tribe, lifted the girl's body nine times and then placed her facing eastwards in a grave containing small green stones, shellfish and snail shells. Any archaeologists who happened to come along later would just find a skeleton, facing east, surrounded by shells. They might speculate that shells were highly valued items, perhaps a form of currency, or a 'food' that the dead person could use in the afterlife. From that sparse evidence, they could not know that actually the Kogi believe they are returning the child to the womb. The lifting of the body nine times symbolizes, in reverse, the nine months of pregnancy. Nor could they know that the shell is intended to be a 'husband' for the girl in the after-life, to prevent her becoming angry and taking the life of a young man of the tribe.[3] To appreciate all this, a visitor would need to encounter the living group and be immersed in their beliefs. Archaeology can give us only an impoverished account.

What beliefs did our ancestors have about the sun? What meaning did they attach to a boar's jaw or a stone tool? We cannot know. We can look at ethnographic evidence from peoples who live in similar ways today – but these are unusual by virtue of the fact that they have remained as they are and have not become like us. All we can do is speculate that, as they are humans rather like us, they may have believed something that we feel we might believe.

Is eating people wrong?

Like speculation, luck is often important in archaeology. A very grisly collection of pre-human bones at Atapuerca, about 10 miles from Burgos in northern Spain, was dis-covered only because of the completely unrelated efforts of one Richard Preece Williams. In 1896 this British entre-preneur undertook to construct a narrow-gauge mining railway to transport coal and iron ore from the Sierra de la Demanda in northern Spain to a junction on the Bilbao line, from where it could be delivered to the iron and steel furnaces in the Basque country. The work required around 1,500 labourers and was finished in 1901. It was the con-struction of this railway, cutting through the foothills of the Sierra de Atapuerca to open a path for the railways, which led to the rediscovery of the Cueva Mayor cave, for part of its entrance collapsed during the work. As it happens, the cave system in these parts had been known about for years and then for the most part forgotten. Inside the Cueva Mayor there is an inscription left by Friar Manuel Ruiz, dated 22 October 1645, and ever since then local people had occasionally visited the entrance, a few braver spirits even venturing further inside.

At a small bend near one entrance appears a hole called 'the silo', formally known as the Sima de los Huesos or 'Pit of Bones'. This is a place where, it is speculated, 250,000 years ago our hominid ancestors dragged several dozen corpses of other pre-humans into the cave and rolled them down a muddy, steep-sloping, 46-foot shaft. The corpses would have landed at the end of the shaft with their arms and legs flailing like rag dolls. From there, they settled into the bottom of a pit. Over the following years, various animals blundered into the hole. They soon died at its base, there being no exit and no means of clambering out. Some of these wild animals survived the fall long enough to gnaw the human remains. In the course of several hundred years, a layer of animal remains accumulated above the jumbled human ones, and then the original mouth of the cave closed over. Graffiti in the cave show that, at some time during the Middle Ages, the pit was rediscovered by local village lads from nearby Ibeas de Juarros. These young men, as young men do, took to proving their manhood by venturing into the Pit of Bones. The idea was to fetch bear teeth for the girls. But getting them wasn't easy. With the original cave entrance closed, reaching that 46-foot shaft required a crawl through a tunnel of around 1,600 feet.

After slow exploration following the Second World War, a PhD student descended into the Sima in 1976. He came out not only with some bear's teeth, but also with what turned out to be a pre-human jaw. Suddenly, Atapuerca became famous, because this jawbone was clearly very ancient indeed. It is now thought that our ancestors have inhabited the area continuously for one million years. More recently it has become obvious that bones from the Sima had not been gnawed by animals alone. Many of the bones show marks and injuries which make it quite likely that both adults and small children were butchered by our early human ancestors.

Similar finds have been made in Britain. In Pontnewydd in Wales, Neanderthal remains appear to have been deliberately placed in caves. Why do caves feature so often in prehistoric activity? Perhaps because caves are dark, echoing and mysterious places; they rumble with strange sounds of wind and water, and the chill air inside feels threatening even to modern humans. Caves are also symbolic of the womb – so, by placing the dead inside them, our ancestors were perhaps expressing their ideas about a cycle of life, death and rebirth.

Most of the Pontnewydd remains are those of young adult males. In his summary of the evidence, Paul Pettit, a research fellow at Keble College, Oxford, makes the observation that perhaps young men were the only ones stupid enough to go caving.[4] But there is clearly a more serious implication, as Pettit points out. Young men would have been the most valued sector of the group – prized for speed, strength and fearlessness, all vital in warding off predators and hunting down prey. Perhaps, as a high-status 'class' within Neanderthal society, the young hunters were the ones given special treatment when they died. However, we should be aware that 'young' is a relative term. We think Neanderthals rarely lived beyond the age of thirty.

At other sites, the use of red ochre to stain bones also suggests some beliefs about how the dead were to be treated, and thus some idea that the dead are not 'just dead'. Neanderthal burial, if that is the right word for it, was communal – the dead were placed, or thrown, on top of other dead bodies. At Krapina cave in Croatia, the remains of up to seventy individuals were found in parts of the cave that would suggest they lived more than 100,000 years ago. Many of their bones bear cut marks, which seem to have been made by stone tools. Some analysts have taken this as evidence that the Neanderthals practised cannibalism. But the practice of 'defleshing' the dead as part of the burial ritual has been recorded among certain modern-day tribes, and the nature of the marks on these bones is arguably more consistent with that ritual activity rather than with the preparation of the bodies for consumption. It is a highly contentious subject – but whatever we make of the evidence, it points to some deliberate treatment of the dead. Neanderthal graves have also been shown to contain flat limestone 'pillows' for the dead, and pollen traces that might indicate the presence of flowers deliberately put into the graves.

Cannibalism is a theme that crops up throughout the history of our species. Great caution must be exercised when interpreting faint marks on human bones so long after their owners' demise. But the skulls found in Croatia – with evidence that they may have been deliberately crushed – has led some to argue that Neanderthals ate the brains of other hominids 100,000 years ago. Many of the criteria which some archaeologists use to identify cannibalized remains – facial mutilation, burnt bones, dismemberment, cut marks, bone breakage and hammer-stone abrasions – are said to be present on bones at this site. Bones found at various other sites have flint marks on them which most closely resemble the marks made by deliberate butchery. But it is not clear why this took place. In Atapuerca there is a suggestion of human flesh-eating from a long way back in time, but this could not have been because meat was scarce. Animal remains indicate that many mammalian species were plentiful, species that our ancestors were well equipped to hunt.

In general, the eating of our fellow humans seems to be

not a matter of physical nutrition but a spiritual activity – that is, related to beliefs about gods or spirits or an after-life – and this has led some to give it the name 'divine hunger'. Marshall Sahlins, Professor of Anthropology at the University of Chicago, suggests that cannibalism 'is always symbolic even when it is real'. Outside situations like disaster or starvation, cannibalism is hardly ever just a matter of food. Most reports of cannibalism in recent centuries suggest that it had a strong ritual element.

The rituals of cannibalism may be better understood by examining practices of more recent societies, about which we obviously know much more than we do about pre-historic humans. Even though they left no proper written records, the Aztecs are perhaps the best 'modern' example. Many of their writings were burned by one of their own emperors, Itzcoatl, and many that survived his bonfire fell victim to the invading Spanish in the sixteenth century. Nevertheless, we have much detailed information about the Aztecs through their extraordinary sculptures, paintings and other artefacts. We also have contemporary and eye-witness accounts of these people, as well as some interviews with Aztecs recorded by the *conquistadores*. Some of these accounts were almost certainly exaggerated to make the Aztecs seem as savage as possible, thus justifying the Spanish conquerors' own brutal behaviour. But overall they probably provide a reasonably accurate, if graphic, account of Aztec religious beliefs and ceremonies.

To understand what influenced the behaviour of the Aztecs we need to look at their antecedents in Central America and the history of the ancient city of Teotihuacan. Two thousand years ago Teotihuacan was the biggest city in the Americas. It stretched over 8 square miles on the central Mexican plateau and housed up to 200,000 inhabitants. It was the seat of a sophisti-cated society, with a priestly caste, merchants and farmers.

It contained palaces, shops and markets, carefully built watercourses and an efficient city refuse disposal system. The wide Avenue of the Dead, which stretches out straight as a runway for 2 miles beyond the Pyramid of the Moon, dwarfs The Mall near Buckingham Palace. This flat roadway was once paved with cement and its neighbouring plazas were decorated with coloured stucco. We have no idea whether any vehicles ever journeyed along it, but it seems that this advanced civilization did not use the wheel. In addition to the massive Pyramid of the Moon at its northern end, there is the even bigger Pyramid of the Sun about half a mile away on its eastern side. After the two biggest pyramids at Giza in Egypt, these are the largest man-made buildings of the ancient world. Either side of the Avenue of the Dead, stretching as far as the eye can see and spaced at regular intervals, are many smaller pyramids, each with steep stone staircases rising to the summit, where in former times an altar would have been sited on the flat, square top.

Some of these pyramids have been excavated, and many have been reconstructed out of the blocks of volcanic lava of which they were originally built. A man's skeleton has been found under nearly every one of the pyramids that archaeologists have studied. In some cases the hands were secured behind the back and the knees bent in a posture of submission. Many of these skeletons wear necklaces made of six or seven human lower jaws. These pyramids are not at all like the funerary tombs seen in ancient Egypt. They were not built to protect a great person after death on the way to the afterlife. On the contrary, the individuals whose remains we find in the Teotihuacan pyramids were killed by the people who lived and worshipped in this city. The inhabitants seem to have made human sacrifices because they believed that the bodies of their victims contained spirits who would protect the pyramids.

Around 700 CE some mysterious, undocumented disaster occurred. This flourishing city was attacked, burned and largely razed to the ground. We do not know who the perpetrators were. Possibly the neighbouring Toltecs were the culprits. But no army occupied or settled the city; it was just left abandoned as most of the surviving citizens fled. Stories of their religious practices, however, continued to be recounted by local people, and they left an enduring record in the stone-carved glyphs and pictograms adorning the city. Much later, around the middle of the thirteenth century, a new tribal people emerged from western Mesoamerica. At first insignificant nomads, the Aztecs crossed bare plain, marshlands and desert. They came believing a legend – that they would eventually find a promised land. They were excellent soldiers and fought off the surrounding inhabitants who were not well-disposed towards them. Initially without a king, they carried their tribal deity, Huitzilopochti, on the shoulders of four priests wherever they travelled. Along the route, it seems their god decided to call them by a new name, 'Mexica'.

Settling the region now dominated by Mexico City, the Aztecs soon came across the mysterious sacred ruins of Teotihuacan. These old buildings impressed them deeply; so they adopted Quetzalcoatl, the feathered serpent god that the earlier inhabitants of the now defunct city had worshipped, alongside Huitzilopochti. They also copied many of the sacrificial rituals of these people. And remarkably quickly, probably because of their military skills and their initial willingness to act as mercenaries for surrounding rulers, they established the Aztec Empire, building their own cities with equally great pyramids. This was soon to become the most powerful empire Mesoamerica had ever seen, with a complex infrastructure, a system of taxation of neighbouring tribes, a rigid legal code, a priestly society and a hierarchy of privileged nobles. Eventually they chose an emperor who was also a god.

The Aztecs had an unquenchable taste for human blood. Bloody rituals had always been a prominent feature of the religious practices of Mesoamericans, but the Aztecs' thirst seemed insatiable. The priests performed autosacrifice, sometimes daily, using their own blood – the most valuable commodity possible – to propitiate the gods. They would mutilate themselves with sharp obsidian knives, repeatedly pierce their ears or arms with thorns or sharpened bones, or make gashes in their tongues or slits in their foreskins. Contemporary descriptions of these priests do not present a very attractive picture. Their long, unkempt hair was often caked in matted blood, their ears, tongues and genitalia were shredded and scarred, their bodies stank of decaying flesh and putrefaction. It seems that the priests viewed autosacrifice as a kind of symbolic death – a payment of the debt for continued life. So perhaps it was inevitable that this came to be seen as an inferior substitute for another ritual.

In the Anthropology Museum in Mexico City there is a very famous artefact – the Stone of the Sun. This sculpted red gritstone disc, 16 feet in diameter, is decorated around the periphery with carvings representing the twenty-day cycle of the Aztec calendar. It is said that young, hand-picked warriors would fight each other in the centre of the disc, rather like the struggle between Japanese Sumo wrestlers contained within their little ring. At the centre of the disc, where the warriors might stand, emerges the carved face of the god Xiuhtecuhtli. On either side, in each hand, he holds a human heart, and his protruding tongue is transfixed by a sacrificial knife.

For the Aztecs, as for their predecessors, the ultimate religious ceremony was human sacrifice. It seems that hundreds or thousands of humans may have been immolated each year. The *conquistador* friars accompanying Cortés on his subjugation of Mexico record that the pyramids were stained deeply red and brown with the blood of sacrifice. The stench was so bad that the Spaniards could smell these cities well before they could see them. Most of the Aztecs' victims were captured warriors; others were well-bred, handsome citizens, and some virgins or children were also subjected to this terrifying death. The Aztecs believed that their sacrificial victims would be transformed into hummingbirds before joining the Sun God in paradise. They were housed in comfort, sometimes for months and occasionally for as long as a year or more, until the time of the ceremony. After ritual cleaning and purification, during the final run-up to the sacrifice, it seems they were treated as gods. There is some evidence that some victims, too, may have seen it as an honour to be the object of sacrifice. But perhaps it was different when the actual moment of truth came. When the victim was held down, immobilized by three or four people, back stretched out in an arc over the curved boulder on the altar, they might just have had second thoughts. The priest, poised with an obsidian knife, would make one transverse cut under the ribs to open the abdomen. A rapid incision was made in the diaphragm and the priest would thrust his hand deep into the chest cavity. The highest act of worship was to grasp the heart from the chest of the living victim, sever the vessels, and then hold it high for the people to see while it was still beating. Once the victim was dead, the body would be sent rolling down the steps of the pyramid. At the bottom, the corpse might be decapitated and the head mounted on the skull rack at the side of the temple.

This method of dispatch was only one of several ways of performing Aztec human sacrifice. Father Diego Duran, presumably an eyewitness, tells of gladiatorial sacrifice, where a prisoner of war was tied to a round carved stone and forced to defend himself against a hand-picked Aztec warrior. The prisoner was armed with a sword made of feathers, while the Aztec was dressed for battle and had the advantage of a large club lacking only the usual studding with sharp obsidian blades. Father Duran observed drily that 'whether one defended oneself well or fought badly, death was inevitable'. Some soldier victims were tied to wooden scaffolds and, like St Sebastian, shot through with arrows until they were dead. Another sacrificial method was decapitation, a treatment sometimes used for virgins. Perhaps the most cruel of all was that reserved for a few captives, who would be repeatedly thrown into a fire and dragged clear, and then forced to submit to the heart sacrifice while still alive.

Many sacrifices were concluded by cannibalism. If the victim was a captive, his or her captor would hold a position of honour and might entertain appropriate guests to a sacrificial meal, where his family might eat a portion of the flesh – perhaps stripped from the thigh of the victim. This was not a normal dinner party. Rather, it was a religious ceremony at which only a small portion of the body would be consumed and sometimes a sample of the victim's blood drunk from a special bowl. It was designed to honour the victim and to give him a place as kin – as a relative of the family sponsoring the meal.

It is difficult, of course, to untangle the ritual significance of such ceremonies. It is known that the Aztecs believed that the gods required human sacrifice to keep the universe going. It is also likely that such occasions offered a chance of expiation for the moral transgressions of the celebrants. The anthropologist Michael Harner has argued that the prevalence of human sacrifice among all these Mesoamerican peoples is attributable to a lack of protein in their diet. However, it is known, for example, that while the Aztecs did not have access to many domestic animals, they almost certainly got adequate supplies from pulses and beans. And many Aztec sacrifices took place at harvest time, when such food was abundant. Moreover, we know from the early Spanish *conquistadores* that the sacrificial victims were eaten exclusively by the nobility, who, presumably, would have had their pick of any available menu. In any event, the amount of human meat consumed would have been an inadequate dietary supplement.

Much of the power of sacrifice probably resided in a combination of politics and religion. What pertained to the Aztecs is likely to have been equally true of prehistoric human societies. Aztec kings ruled only because the gods wanted them to. The king acted as protector of the priests and a guardian to the temples. Human sacrifices, in addition to striking abject terror into the minds of all who saw or heard about them, were a powerful way to show the conjunction of gods and the state. These large public spectacles confirmed the power invested in the king. They must have provided a very effective way of deterring both the state's citizens and its enemies from any form of resistance or aggression towards the ruling class.

The gruesome details of the Aztecs' sacrifices are particularly remarkable when we consider that these rituals ended only comparatively recently – when Hernán Cortés invaded Central America in 1519, over ten years after Michelangelo had completed his paintings in the Sistine Chapel. While the European Renaissance was already in full swing, the Aztecs were still wedded to the pagan rituals of Baal and Moloch. Could it be that western civilization owes more than it generally recognizes to the rejection of the Golden Calf at the time of Moses?

Human cannibalism seems most easily explained as a kind of symbolic communion, an attempt at acquiring the dead person's powers, a release of the soul, or an act of revenge or punishment inflicted on the ghost of the dead. Although it seems abhorrent to us, cannibalism is also clearly a tool for maintaining and regenerating the social order.

Because cannibalism often involves eating relatives, rather than enemies or people from other families, some anthropologists think that, when humans engage in cannibalism, they are trying to reinforce the unity of the group. In the act of consuming kin some 'essence' of the ancestors is handed on and absorbed, keeping the group together and linking the present to the past. Until relatively recently, Gimi women in Papua New Guinea ate the flesh of their dead menfolk to prevent it from rotting in the ground. They thought that by doing this, they were freeing the dead men's spirits so that they might rejoin those in the ancestral forest.

Beth Conklin from Vanderbilt University studied the Wari tribe of Guyana in South America. Until pressure from government officials and missionaries forced them to desist about forty years ago, the Wari ritually ate parts of their dead relatives. This ritual seems to have been an integral part of the emotional recovery of the living from death. Dr Conklin is quoted as saying, 'It marked a distance between the people doing the eating and the person who is eaten. The Wari believe you need to gradually create emotional distance between the living and the dead, because in a small society, the ties of love and affection to your family are your strongest bonds, and they don't dissolve or loosen with death.' Some people have suggested that eating part of a dead person allowed the spirit of the deceased to be absorbed by the entire tribe, and that this was therefore considered by the Wari the most respectful way of treating a corpse.

Cannibalism was not an unknown practice at the time of the ancient Hebrews – indeed, it has been suggested that the word 'cannibal' is derived from the Chaldean *cahna-bal*, meaning 'priest of Baal' – *cahna* being an emphatic form of *cahn*, 'a priest' (from which comes *cohen* in Hebrew). But at some point in human history, cannibalism largely ceased to be the fulfilment of a religious impulse, and came to be regarded as an irreligious and horrific activity. Nevertheless, the ritual of cannibalism has occasionally persisted into relatively modern times. The recent case of Arwin Meiwes, currently serving eight years in prison in Germany, is highly disturbing partly because his perverted appetites were by no means unique. It will be recalled that he searched the internet for a willing victim to cannibalize: evidence submitted to the German court alleged that Meiwes had responses from no fewer than 430 other men who indicated an interest in sharing his practice. Equally remarkable is the ritualistic manner in which Meiwes and his ultimate victim shared parts of the victim's own body, the genitals.

Modern horror at cannibalism reflects a longer-standing revulsion. Surely this is why rabid anti-Semites have repeatedly accused Jews of cannibalism. The so-called blood libel – the myth that Jews consume Christian blood during the Passover feast – has led to untold suspicion and hostility over the centuries. Even before the time of Christ, Jews were accused of ritual murder – in spite of the fact that the Jewish religion expressly forbids the consumption of any kind of blood. Moreover, once they had left Egypt, the Israelites were allowed to perform animal sacrifice only within the confines of the Jerusalem Temple under rigidly controlled conditions. A particularly malignant rumour was promulgated by Apion, the Greek rhetorician and rabble-rouser. In the second century BCE, during the reign of the Seleucid king of Syria Antiochus Epiphanes, he alleged that Jews had captured a Greek foreigner, fattened him up for a year, and then conveyed him to a wood where he was slaughtered. He claimed that during the actual murder, those standing around swore an oath of hostility to the Greeks. This defamation became one of Antiochus' justifications for forcing the Jews to adopt Greek culture (so-called 'Hellenization') and for the subsequent desecration of the Temple in Jerusalem.

It wasn't only the Jews: early Christians, too, were accused of human sacrifice. In the second century, the eminent theologian Father Tertullian wrote: 'We are said to be the most criminal of men, on the score of sacramental baby-killing, and the baby-eating that goes with it.' It is probable that cannibalism was a 'catch-all' accusation slung at anyone practising a new and unfamiliar religion, rather than based on any solid evidence. In more recent times, Satanists have provoked unsupported rumours and have been accused of murdering babies and drinking their blood. In many, if not most cases, it seems society is ready to accuse breakaway groups of violating their most precious taboos as a way of justifying their persecution of them.

Judaism and Christianity perhaps lend themselves rather readily to accusations of cannibalism. Animal sacrifice was once a central part of Jewish worship; and in the book of Genesis the father of the Jewish people, Abraham, shows his loyalty to God by being prepared to sacrifice his son Isaac. Many scholars have written commentaries about this extraordinary episode, one of the most puzzling human dramas in the whole of the Bible. Part of what seems to be happening here is that God is testing Abraham. What He actually demands is faith and obedience, *not* blood. Even so, analogous traces of a somewhat similar tradition survive into Christianity – so that the central ritual of the faith to this day is the Mass or Eucharist, in which people symbolically consume the body and the blood of Christ.

Sadly, Christians did not learn from the early experiences of their own persecution. Throughout the Middle Ages, Christians turned on the Jews with false accusations of this sort. One famous early blood libel was in 1255, when the story of 'Little' St Hugh of Lincoln was related by the chronicler Matthew Paris: 'The Child was first fattened for ten days with white bread and milk and then . . . almost all the Jews in England were invited to the crucifixion.' As a result of this fabrication, ninety-one members of the Jewish community were sent to London for trial, and nineteen of them were executed. It may seem implausible, but the blood libel, with its hints of ritual cannibalism, continued in Russia and elsewhere until modern times. The Nazis used it repeatedly, giving the 'Jüdischer Mordplan' (Jewish murder-plan) front-page headlines in their newspaper, *Der Stürmer*, on several occasions.[5]

The religion of the painted caves

Chris Henshilwood of the Iziko Museums of Cape Town, Director of the African Heritage Research Institute, recently described the remarkable find of some unusual pieces of red ochre in the Blombos Cave in South Africa. At least two of these pieces, measuring just a few centimetres in length, were carefully flattened and ground along their edges. Both also have very clear repeated diagonal lines, a kind of evenly repeated cross-hatching, engraved across the ground surface. Most intriguingly, the patterns are very similar on the two pieces. The evidence – careful preparation of the surface to be engraved, the geometric patterns produced, the technique of engraving and the fact that the final design is similar on both pieces – points to deliberate human activity. Various archaeological and chemical tests place the age of these relics as approximately 77,000 years. This suggests that, though they lacked writing, our early ancestors nevertheless had a symbolic way of expressing themselves.

Cro-Magnon man, an early human, emerged between

30,000 and 40,000 years ago. Bits of skeleton and personal belongings dating from that period were first found in France in the 1860s, and since then a number of their caverns in western and central Europe have been discovered. Cro-Magnons seem to have been natural artists, producing finely crafted stone and bone tools, and jewellery made from shells and mammoth teeth.

Among the Cro-Magnons' greatest cultural legacies are their remarkable cave paintings. Those at Lascaux in south-western France, around 17,000 years old, are glorious: animated, detailed and multicoloured. Similar art, though not always in such good condition, has been discovered across Europe, from the Asturias to the Don River, dating from 30,000 down to 9,000 BCE. The fact that these paintings involve similar techniques, and depict similar scenes based on hunting, has led some to suggest that there was an early, widespread 'religion of the caves'. This may be pretty fanciful; but nevertheless, the very use of painting to make representations suggests a symbolic faculty of the mind, which is certainly the bedrock of belief in deities and worship.

Some years ago I visited one of the French caves – the Grotte de Gargas. Gargas does not have so many of these enigmatic paintings of animals, though there are images of bison and antelopes. But here visitors see something far more mysterious. Low down on the walls of this vast cavern are numerous painted stencils of human hands: around 250 of them, both of children and of adults. Many different pigments (pulverized charcoal seems to have been one medium) were used, so that the designs are white or black, red or yellow. It seems that these people, in making their paintings, placed a hand on the wall, and then, with a mouthful of pigment, blew a spray of colour around it. The technique of carbon-14 dating puts one of the hands at about 27,000 or more years old. A similar find, of fifty-five hand-prints, has been made in the

Cosquer Cave, also in France. Some of these are stencils and others are 'positive prints', made by the technique widely used in nursery schools today: a hand was dipped into pigment and then applied to the rock wall.

What is extraordinary is that, in both caves, most of the hand-prints seem to have one or more fingers missing. Nobody knows why this is. Do these hands bear witness to particularly cruel weather during an ice age, when frostbite would have been common? This seems unlikely, because the thumb in these prints is nearly always intact. Could it be the result of some bizarre mutilation, possibly some sacrificial ritual? Certainly many human groups across the world have engaged in deliberate scarification and mutilation – recall the Aztec priests. But no skeletons from this region have been found with digits missing.

Perhaps the most likely explanation is that the fingers were not missing: rather, the hands were deliberately drawn with one or more fingers bent, so that they did not leave an image. It is possible that this was some kind of greeting, or a coded language associated with hunting or other rituals – similar to the silent language used by hunting tribes like the African Bushmen and the Australian Aborigines. The mysteries do not end there, however. Splinters of bones have been found by French archaeologist Jean Clotte in cracks in the rock, near to certain handprints. Where were they from, and why were they there? Perhaps, rather like the notes that religious people stick in the cracks of the Western Wall of the ruined Temple in modern Jerusalem, they represent prayers to a deity asking for peace, health or some other form of help.

These traces will always be a source of controversy, and archaeologists have fought each other ferociously about the issues raised by cave paintings. The inaccessibility of the caves used – the one at Cabarets, in France, for example, takes several hours to reach – may indicate that the paintings had a ritual purpose. Perhaps the ordeal of the painter, crawling alone through dark tunnels for hours, was a means of giving thanks for a successful hunt, or a means of securing one in the future. Religious paintings by modern humans can be objects of adoration or meditation. But in other systems of belief, a drawing may be a magical way of making the events depicted in the drawing happen. Similar symbolism is seen frequently in some tribal societies today: for example, a traditional healer may put sacred stones into a pot in order to 'make' a barren woman pregnant.

Religious practice also often involves an ordeal of some kind, either to 'make' something happen, or at least to induce the comforting belief that something will happen. In parts of Africa, pubescent boys are subjected to bewildering and painful ordeals, such as hunting rituals and circumcision, before they are acknowledged as men. Sometimes there's a sense of a 'trade' going on with the supernatural world: I go through X ordeal so that you (God, gods, spirit) will give me Y in return. Maybe these paintings, in their dangerous and inaccessible locations, were a part of some deliberate ordeal – a test of skill and bravery for young men about to become hunters. Or possibly they were part of a 'deal' with the spirits: human effort and pain expended in return for a successful kill to follow.

Magic, shamanism – and the myth of spiritual evolution

Humans almost certainly believed themselves to be in a spiritual relationship with nature 15,000 years ago. The evidence was discovered in 1914, in the Trois Frères Cave in the south of France. Here a long, confined shaft at the entrance finally gives way to a vast, cathedral-like space. The walls and ceiling are covered with pictures of animals

– mammoths, bison, stags, bulls and ponies. The inhabitants of the caves have drawn many spears flying towards these animals.

A really striking figure dances about 14 feet above you. He is painted some 3 feet high and overlooks the hunt. He is bent at the waist, in a jumping pose which quite resembles the gait adopted by Africans dancing in celebration of a successful chase. This picture – known as the 'Animal Master', or 'God of the Cave', is the most arresting image in all the prehistoric art I've seen. It requires little imagination to agree that he appears to be wearing some kind of ritual mask: the eyes, ears and face are almost owl-like. He wears large antlers like a deer, his body, belly, shoulders and forelegs resemble a lion, and the tail is that of a groomed horse. But his beard, posture and expression, as well as his feet, lower leg musculature and penis, are unmistakably those of a man. He resembles an all-powerful sorcerer, with hunched shoulders and head tilted to one side, quizzically supervising the hunt and ensuring plenty of kills.

Sadly, this cave has now been closed to protect the strange paintings. The influx of modern human visitors has led to fungal overgrowth on the cave walls and now – having survived almost completely intact for fifteen millennia – the whole remarkable cavern is at risk of ruin. But if we cannot see the paintings, nor can we forget them. What were the motives of those who painted this masterful scene?

The great American expert on mythology Joseph Campbell once suggested that these primitive men were trying to understand and accept a strange world where people needed to kill to live. Surely, like us, early man would have had dreams and indeed nightmares. Perhaps – though of course this can only be speculation – Cro-Magnon man, living so close to nature, had something of a guilty conscience when the animals he slaughtered, with the blood and the squeals, came back in his dreams, maybe threatening revenge. It's possible that a mythology emerged to explain and resolve this anxiety. One form it could have taken was the belief that the beasts would submit themselves to being killed if the proper ritual were undertaken. Campbell argues that the hunters may have communicated with the spirits of the game in their rituals. If they thought as the animals thought and paid tribute to their gods, and thanked their prey for the sacrifice they had made, they would gain expiation from the guilt of killing. So the animal songs and animal dances needed to be performed with reverence, and their meat and skin taken carefully and respectfully to avoid violating the souls of the animals.

Another, similar image is seen on a slate slab at Lourdes. It depicts a man wrapped in a deerskin, with a horse's tail and a head surmounted by antlers. While we can never be anything like sure what these images meant for the people who painted them, we know that many of the supernatural beings humans have envisaged over the centuries are 'composites' of human and animal features. Ganesh, the elephant-headed god of the Hindus, is a classic example, as are the half-equine centaurs of Greek legend. So these figures in the caves might have been depictions of animal gods.

The Victorian British were among the first to study religious beliefs in so-called 'primitive' societies. Although their conclusions were often drawn from rather scanty evidence, they are to be praised for challenging the earlier view of Christian missionaries, who thought that, until Europeans came along with their bibles, tribal people had no religion at all. Many Victorian authorities thought that all religions began with animism. The term 'animism' refers not to animals but to the idea that principles, elements or features of nature – for example, the wind, trees, certain creatures – possess an *anima*, or soul/spirit. Nowadays, we might call these beliefs paganism. The evolutionary view of men like E. B. Tylor (1832–1917), regarded as the father of anthropology, was that animism and magic gave way to organized religion and eventually science – a neat, if desperately simple, view of human history.[6]

We can see how certain aspects of our distant ancestors' lives would have made animist beliefs useful, attractive, even necessary. For a start, as hunter–gatherers dependent upon the climate and the movements of migrating herds, prehistoric people lived an existence very close to nature. This nature would have inspired feelings of awe, as at the majesty of a waterfall or a sunset, or fear, as at the sight of an angry bear towering several feet above the height of a man. Our ancestors might have gone on to attribute human characteristics to these experiences. We do it ourselves, all the time. We make statements like 'Germany has revoked her decision to back the Maastricht Agreement', or 'the committee expressed its regret that Professor Winston did not attend the meeting'. We call countries and boats and cars 'she'. In these expressions, we attribute human characteristics to entities that plainly do not have them. Why? This is possibly a means of rendering things bigger than us comprehensible to our brains, reducing them to the more familiar framework of human relations. In our palaeolithic past, animals like bison and bears had a dual aspect – they had the power to kill us, but also to feed and sustain us. So the idea emerges that when we kill and eat the flesh of a bison, we pay respects to its departed spirit, or to the spirit that represents all bisons. We also hope that we will not be injured or killed by them, and may obtain more of them to eat.

There is another explanation for these weird half-man, half-beast figures. At Starr Carr in England archaeologists found an antlered skull with holes drilled through it, as if it had been worn as a head-dress. This is similar to head-dresses worn by modern-day shamans in Siberia. The shaman is a specific kind of religious 'officer' found in a number of traditional societies. Shamans often enter trance-like states, sometimes through the ingestion of drugs, in order to have dialogue with the spirits and divine the causes of misfortune, such as illness or accident. Their 'solutions' often include some ritual or offering to an angered spirit; in some cases, they are believed to heal the sick, by doing battle with malign spirits or enticing them away from the body they have decided to occupy.

We cannot infer too much from the similarity of one possible head-dress in a painting and another possible one found in an excavation to similar features in surviving societies: that would be like assuming Nazi beliefs everywhere we found a swastika design, or Christianity everywhere we found a cross shape. But shamanistic beliefs are common to societies of the hunter–gatherer type – such as the Inuit people who today inhabit the far north of Canada – so we do have one further explanation for the images in these ancient paintings, and a more definite indication of religious beliefs. The human–animal figure at Trois Frères could have been a shaman, who brokered the deal between the hunters and the hunted prior to the kill.

Prophets, too, sometimes used magic – at least, it would have appeared to be magic to the onlookers they wished to convert. There are also many examples of possession or visions in the Bible. Gideon is reported to be clothed or possessed 'with God's spirit' (Judges 6: 34); and the divine spirit came suddenly upon Saul, transforming him and provoking great fury and ecstatic frenzy (I Samuel 10: 6ff.). One of the great stories of 'magic' in the Old Testament is set in the time of a great drought in Israel, described in the first book of Kings. The story tells how Elijah competed with the priests of Baal, the local god to whom children were sometimes sacrificed. He stood, a lone figure at the summit of Mount Carmel, pitted against

450 prophets of Baal and 400 prophets of the goddess Asherah. His opponents slashed their flesh and danced around, according to their rituals. But nothing happened. Elijah mocked the priests. In one of the earliest examples of humour, the Bible recounts how Elijah told them to shout louder – perhaps their god was 'asleep'. Then, with a piece of stage-magic, Elijah drenched his altar with water and prayed to God – and lightning struck. The altar fire was lit – and torrents of rain ensued. Elijah was, in short, acting quite like a traditional shaman, even if his God was a very different one from the nature-and-fertility deities worshipped by his competitors.

In twenty-first-century America, evangelist preacher Benny Hinn conducts televised and highly choreographed prayer meetings at which he allegedly cures HIV, frees the disabled from their wheelchairs and even claims to give life to the dead. Once again, these are not unlike the activities carried out by the shaman in a hunter–gatherer society – even if they have now become a multi-million-dollar enterprise and a form of mass entertainment into the bargain. So we should be sceptical of any evolutionary notions that suggest that 'this type of religion turned into this'. Such ideas are arrogant, because they carry an assumption that the routes to God of other societies and ages are 'simpler', 'more primitive' than our own. They are also wrong – because all the ideas humans have ever had about God are present throughout the world, throughout history.

We know little about the ideas of our palaeolithic ancestors, but we know something about how they lived. And it was a life that must have favoured quite a lot of sitting around and pondering. We have a false notion of our ancestors' lives as being, as the seventeenth-century philosopher Thomas Hobbes put it, 'nasty, brutish and short'. It is equally silly to imagine them living in some sort of innocent bliss, at one with nature. But the lifestyle of hunter–gatherer societies contains an enviable amount of free time. Look at anthropological films we see on television. Notice how much time is spent in the camp, whittling sticks, chewing the fat, exchanging jokes. By comparison, agriculture demands daily drudgery in accordance with the fixed rhythms of the seasons; and life in industrial societies is even more rigidly controlled by the clock. But hunter–gatherers, in the optimum conditions of good weather, peace and plenty, have plenty of time. They have fire, so they also have a source of light and warmth – they are freed from the timetable set by the rising and setting sun. It's not hard to imagine that, for social animals with language (and blessed by a lack of television or MP3 players), a form of speculative story-telling became a major pastime. That in turn would have encouraged our ancestors' minds to wander from the particular to the general, to see links and similarities between their lives and the world around them. Someone notices that women's biology is like the phases of the moon; that the sun of the day-time is 'opposite' to the moon of night-time, as men are 'opposite' to women. The moon then becomes female and the sun male. A whole range of other classifications can then be made – success in women's activities, like childbirth, depends on the favour of the moon goddess; men's, like hunting, on the sun god. From the easy rhythms of their existence, palaeolithic humans had a facility for enquiring about their world, and seeking to explain it.

Consciousness and consolation

In the 1920s a British anthropologist called E. E. Evans-Pritchard was conducting research in a Sudanese village when a wooden grain store suddenly collapsed, killing a member of the tribe. He was able to use this regrettable event to confirm his own ideas about religion in traditional societies. Evans-Pritchard was convinced that the idea of magic evolving into religion and then science was flawed. From his experiences in Africa, he felt that people in traditional societies were no less rational than westerners, and were certainly no more or less stupid. There had to be another reason why people believed in magic, spirits, witchcraft and the like.

The villagers knew, from a cursory examination of the pitted posts of the fractured granary-house, that termites had eaten through the wood. But we humans are never happy with that kind of explanation. It catapults us into a random universe in which anything, bad or good, can happen at any time. And this is pretty unbearable. What we want to know is *why* the granary collapsed, *at this time*, and *on this particular person*. For the Azande people being studied by Mr Evans-Pritchard, the answer was sorcery – and a whole range of rituals and beliefs were in place to work out who the sorcerer was, and to neutralize their powers. Supernatural ideas existed to explain what wasn't readily explicable.

This kind of explanation is not found only in African villages a century ago. After the tsunami that afflicted south-east Asia in December 2004, the Cambodian king claimed that his country had been spared because he had prayed to the goddess Indra. In ancient Israel, the prophet Amos interpreted hostile incursions by foreign nations as God's wrath with the sinful Israelites, who had forgotten the covenant they had made with him in the desert. In the modern world, certain Christian leaders still declare that AIDS is a divine punishment for promiscuity or homosexuality. Which of us, when faced with some personal misfortune, hasn't looked upwards and asked, 'Why me?'

Consciousness, which separates us from our animal counterparts, is both a blessing and a curse. It frees us from the simple following of instincts, frees us from the moment and allows us to plan ahead. Understandably, it also gives us a great sense of power and capability. But only so much. However powerful and capable we are, some things are beyond our powers and always have been. Babies die – even in the face of 'powerful' western medicine. Misfortune strikes when we least expect it. This induces an overwhelming feeling of bewilderment and hopelessness. This can take us in either of two directions. We can lie down and die. But our genes urge us not to. Or we can find a way of living alongside that uncertainty – by explaining it, by inventing certain ways of acting that give us a comforting sense that we can control it. In many cases, throughout our history, these ideas have involved a relationship with supernatural forces – a god, for example, or spirits residing in nature, or angry ancestors, or human individuals with malicious magical powers.

Yet ideas about the supernatural do not exist merely to account for misfortune and suffering. In many societies they have an *aetiological* purpose – that is, they explain in a broader sense why things are as they are. We can see a lot of the references in the book of Genesis in this light. The tale of Adam and Eve's eviction from the Garden of Eden could be read as an attempt to explain the divisions and differences between men and women, why serpents slither on the ground, why women and not men suffer the pain of childbirth, why we all have to die. It also accounts for the whole question of how we got here, and explains the origins of certain place names – for example, the Bible tells us that Peni-El (face-of-God) is so named because it is where Jacob met God face to face and wrestled with him.

Understanding why things are as they are makes them easier to accept. The loss of a loved one, perhaps on the battlefield or in a foreign country, when the body may not be recovered and the death confirmed, is one of the most difficult events to deal with. When a love affair breaks down, the injured party often talks of wanting 'closure', which means, really, that they want to know why their partner left or stopped loving them.

The need to understand what is happening is integral to the way the human brain makes sense of the world around us. In our tree-dwelling past, the sudden crack of a branch or an unfamiliar animal cry would have alerted us to the possibility of being attacked and eaten. So we developed the kind of brain that seeks out answers for what we can't immediately understand. If a weird noise is keeping us awake, we may find ourselves getting up and pacing around, trying to find the source. Even if we cannot stop it, knowing from where the noise comes gives us the mental peace needed for a few hours' bliss before the alarm disturbs us. For this information-hungry device, the human brain, it's not enough to know 'my boyfriend has left me' or 'there is a noise' – our brain demands to know *why*: an extra level of knowledge. Small children playing the interminable 'Why?' game are enslaved to this basic feature of brain function. Understanding the causes of things feeds this powerful human need. And the supernatural can provide some persuasive answers in situations where no other answer presents itself.

The Venus of Willendorf is a remarkable statuette of a female figure, carved in limestone. It dates from perhaps 25,000 BCE. The more one gazes at it, the more one senses the presence of the ancient sculptor who made it. It is not as intricate as some of the more detailed cave paintings, but it leaves an indelible impression none the less – perhaps because it speaks to something so basic in our make-up. This woman is obese, with bulging flanks, and she has two pendulous breasts. Her belly, which is fat and swollen, has a prominent navel. This navel is sufficiently inverted to suggest that she is not pregnant, though the shape of her body strongly suggests that she has had babies in the past. The vulva and labia are clearly carved.

In contrast to the detail of her genitalia, her head is very basic and unrealistic. Where her face should be, there is just a series of lines and protuberances at regular intervals. It is almost as if she has covered her face with a curly hairpiece. Her thighs are short, ending in two well-proportioned knees – but she has no feet so, unless these have been broken off at some time in the past, this statuette could not have stood unsupported. Some scholars have suggested that she might have been carved so that she could be stuck in the ground. Similar effigies have been found at various archaeological sites across a wide region, from Lake Baikal in Siberia to Liguria in Italy. These 'Venus figurines' seem not to have been deliberately buried, but may have played a role in daily life – certainly many have been found in domestic settings, as if they were household objects.

We can't ever know what true purpose these squat figurines served, but we can make some observations about the beliefs that lay behind them. The drive to make babies is one of the most powerful human imperatives. In many societies even today, an infertile woman is a complete outcast, with no status and no secure future. The imperative to reproduce would have been even more potent for our ancestors. In palaeolithic times, and for long after that, children *were* the future – ensuring the survival of the species, and the survival of immediate relatives. Before the concepts of pension plans or savings accounts or the welfare state, having children was the only means by which we could ensure our own survival once we had become too feeble to get our own food. In many parts of the world, this remains a rationale for having a large family. The world population boom has been largely the result of poverty. In poor countries, where infant mortality is high, there is huge pressure to have large families to improve the chances of at least some children surviving to ensure their parents' well-being. This is why improvements in infant and child health are so much more effective than massive contraceptive programmes in limiting population expansion.

Throughout our society, women wish to have babies because of a biological need. Most suffer silently, but often immeasurably, when they cannot fulfil this demand of their minds and bodies. Men, too, often feel a powerful instinct towards fatherhood – though many of them simply wish their partners to be happy and will do whatever it takes to facilitate it. But both men and women, for the most part, express their desire to have a family in quite non-practical terms: 'I'd be a good Dad,' 'It'd make our lives complete,' 'I just know that's what I'm here for.' No-one I have yet encountered in twentieth- and twenty-first-century London hospitals has said to me that having a baby was necessary for their own survival.

Our ancestors must have seen the matter very differently. Of course, early humans had the same genes we have, and similar emotions. Many a young woman must have dreamed longingly of the family she would bear; many an infertile one must have gazed with sadness on the exuberant family of her neighbours; many a young man must have larked around with someone else's kids and realized he would enjoy fatherhood. Of course, we were not slaves to those genes. Cro-Magnons had an imaginative, emotional life not so dissimilar to that of modern man – the skilful and beautiful paintings of Lascaux tell us that. But survival was precarious. Life was generally shorter and subject to constant hazards – attack, famine, disease. The more assailed we were by dangers, the more our genes would have driven us to reproduce. But pregnancy and childbirth were pretty precarious, too. Disease, stress and hunger affect fertility. Infections can whisk life away. The whole business of having children may have seemed like one of the mysteries we could not control. So it is not surprising that prehistoric religion was directed in large part towards explaining and controlling it.

The widespread presence of figurines like the Venus of Willendorf has led some scholars to argue that an early 'Mother Goddess' cult existed throughout the palaeolithic age. Some have argued, more fancifully, that all early religion centred on the worship of the feminine in humanity and nature. This was reflected, so the theory goes, in patterns of matrilineal descent – in other words, tracing your ancestry through your mother rather than your father. Perhaps, as societies became more complex and wealthy, the male principle – expressed in strength, war, 'male' tasks like hunting and metal-work – replaced the female, and so a masculine God or gods supplanted the original Earth Mother. This theory was very popular at one time, and it attracted the literary imagination of novelist Robert Graves and poet Ted Hughes, among others.

It's a neat idea; but others have pointed out that matrilineal descent is also found in many male-dominated, warlike, technologically advanced societies, such as ancient Israel. Modern-day Jews trace their descent matrilineally: I am a Jew because my mother was. Yet the God of the Israelites, always spoken of in the masculine form in Hebrew, if not male, was certainly not female.

To argue that at some point in the past male gods replaced the female goddess is, in short, a sloppy kind of evolutionary thinking. But it is certainly true that people across the ages have worshipped certain feminine attributes – fertility, nurturing, and cyclical patterns related to the phases of the moon. The Greek poet Hesiod, writing in about 700 BCE, says that when the god Kronos had cut off his father's penis and testicles, he lobbed them into the sea. The circle of white foam that resulted (who knows what his members contained) created Aphrodite, the beautiful goddess of sexuality. Zeus, the father of the gods, gave her some fairly sound, if sexist, advice: 'No, my child, not for you are the works of warfare. Rather concern yourself only with the lovely secrets of marriage.' Her attributes were certainly worthy of veneration. The adoration of Mary, 'Our Lady', 'Mother of God', in Christian Europe points to the notion that our predecessors were highly receptive to the idea of worshipping the feminine. As the creators of life, women's bodies are quite obviously a source of wonder; so it's not hard to imagine our ancestors seeing parallels with the creation of the world itself, and with the observable rhythms of life and death in the natural environment that surrounded them.

God settles down

Around 9,000–8,000 BCE there were major changes in the climate and landscape of Europe and Asia. The ice retreated from the steppes, to be replaced by forested areas, and animals began moving north. Diminishing amounts of game may have compelled humans to diversify their diet, and to settle increasingly near lakes and rivers where they could live on fish. And, as we have seen at Abu Hureyra, humans took a further leap. They established more permanent settlements, harvesting crops and domesticating animals. These changes in the patterns of human life inevitably led to changes in the Divine Idea.

At Stellmoor, near Hamburg, people were making sophisticated bows and arrows 10,000 years ago. Modern experiments show their arrows would have been accurate at 50 metres and, with their flint points, more lethal and destructive than a modern metal-tipped arrow. At the main excavation, archaeologists made an interesting find: the bodies of twelve whole reindeer, preserved in a peat bog. Controversy rages about this – some people have pointed out that people might have buried the corpses to keep them fresh, as groups living in polar regions do nowadays. But others have suggested a definite religious purpose to this forgotten cache of valued meat and assert that it was a sacrifice. They point to the finding of a reindeer skull on top of a wooden pole to indicate ritual activities, possibly associated with a reindeer-spirit.

As we have seen already, the Divine Idea quite often has a 'commercial' element, in that people undergo pain or hardship in order to procure something from the supernatural power. From this period of prehistory onwards, we find considerable evidence of humans deliberately giving up things that were valuable to them – tools, pots, trinkets: throwing them into pits and streams, even sometimes breaking them beforehand. The Jewish wedding ceremony involves breaking a glass, and while a whole range of symbolic meanings are attributed to this (the destruction of the Second Temple or breaking the bride's hymen, for example), it nevertheless contains an element of sacrifice echoing the reindeer sacrifice of Stellmoor. You relinquish something of value, but gain something greater.

While our European ancestors might have been lining up their valuables in ditches, in the Near East people were engaged in more practical endeavours. At Wadi en Natuf in modern Israel, archaeologists have found some of the earliest evidence for settled patterns of living. The Natufians inhabited semi-subterranean caves and open-air sites, and harvested wild crops with stone sickles before grinding them with phallus-shaped pestles. This era marks the beginnings of agriculture. At sites across the Near East, such as Catal Huyuk in Turkey, we find evidence of an array of related changes: permanent dwellings, sometimes including hundreds of family units, gradual domestication of animals and crops, skilfully made tools.

Farming is a more efficient method of survival than the hunter–gatherer life: even though there was drudgery, there would have been greater returns for less effort, and more food for a greater, settled population. In what used to be the fertile plains of the Near East, villages became towns and eventually major cities. The biblical description of Nineveh, for example, says that Jonah took three days just to cross the city. While this must have included 'suburbs', it does indicate the emergence of a substantial population and necessary changes in the way humans communicated.

As farming techniques increased in efficiency, some individuals were able to acquire a surplus of food, making them 'richer' than others. So we begin to see the emergence of more pronounced differences in wealth and status: the beginnings of 'class'. The presence of surplus allows some individuals to disengage from the business of producing food and concentrate on other tasks. Pottery, house- and boat-building, tool manufacture and art become complex skills, in which specialized individuals acquire greater ability. Luxury goods – the ancient equivalents of the Rolex or the Rolls-Royce – emerge as a means by which people demonstrate their power and status; and we begin to find these in graves, once again indicating that our ancestors believed in an afterlife.

But agriculture is not merely related to a change in the patterns of social life. It represents a new way of looking at the world. Humans no longer find food; they *make* it – inspiring a fresh degree of confidence in human abilities and understanding of the natural world. Prompted by the demands of this new occupation – when to sow and plough and thresh – they begin to measure time, developing systems of notation for marking the passage of the days and the seasons. Rather than being part of nature, humans are now somewhere outside of it, shaping and changing it, bringing it under their mastery.

We did not dispense with the Divine Idea once we began to farm. Agriculture merely brought about a difference in the way we approached it. For a start, farming still involves an element of 'luck' – the right weather, the absence of pestilence – and these things are still beyond human control. In the plains of the Near East, human efforts could easily be thwarted by the regular flooding of great rivers like the Tigris and Euphrates. These uncertainties must have intensified the anxiety felt by hunter–gatherers. Nomads can at least move on if a natural disaster strikes an area; but for people who have invested time and resources in building homes and other structures, rather more is at stake. The Divine Idea is still very much needed to provide consolation and explanation. Meanwhile, as the business of being bound to a formal cycle of ploughing, planting and harvesting regularized our existence, so it also regularized our worship.

That is part of the reason why, in sites across the Near East from this time, we find evidence of the first sanctuaries: that is, sites with a specifically religious purpose. The earlier beliefs centred on a Mother Goddess are still much in evidence, as the wall paintings at sites like Catal Huyuk testify. The dead are buried underneath houses, with quite lavish arrays of grave-goods, reflecting differences in wealth and status. These often included multiple female statuettes, suggesting that beliefs about the afterlife were tied into worship of the feminine principle and a cycle of birth and rebirth.

The presence of primitive places of worship may indicate a new way of thinking about the supernatural. The very nature of the Venus figurines suggests that the Divine Idea now involved worship of a human form. Our farming ancestors seem to have been applying a certain anthropomorphic tendency to their gods, giving them human attributes. We made them 'settled', like us, by building specific sites at which they were worshipped, almost implying that these impersonal forces of nature should live in houses, dwelling-places that we ourselves have built. God is no longer all around us, in the wind and the storm clouds or the migrating herds of bison, but located in a specific habitation. We see a development of this idea in the notions of the ancient Israelites, whose God becomes attached to a specific people and place: first of all a tabernacle carried around by desert pastoralists, later a vast temple complex built at the heart of a great city. We could say that, in this age, humans still worshipped the supernatural in order to secure what they needed to survive, but the two – human and God – had taken a step nearer to one another.

Throughout the ancient Near East, this trend of increasingly complex human groups living by farming reached its apogee in the city-states of Mesopotamia, which began to establish themselves around 3000 BCE. Differences in wealth and status became more pronounced as metalworking skills developed. Individuals emerged with the power to acquire goods from far away, and we find elaborate displays of this power in the archaeological record – ornaments, armour, swords and helmets, featuring skilled work and materials that can only have been procured by people of kingly rank.

Once humans had settled, developed agriculture and amassed property – whether in the form of grain, land or animals – methods of keeping records became more important. Perhaps the earliest were tally sticks: pieces of twig, bark or hardwood on which our ancestors carved notches. Certainly something of this sort persisted right up to the Middle Ages in parts of Britain. But wooden tallies of this kind were easily destroyed, and stone or clay would have been a more permanent alternative. The earliest incised 'counting tokens' date from about 7000 BCE and were found in the region of Mesopotamia fed by the Tigris and Euphrates rivers. Judging from the archaeological record, around 4100–3800 BCE these tokens evolved into stamps that could be impressed on wet clay – almost certainly the beginning of written language.

While some scholars argue that writing may have started in ancient Egypt or even in the Indus valley, in what is now Pakistan, Mesopotamia is almost certainly the first place where humans developed written language – not just pictures carved in clay. Interesting examples of early writing are found at the excavations of Uruk in the area of southern Mesopotamia called Sumer, now Iraq. Uruk, recorded in the Bible as Erech, was an impressive city. According to legend, its walls, built by the king, Gilgamesh, were about 6 miles in length and enclosed nearly 1,000 acres; the city housed some 50,000 inhabitants. An organized social structure was needed because the inhabitants built, and shared, an irrigation system.

My good friend Professor Mark Geller, of University College London, has translated a quantity of prayers from ancient Sumerian which are perhaps 4,500 years old. The older ones, on fragments of clay tablets, ask the gods for health and for food, and provide incantations to protect against attack by demons. Professor Geller notes that these bits of tablet, with their inscriptions written in cuneiform, are found everywhere, not just in temples or palaces. Many – like modern-day text messages – contain a record of extremely trivial transactions.

This suggests to Professor Geller writing was not just a secret guarded by a privileged minority. In Sumer, many ordinary houses had a room set aside for archives such as commercial records, records of ownership, bills of sale, a last will and testament – a kind of personal library. One excavation revealed a house that burned down about 2,000 years ago. One room contains a large number of tablets; in the other there is the skeleton of a man, almost certainly killed in the fire, accompanied by just one tablet: the index to the rest of the tablets in the next room. Professor Geller feels that levels of literacy were related to the cost of the writing materials. Clay would have been extremely cheap, and that is why tablets written in cuneiform are scattered everywhere throughout the region. By comparison, humans in later civilizations may have been less literate because the medium was far more expensive. Until the printing press was developed in the fifteenth century, Europeans were restricted to writing on costly parchment made from animal skins. While I find Professor Geller's view very interesting, other Assyriologists disagree. Dr Irving Finkel of the British Musuem in London feels that literacy was rare in Mesopotamia – even some kings, he believes, could not read or write.

Babylonia, the southern region of Mesopotamia, was effectively two regions: Sumer in the south and Akkad to the north. The language in Akkad eventually gave rise to Hebrew, Aramaic and Arabic. The Sumerian language died out after about 2000 BCE, but it is almost certainly the people of Sumer who invented writing. Now that we can translate much of the Sumerian language, we know a great deal about these people – including a great deal about their religious beliefs. They had a huge number of gods, related to the planets, animals and the weather. For example, in the south there were city gods closely related to marsh life and to fishing and hunting. Enki was the god of fresh water and of marsh life in Erred in the west, and Nanshe the goddess of fish in the east. The presence of written records means that we are no longer restricted to looking at the traces of what people did; we can also examine direct accounts of what they believed.

Order! Order!

When in the height heaven was not named,
And the earth beneath did not yet bear a name,
And the primeval Apsu, who begat them,

> And chaos, Tiamut, the mother of them both
> Their waters were mingled together;
> And no field was formed, no marsh was to be seen;
> When of the gods none had been called into being,
> And none bore a name, and no destinies were ordained;
> Then were created the gods in the midst of heaven,
> Lahmu and Lahamu were called into being . . .

These lines are from the opening of the *Enuma Elish*, 'From on High', the Babylonian or Mesopotamian creation story. It tells of the conflict between an ordered cosmos and the chaos which preceded it. This particular version (older variants exist) was written in the Akkadian language some time in the twelfth century BCE on seven clay tablets.

The initial chaos described in the *Enuma Elish* – 'when sweet and salt water mingled together', in one earlier translation – gives birth to the gods, who emerged two by two from each other, in greater and greater complexity of form and nature. The subsequent struggle among the gods culminates in the sculpting of the universe by Marduk, the sun god, who splits the body of Tiamut, the sea goddess, into two, thus creating the sky and the world.

Marduk, who was regarded as the local god of the city of Babylon, went on to create man almost as an afterthought, from a mix of dust and the blood of a particularly thuggish god called Kingu. So in Babylonian terms, man, though divinely created, had nothing much to be proud of, for all his mastery of agriculture and architecture. And in Marduk, by far the most powerful in a whole pantheon of Mesopotamian gods, we perhaps have a glimpse of something approaching a monotheistic figure.

Creation stories across the world share elements in common with the *Enuma Elish*. The origins of the world as an act of separation by a god are particularly widespread.

God splits the sky from the waters in the book of Genesis. He also, in chapter 2, creates Adam from dust. In Greek mythology, Prometheus similarly creates man from mud. This does not necessarily imply that one society 'gets' its myths from another, even if they are geographically close. It's more likely that, evolving as we did with the same brains and largely similar needs, humans in separate places and societies developed similar responses to the questions of origins. Plants observably spring from the earth, so it's not difficult for people to surmise that that is where they came from too.

Although, unlike the book of Genesis, the *Enuma Elish* contains little observation on the moral conduct of humankind, it is not just a story of origins. It was a ritual object in itself, recited every year at the Akitu, the Babylonian New Year festival, which took place in the month of Nisan, or April. (Nisan, incidentally, became the name of the Hebrew spring month.) In Babylonian belief, the world of men was seen as a model of the world of the gods. The city of Babylon was thought to be a model of how heaven looked – its name, Bab-il-Ani, means 'Gate of the Gods'. The river Tigris was reflected in the star Anuit, and the Euphrates in the Star of the Swallow. We also see this idea reflected in early Persian beliefs, which stated that every aspect of the world (*menok*) had a parallel in the heavens (*getik*). In Babylon, this link with the divine world was celebrated and perpetuated, through the king, in a twelve-day festival which assured him of continuing kingship and his society of continuing fortune.

This festival was believed to have power because it re-enacted the activities of the gods in creating the universe.[7] This re-enactment is a common feature of many religious rituals. When Christians wash one another's feet, they do so in direct imitation of Christ's own actions, as described in the New Testament; and this in turn stems from the earlier occasion on which Abraham washed the feet of the angels who came to visit him. The same is true in respect of the bread and wine consumed in the Mass or Communion, a repetition of Christ's actions at the Last Supper. In the religious texts of the people who settled the Indus valley, the founders of Hinduism, we find lines such as 'We must do what the gods did in the beginning' (Satapatha Brahmana, bk VII, ch. 2, para. 4) to explain the purpose of rituals.

The symbolism of the twelve-day Akitu festival was closely bound up with ideas about death and rebirth. The king was ritually mocked and humiliated, even slapped in the face by the priests, symbolizing the death of his power. A scapegoat, believed to be the embodiment of the city's sins and misdeeds, was ritually slaughtered. A mock battle was enacted, replaying the original struggle between Marduk and Tiamut. And finally, after the recitation of the *Enuma Elish*, the fate of the following year was decided by oracles. The New Year ceremony, in essence, cancelled out the year that had gone before. By re-enacting a series of events that were believed to have taken place before time began, the Babylonians were taking themselves out of the calendar, resetting the clock to zero, and causing the whole of creation to be renewed.[8] To an extent, the Akitu was a large-scale vegetation rite (a scholar's way of saying 'harvest festival'), but with the added significance that, by re-enacting the creation story, the Babylonians saw themselves as guaranteeing good harvests for the future. Future success depended on revisiting and stressing continuity with the mythical past.

There are some clear parallels between these Babylonian beliefs and those of ancient Israel. The 'scapegoat', a real goat ritually invested with the sins of the people, was a feature of Israelite ritual. In Leviticus 16: 21, Moses' brother Aaron is instructed to take a goat and abandon it in the wilderness, as a symbol of Israel casting out its sins. The *Enuma Elish* also has many similarities to the Israelite creation myths we find in Genesis. Both texts have a seven-day creation scheme, beginning with light on the first day, moving on to the firmament on the second day and dry land on the third. In fact, they mirror each other right up to the seventh day, when both God and the gods take a breather.

The themes of the *Enuma Elish* are also found in the legends of the Israelites' reviled neighbours, the Canaanites. In their legends, Baal, the god of fertility and storms, engages in a battle with Yam, the god of the sea. In this we can clearly see the concerns of a people whose lands were under constant threat of being flooded by hostile rivers and seas, and who yet depended on rain to keep their crops growing. Central to the Canaanite story is Baal's death and descent into the underworld, his retrieval by his lover and sister Anat, and his eventual return to the divine realm. This theme, the death and rebirth of the gods, is central to much of human religious thought, and is probably related to the cyclical renewal of the natural world.

If scapegoats and re-enacted battles seem a million miles from our own staid patterns of worship, it's worth remembering that millions of people believe that Jesus Christ died and was born again, and that Christians celebrate this event in the springtime, just as Near Eastern peoples engaged in their New Year renewal rites. Across England, the rich symbolism of 'green man' festivals and maypoles provides more than quaint carnivals for tourists – these are the vestiges of older belief systems which celebrated fertility, death and rebirth.

Babylonian religion, with its many gods personifying the forces of nature, showed a strong link to the sort of beliefs our prehistoric ancestors must have had. In Israel, and elsewhere in the world, people were coming up with a fresh version of the Divine Idea, replacing the many Gods with just One.

A Question of Survival

Few people will have heard of Grangecon, County Wicklow – it isn't particularly easy to find on any map, and there is next to nothing on the World Wide Web about this tiny Irish village. Yet spectacularly, in December 1994, it was invaded by thousands of people. These visitors were not there to see the village; they all flocked to its small, rather insignificant post office. But most of them didn't even buy a stamp. The attraction was an extremely weird event which had taken place in the small back room of this building. Following announcements on the local radio and reports in the newspapers, it had become public knowledge and caused widespread excitement. The Virgin Mary had wept.

Mrs Mary Murray, the village's retired postmistress, and her daughter had noticed that there was something seriously wrong with their 12-inch-high ornamental statue of the Virgin. Both eyes seemed to have spontaneously filled with tears and little drops of blood had oozed from the left eye, leaving a brown stain on the Virgin's cheek. As the news spread, an initial trickle of interested visitors became a tidal wave. Ever since that Christmas,

Mrs Murray's visitors have been convinced that they have seen the statue's eyes water; and many of them say that when they were in the back room with the Virgin, they experienced a deep, abiding sense of peace.

Mrs Murray kindly welcomed travellers, but only between eight in the morning and eleven at night, Sundays included. Then, to cope with the increasing crowds, Mrs Murray arranged to have the statue moved temporarily (sealed in a glass case) and placed in the village. Now, apparently, at 3 p.m. every day, the glass case with its holy contents is ceremoniously lifted up and taken from the post office and carried in procession, to the accompaniment of many Hail Marys, to the main street of the village, where it is reverently placed next to the shrine of the Madonna that stands there. Pilgrims have come from all over the world to pray here. Devout Catholics believe that the Madonna herself uses this statue's ability to cry to encourage Christian prayer and devotion. To this day, small groups of the faithful congregate regularly in Mrs Murray's back room and, in the presence of the statue, recite the Rosary and give each other spiritual support.

Miracles in the mind?

These events in County Wicklow are by no means unprecedented. In Catholic communities all over the world in recent years quite similar miracles have happened. Statues of the Virgin Mary have lachrymated in New South Wales (where they had to place cotton-wool balls under the eyes to catch the tears), in Benin in Africa, in Italy near Rome, in Naples and in Calabria, and as far afield as Mexico, Puerto Rico, Barcelona, Trinidad, Las Vegas, Kansas and Virginia.

A variety of different reasons have been suggested for these phenomena. Admittedly, sometimes priests in the

Catholic church have been reluctant to describe such events as miraculous, and investigations into them have generally offered rather simple, often mundane explanations. The manufacturers of the Grangecon statue say that the adhesive used to fix its eyes might become moist in certain temperatures. Such statues, it has been pointed out, are often made with various resins and glues that can liquefy, and pigments that change colour, particularly if there is a change of temperature or an increase in humidity. Sometimes these adhesives are hydroscopic – that is to say, they attract water in the appropriate environment – and this aids the process of liquefaction. It is further said that lighting a candle close to one of these statues might cause fluctuations of air currents or temperature; and that the flickering light produced by a candle could give an observer the impression that the statue was crying or even bleeding.

Devout Catholics are by no means the only people to have their religious zeal excited by such curious occurrences. During the night of 21 September 1995, the dream of a man in New Delhi initiated a worldwide phenomenon. When this man awoke after his fitful sleep, he remembered that Lord Ganesh, the elephant-headed god of wisdom, had developed a severe thirst in his dream. At three in the morning, when it was still totally dark, he rushed around to his local Hindu temple. After a rather difficult conversation, he managed to persuade the sceptical priest there to let him give a spoonful of milk to the small stone image of Ganesh. Both watched with increasing astonishment as the milk disappeared up the trunk of the statue. By dawn, news of the thirsty god had spread like a conflagration. A little later that morning, people right across the city and then across India were offering spoons of milk to their own local statues of Ganesh and observing that the little elephant-god was suddenly accepting their offerings. Within a few hours, tens of millions of the faithful had stormed the nation's temples. In New Delhi, banks halted transactions, food markets closed and most work was brought to a standstill. By the time the morning was well advanced, it was impossible to buy a pint of milk anywhere in the city. Then, just as dramatically, some twenty-four hours later, all the statues in India suddenly stopped drinking.

But by then, many people in India had made telephone calls to family in different parts of the world. Pandit Manilal, the priest of the Ram Mandir Temple in Nairobi, told a reporter how two women had arrived at his temple at about three that afternoon. Excited calls from India had told them about the miracle, and they wanted to see if it could happen in Nairobi. So Pandit Manilal and the two women first offered milk to the temple's image of the elephant-god, and then to the dancing statue of Shiva, the Destroyer, then to Shiva's servant, Nandi, and lastly to Naga Devata, the serpent idol. Each drank the milk thirstily. Within an hour or so, people were thronging through the temple and making offerings. In the nearby private temple of Jyotin Patel, the miracle had begun by 4.30 p.m. that Thursday and continued till 9.30 a.m. the following Sunday. Mr Patel and his sister-in-law, Minal, first tried with a small spoonful of milk, which rapidly disappeared. Minal even recalls hearing a sucking sound. Pandit Manilal, like many other Hindus, believed that these miracles were a sign that an avatar, a kind of super-human Messiah, had just been born and that people would recognize his appearance once he was old enough to announce himself.

Over the next day or so, Hindus crowded their temples in every part of the world – and the miracle was repeated in places as far apart as New York, Chicago, New Jersey, Los Angeles, London, Leicester, Copenhagen, Bangkok, Katmandu, Singapore, and Edmonton in Canada. The world press coverage was immense – in Britain, broadsheet newspapers such as the *Financial Times*, as well as tabloids like the *Mail* and the *Sun*, ran the news on their front pages. CNN and NBC, the BBC, ITN, and many other channels gave this Hindu miracle the most thorough coverage of any news item.

Of course, there were sceptics. Numerous scientists tried to explain the phenomenon as 'mere capillary action'. Others suggested that the porous nature of the statues could explain why one larger figure of Ganesh could absorb several litres of milk in a matter of hours. Many people argued that these demonstrations were simply the mass hysteria of many very gullible people around the world who desperately wanted to believe in a miracle. But some individuals whom one would have thought to be entirely rational, normally quite reliable witnesses seemed convinced that statues everywhere had indeed suddenly started drinking milk. One scientist, Dr Aparna Chattopadhyay, was moved to write to the *Hindustani Times* in New Delhi: 'I am a senior scientist of the Indian Agriculture Research Institute, New Delhi. I found my offerings of milk in a temple being mysteriously drunk by the Deities.' And it is reported that one of Malaysia's leading barristers (surely, after prostitution, the most sceptical profession in the world) was hardly able to summon up strength to drive his car after a small solid metal elephant attached to the dashboard of his vehicle consumed six teaspoons of milk shortly before he attempted to negotiate the morning traffic in Kuala Lumpur.

Jews generally are reluctant to accept the supernatural, and prefer rational or scientific explanations for miracles. But a group of Orthodox Jews in Israel have been trying for a number of years to breed cows in order to produce a pure red heifer. Even though they have enlisted the expert help of cattle-breeders from Nebraska, all their attempts have failed. They embarked on this project because they believe that the birth of a red heifer would enable them to re-enact the ancient rites of purification mentioned in the Bible. If such an animal were ritually sacrificed, with its ashes properly employed they would be able to walk across the Temple Mount in Jerusalem without themselves becoming ritually unclean. This, they believe, would be the first step in building a new Temple in Jerusalem on the site of the original Temple – now occupied by the Al-Aqsa Mosque, a very holy place for Muslims.

But these would-be cattle breeders were pipped at the post on 15 April 1997, when such a calf was born spontaneously in the small farming community of Kfar Menachem, in the north of Israel. This calf with red-coloured hair was born to ordinary-looking parents from a black-and-white herd. The calf was called Zaleel (Melody) and was immediately given special treatment, being separated from the other cattle so that she would not be accidentally kicked and thus rendered ritually unfit. Within weeks of the birth, Orthodox rabbis from all over the country had come to check this little heifer, to see if it complied with all the conditions laid down in the book of Numbers (chapter 19) and the Mishnah (rabbinical scriptures containing the Jewish Law). To qualify as a *parah adumah*, the perfect red heifer described in the Bible, it is necessary for such an animal to have no more than two hairs of the wrong colour on its body. The rest must be entirely red. Moreover, such an animal must have no defect, nor injuries of any kind, and never have been put under the yoke.

The complex rules about the red heifer are extremely puzzling. Rabbinical authorities worldwide have always found them difficult to explain. Whole books, most of them barely intelligible, have been written about them. In Temple times, the remains of such an animal, after it had been sacrificed, were seen as being very powerful. Once a red heifer was over three years old, it would be slaughtered and burned to a cinder, together with any of its dung. The resulting ashes, when mixed with a quantity of spring water and then sprinkled appropriately, had the remarkable property of making a defiled person clean – for example, after contact with a dead body. Equally inexplicably, if a clean person touched this 'holy water', it defiled them. According to ancient rabbinical tradition,[1] only seven (or at most nine – it depends a bit on who's counting) red heifers were ever born and sacrificed, and none seem to have materialized in the last two thousand years. So the arrival of this little calf in Kfar Menachem was more than just a red-letter day. Some saw it as a divine sign. Quite a number of strictly Orthodox Jews believed that such a rare birth in modern times must signal nothing less than the imminent arrival of the Messiah.

The news of this birth did not please all Jews, however. Worried at the thought that Jewish zealots trying to build anything in the Temple area would cause more than merely a riot, the broadsheet liberal newspaper *Ha'aretz* stated: 'The potential harm from this heifer is far greater than the destructive properties of a terrorist bomb.' In the interests of peaceful co-existence between Muslims and Jews, the paper recommended that Melody be shot immediately. As it happened, the need for such drastic action turned out to be unnecessary. Before she was three years old, Melody started to develop white hairs on her tail – more than two of them.

Shortly before this time my son, Joel, visited Kfar Menachem with some of his observant friends, curious to see this creature for themselves. They reported that it looked a very ordinary animal – just a small brown cow. So after all the fuss, this bovine was not divine.

These three examples of belief in the Divine Idea are all very different, but they have certain important aspects in common. First, there is obviously the belief, in each of these three very different communities, that something supernatural has occurred or will occur. Each of them, too, carries some ideas about sacrifice and offering. Throughout the centuries, Catholic imagery has been suffused with images of blood – the 'wounds of Christ' were used as a focus for prayer and meditation; crucifixes vividly depict the blood around Jesus' forehead from his crown of thorns, and trickling down his side from a wound inflicted by the spear of a Roman centurion. Throughout the Middle Ages in many parts of Europe, people flocked to pray in front of phials allegedly filled with the blood shed by Christ. To this day, in the Mass, Christians symbolically consume the blood of Christ. And central to the notion of Christianity in all its forms is the idea that the death of Jesus is a sacrifice on behalf of all of us, and that this sacrifice atones for the sins of all humankind.

The ritual of the red heifer also concerns a sacrifice. Many sacrifices in Judaism are associated with some form of atonement and purification. 'For the life of the flesh is in the blood; and I have given it for you to make atonement for your souls; for it is the blood that makes atonement, by reason of the life [which is in it]' (Leviticus 17: 11). The people were forbidden to eat the blood of any animal because life, of course, belonged to God only. The *hattat* (sin offering) was a regular propitiatory sacrifice in the Temple. Another form of atonement was via the *asham* (guilt offering), used after various civil violations and also after leprosy and certain other forms of ritual impurity. Yet another Temple offering, the *olah*, usually a bull, was made twice daily and burned (with extra lambs on the Sabbath), and these offerings too might also be used for the rituals involving purification. The sacrifice of the red heifer was different from most of these ritual offerings in that it was performed outside the camp. In this respect it is like the scapegoat mentioned in chapter 1, which once a year was driven out into the rocky wilderness with a red ribbon around its neck. Eventually it would fall down a precipice and break its neck – in atonement for the sins of the people.

Sacrifice is also an element in the Hindu miracle. The offerings that were given to Ganesh were of milk – a precious, life-giving liquid. It's also significant that the desired offering came from a cow. Across the world, religions single out certain types of animals, insects, plants and trees as being 'special' or 'sacred'. The cow is sacred to Hindus; indeed, in the events of 1995 described above, one of the milk-drinking gods was Nandi, whose effigy resembles a bull. Bulls featured prominently in many religions: in ancient Assyria the winged bull may represent the god En-lil; in Egypt, the cult of Apis regarded the bull as sacred; the bull featured prominently in ancient Greek mythology, and the cults of the Mycenaean civilization and Mithras are centred on this animal.

The examples above suggest that many religious beliefs have a great deal in common. Some people believe that this is because all religions are different ways of expressing a basic truth. But there's an alternative view. Do all religious ideas have common elements simply because all humans have the same brains, which evolved for the common purposes of surviving and passing on our genes?

Consider the following pairs of sentences for a moment:

(a) We worship this woman because she is the only one ever to conceive a child without having sex.
(b) We worship this woman because she gave birth to seventy-three children.

(a) We pray to this statue because it listens to our prayers and gets us what we want.
(b) We pray to this statue because it is the largest artefact ever made.[2]

If you have a brain roughly like mine – and you do – you will have noticed that the type 'a' sentences share something in common, and the type 'b' sentences share something in common. The type 'a' sentences seem typical of the beliefs one encounters in a religion. Christians revere the Virgin Mary. In south-east Asia, Buddhists place offerings on shrines dedicated to their ancestors. Without being an anthropologist, I feel fairly confident that the type 'b' sentences are not the kind of thing one encounters in religious belief. But why do I have this feeling?

Let's look at another pair of sentences for a moment:

John hits Mary in Shoreditch.
John brews Mary between Shoreditch.

Both of these sentences are made up of real English words. In both cases, the form of the sentence follows the basic scheme of Subject Verb Object, which is how English sentences are constructed. But the second one is nonsense – and that was apparent from the moment we saw it. Some inbuilt 'grammar detection' device in our brains, active from our earliest months, allows us to know what is meaningful and what is not. Maybe the same is true of the religious examples. Why do some concepts seem to make sense only as part of a religion, while others seem inappropriate in that context? Could it be because there is some inherited 'religious' unit in our brains? If so, how and why did it come about?

Nature, nurture and intrinsic religion

E. O. Wilson, Professor of Zoology at Harvard University, is one of the giants of modern genetics. His views on Darwin and his understanding of human behaviour brought him a considerable reputation in the 1970s. He has received many accolades for his work – the prestigious Craaford Prize in Sweden, and two Pulitzer Prizes and the National Medal for Science in the United States. But sociobiology, the discipline he established, comes so close to undermining religious belief and many moral convictions that his work has also sparked huge controversy. His central view, that most of our psychological traits may have been selected for during evolution, prompted various people to vilify him as a racist, a Nazi eugenicist and a misogynist.

Edward Wilson was born in the American Deep South and brought up in a strongly Baptist community. But by the time he was a teenager he had become a convinced atheist. Of religiousness, he wrote: 'The predisposition to religious belief is the most complex and powerful force in the human mind and in all possibility an ineradicable part of human nature.'[3] Of course, many others well before Professor Wilson have argued this; but it is only comparatively recently that psychologists have found ways of testing it.

Leading the fray into this highly controversial area has been Thomas Bouchard, Professor of Psychology at the University of Minnesota. Bouchard had a rare opportunity to study human twins who had been reared apart since birth. This opportunity presented itself because in the United States during the 1950s and 1960s it was considered best to separate at birth twins who were to be adopted. This led to a number of these children being brought up by families who did not even know that their adopted baby had a twin; and sadly, the children themselves were brought up in total ignorance of their 'lost' twin brothers and sisters. Tom Bouchard was able to trace many of these families and establish the fact that many of the twins were identical. Identical twins, of course, are formed in the uterus by the embryo splitting; so identical twins have exactly the same DNA. Non-identical twins – growing from two separate eggs fertilized by different sperm – do not have identical genes, but will just share many general aspects of their genetic inheritance, as do any other brothers or sisters in one family unit.

The existence of many twins, both identical and non-identical, all reared apart, provided Bouchard with an extraordinarily felicitous situation for scientific investigation. He recognized that these twins, if compared with each other as they grew up, should provide an important way of measuring genetic and environmental influences. His groundbreaking and painstaking work in what became known as MISTRA (Minnesota Study of Twins Reared Apart) and his novel systems of data analysis, produced and applied over the 1980s and 1990s, gave rise to some extraordinary insights into which aspects of the human condition are more likely to be due to nature, and which to nurture. And in doing this research, Boucher provided some hard evidence to support E. O. Wilson's view that religiosity was inherent in human nature.

In one famous study, Bouchard concentrated on seventy-two sets of twins who already had reached adulthood. He first established which of the twins (thirty-five sets in all) were genuinely identical by getting data on some of their specific individual proteins – and thus their genetic make-up – from the Minneapolis Blood Bank. The twins were then invited to complete personality questionnaires of a type whose use had been validated in other psychological studies. Such questionnaires, which are widely used by psychologists, are quite detailed and investigate issues including responses to family values, and various aspects of personality and psychology. Questions take the form of statements, to which the respondents have to rate their level of agreement on a scale of 1 to 8. The following is a small sample of the many statements relating to religion:

I enjoy reading about my religion.

My religion is important to me because it answers many questions about the meaning of life.

It is important to me to spend time in prayer and thought.

It doesn't matter to me what I believe as long as I am good.

I go to my (church, synagogue, temple) to spend time with my friends.

I pray mainly to gain relief and protection.

Although I am religious, I don't let it affect my daily life.

When Bouchard and his team compared the answers to these and other personality questions, they found strong statistical evidence that identical and non-identical twins tended to answer differently. If one identical twin showed evidence of religious thinking or behaviour, it was much more likely that his or her twin would answer similarly. Non-identical twins, as might be expected (they are, after all, related), showed some similarities of thinking, but not nearly to the same degree. Importantly, the degree of religiosity was not strongly related to the environment in which the twin was brought up. Even if one identical twin had been brought up in an atheist family and the other in a religious Catholic household, they would still tend to show the same kind of religious feelings, or lack of them.

Bouchard's research, which has generated considerable controversy, went beyond just measuring religiosity. His questionnaires supported the idea that there are two types of religious attitude. Gordon Allport, a Harvard psychologist who died in 1967, had done some key research on various kinds of human prejudice in the 1950s and had come up with a definition of religiosity still in use fifty years later. He suggested that there were two types of religious commitment, extrinsic and intrinsic. Extrinsic religiosity he defined as religious self-centredness. Such a person goes to church or synagogue as a means to an end

– for what they can get out of it. They might go to church to be seen, because it is the social norm in their society, conferring respectability or social advancement. Going to church (or synagogue) becomes a social convention. Some cynics would say that this sounds a bit like many of my Church of England friends in the public school where I was educated. Dr Allport thought that intrinsic religiosity was different. He identified a group of people who were intrinsically religious, seeing their religion as an end in itself. They tended to be more deeply committed; religion became the organizing principle of their lives, a central and personal experience. In support of his research, Dr Allport found that prejudice was more common in those individuals who scored highly for extrinsic religion.[4]

Bouchard has consistently found in many of his studies that intrinsic religiosity – which seems to incorporate a notion of spirituality – is much more likely to be inherited. Extrinsic religiosity tends to be a product of a person's environment and direct parental influence. Bouchard also found that tendencies towards fundamentalism were also rather more likely to be inherited. It is of some interest, too, that, in the populations that Bouchard and his colleagues have studied, women tend to have inherited rather more religious attitudes than men.

Work by several other scientists has inclined to confirm Bouchard's findings. One study conducted by an international team at the Institute of Psychiatry in London under Dr Hans Eysenck (himself no stranger to controversy) looked at information from twins living in the UK and Australia. This team gleaned data from identical and non-identical twins who had been brought up together. The authors looked at their attitudes to matters as diverse as whether they believed in the death penalty, observed the Sabbath, liked modern art, believed in moral training, were in favour of student rags, chastity, nudist camps, legalized abortion, mixed marriages and pyjama parties, and believed in the truth of the Bible. This study was published in one of the world's leading scientific journals under the lead authorship of Dr N. G. Martin of Virginia Medical College in 1986 (one wonders whether pyjama parties were still in vogue so recently . . .).[5] The researchers found that attitudes to Sabbath observance, divine law, church authority and the truth of the Bible showed greater congruity in identical rather than non-identical twins – again supporting the idea of a genetic influence.

Needless to say, these data and the views these scientists have expressed have caused huge argument from the moment they appeared, and many other scientists have disputed them. Research conducted by Jeremy Rose of the University of Texas at Austin found a strong relationship between religiosity and a child's environment. He certainly is not convinced that there is a major genetic component to human religious belief. He sums up: 'Behavior genetic research should attempt to more precisely delineate the components of moral and religious attitudes, beliefs, and behaviors and their sources of variation.'[6] If we want to know whether the Divine Idea is carried in the genes, we need to understand *why* this should be the case. What benefits could it offer us?

Religion – the costs and benefits

I recently sat through a fundamentalist church service in Florence, Kentucky. The pastor, Brad Bigney – admittedly an amazingly charismatic speaker – spoke for over an hour to his congregation, who seemed to be deeply committed Christians most of whom believed in the literal truth of every word of the Bible. Most of these people, for example, believe that the Grand Canyon was created by Noah's Flood, and that God put fossils of dinosaurs in place to fool gullible men into believing in evolution. The fact that they were card-carrying fundamentalists was pretty obvious – let any preacher, no matter how charismatic, speak for more than twelve minutes in my synagogue and see what happens.

At the end of the Reverend Bigney's dramatic sermon, he summed up. 'No-one in this Church need ever worry,' he declaimed. 'Worry is unnecessary. If any of you worry,' he went on, 'you are nothing but an atheist!' This notion that a belief in God solves all our problems is initially very satisfying – but, on reflection, it can also be terrifying. The certainty it implies seems to falsify the human condition completely, robbing us of free will and negating our capacity for choice. It has also led religious people to commit terrible acts. But presumably, to the intrinsically religious, such certainty must make many of life's difficult decisions much easier.

If religion provides this kind of comfort, then it's possible to argue that humans evolved a propensity for religion because it promotes mental and physical health. The evidence generally is that intrinsic religiosity seems to be associated with lower levels of anxiety and stress, freedom from guilt, better adjustment in society and less depression. Extrinsic religious feelings, on the other hand, where religion is used as a way to belong to and prosper within a group, seem to be associated with increased tendencies to guilt, worry and anxiety.

The Dolley Pond Church of God With Signs Following was founded in Tennessee in 1909 by one George Went Hensley. This former bootlegger took to the pulpit in a rural Pentecostalist community in Grasshopper Valley. One Sabbath, while he was preaching a fiery sermon, some of the congregation dumped a large box of rattlesnakes into the pulpit (history does not record whether they were angry or just bored). Without missing a beat, in mid-sentence, Hensley bent down, picked up a

3-foot-long specimen of this most venomous of snakes, and held it wriggling high above his head. Unharmed, he exhorted his congregation to follow suit, quoting the words of Christ: 'And these signs will follow those who believe . . . in my Name . . . they will take up serpents.'

News of Hensley's sermon spread through Grasshopper Valley; others joined him in handling snakes, and the practice caught on. There have since been around 120 deaths following snake-bite in these churches, but most of the congregants tend to refuse medical help if they are bitten, preferring to believe that divine intervention will be more efficacious. Sadly, Hensley himself perished from a snake-bite in 1955, and shortly afterwards the US government wisely acted to prevent the practice – although it is still legal in parts of the States. Today snake-handling continues mostly in small communities in rural areas of Tennessee and Kentucky, as well as pockets in other Southern states. Participants feel that 'the spirit of God' comes upon them as they open the boxes containing the snakes. Often lifting three or four of them up simultaneously in one hand, holding them high and allowing the creatures to wind around their arms and bodies, they praise God ecstatically.

In October 1998 the *Knoxville News-Sentinel* ran the following story:

One of the prominent leaders of snake-handling churches in the Southeast died October 3 after being bitten by a rattlesnake during a church service at the Rock House Holiness Church in rural northeastern Alabama. John Wayne 'Punkin' Brown, Jr., of Parrottsville, Tennessee, was preaching with his own 3-foot-long timber rattler in hand when the reptile sank one fang into his finger. Mr. Brown's wife, Melinda, had died three years earlier from a rattlesnake bite received at the Full Gospel Tabernacle in Jesus Name Church in Middlesboro, Kentucky. The Browns left behind five young children. The Browns had been bitten dozens of times before the fatal bites.

To many of us, religious or not, this type of activity seems little short of outright lunacy. And it's certainly the case that religion and mental ill-health have long been linked. The disturbed individual who believes himself to be Christ, or to receive messages from God, is something of a cliché in our society. Ever since Sigmund Freud, many people have associated religiosity with neurosis and mental illness.

Many years ago, a team of researchers at the Department of Anthropology at the University of Minnesota decided to put this association to the test. They studied certain fringe religious groups, such as fundamentalist Baptists, Pentecostalists and the snake-handlers of West Virginia, to see if they showed the particular type of psychopathology associated with mental illness. Members of mainstream Protestant churches from a similar social and financial background provided a good control group for comparison. Some of the wilder fundamentalists prayed with what can only be described as great and transcendental ecstasy, but there was no obvious sign of any particular psychopathology among most of the people studied. After further analysis, however, there appeared a tendency to what can only be described as mental instability in one particular group. The study was blinded, so that most of the research team involved with questionnaires did not have access to the final data. When they were asked which group they thought would show the most disturbed psychopathology, the whole team identified the snake-handlers. But when the data were revealed, the reverse was true: there was more mental illness among the conventional Protestant churchgoers – the 'extrinsically' religious – than among the fervently committed.

When I was a medical student I remember spending a great deal of time with a seventy-year-old woman, a widow called Joan. She had been transferred from a country hospital in Essex to a hospice for the terminally ill in east London. Joan was dying. She had excessive wasting and weakness because of an extensive cancer of the ovaries. The cancer had migrated to the bones of her back and her legs, and caused severe pain every day. There were also painful metastatic deposits in her liver. She found it difficult to eat because she felt very sick much of the time, and even when she wasn't nauseated she found it difficult to swallow. I remember one particular evening spent with her when she made a deep impression on me. She typified, I think, what is meant by someone who is intrinsically religious.

'I've had a wonderful life,' she told me; 'although I've never been a regular at church, I've always believed in God and feel now that my spiritual beliefs are increasing. But I've been so fortunate – I've always been happy. Now I pray every morning and evening, and I am very much at peace. Of course, I don't relish dying,' she said, 'but I have very little fear of it.

'What a great place this hospice is – it is even decorated with my favourite coloured wallpaper. And the people here have been incredible. Yesterday, they took me outside [this was the East End of London – hardly a salubrious area] and I could see the trees and the little green park. Now I know why it's called Bethnal Green. Wonderful to see God's work growing afresh this spring.'

Joan seemed to have few regrets about her life – 'I just take things day by day, and am pretty content with everything that I have. And I've been very lucky – the pain – even at night – has been quite bearable, especially with the injections of morphine. I do wish I felt more like eating; I know I shall die soon, but I feel ready to leave – I feel very tired.' Three days later – it was my weekend off and I did not see her – she died in her sleep.

So it is possible that strong levels of belief in God, gods, spirits or the supernatural might have given our ancestors considerable comforts and advantages. Many anthropologists and social theorists do indeed take the view that religion emerged out of a sense of uncertainty and bewilderment – explaining misfortune or illness, for example, as the consequences of an angry God, or reassuring us that we live on after death. Rituals would have given us a comforting, albeit an illusory, sense that we can control what is in fact ultimately beyond our control – the weather, illness, attacks by predators or other human groups.

However, it is equally plausible that the Divine Idea would have been of little use in our prehistoric rough-and-tumble existence. Life on the savannah may have been in the open air, but it was no picnic. Early humans would have been constantly on the look-out for predators to be avoided, like wolves and sabre-tooth tigers; hunting or scavenging would be a continual necessity to ensure sufficient food; and the men were probably constantly fighting among each other to ensure that they could have sex with the best-looking girl (or boy) or choose the most tender piece of meat from the carcass. Why would it be necessary, in the daily scramble to stay alive, to make time for such an indulgent pursuit as religion?

Richard Dawkins, our best-known Darwinist and a ferocious critic of organized religion, notes that religion seems to be, on the face of it, a cost rather than a benefit:

Religious behaviour in bipedal apes occupies large quantities of time. It devours huge resources. A medieval cathedral consumed hundreds of man-centuries in its building. Sacred music and devotional paintings largely monopolized medieval and Renaissance talent. Thousands, perhaps millions, of people have died, often accepting torture first, for loyalty to one religion against a scarcely distinguishable alternative. Devout people have died for their gods, killed for them, fasted for them, endured whipping, undertaken a lifetime of celibacy, and sworn themselves to asocial silence for the sake of religion.[7]

It seems at first glance as if Dawkins is arguing that religion is an evolutionary disaster area. Religious belief, it seems, would be unlikely, on its own merits, to have slipped through the net of natural selection. But maybe that interpretation of what Richard Dawkins is saying neglects some of the further benefits that religion might well offer in the human quest for survival and security.

I was at St Peter's in Rome for the Easter celebrations just days before the death of Pope John Paul II in spring 2005. Many readers will recall the mass pilgrimages that converged on the city during the following weeks as his body lay in state, uniting up to a million visitors from different ages, backgrounds and cultures in an outpouring of affection and grief. Considering that he was such a deeply conservative leader, it was surprising to see how many young people came from far afield to pay their respects. As this dying man, from a bedroom window in the Vatican, struggled to make the sign of the cross that Easter to bless the crowds below, there was a round of gentle applause from the watching crowds. Inevitably, most people there at such a moment felt moved to applaud – myself included. Ostensibly an outsider (and one with, you might argue, long-established reasons to be indifferent to a pope), I felt part of a common spirit that day. Perhaps, in essence, that is what religion does: it joins people together. In his book *Darwin's Cathedral* David Sloan Wilson, Professor of Biology and Anthropology at Binghamton University in New York State, says that religiosity emerged as a 'useful' genetic trait because it had the effect of making social groups more unified.

This is not, in itself, a new idea. Although they study human cultures, not genes or psychology, anthropologists have been proposing that religion acts as a form of 'social glue' for over a hundred years. Born in France in 1858, the sociologist Émile Durkheim was the son of a rabbi. Although he experienced a brief bout of mysticism under the influence of a Catholic teacher, for most of his life Durkheim was a devout atheist. He nevertheless was quick to see what benefits the Divine Idea gave to human societies. He defined religion in an enviable soundbite: 'a unified set of beliefs and practices relative to sacred things, that is to say, set apart and forbidden, – beliefs and practices which unite one single moral community – all those who adhere to them'.[8]

Durkheim was not the first to see religion as a powerful collective force. The ancient Chinese Book of Rites, from the third century BCE, noted that 'Ceremonies are the bond that holds the multitudes together, and if the bond be removed, those multitudes fall into confusion,' demonstrating that the Chinese aristocracy of the time were well aware of the social value of rituals. A century or so earlier the Greek philosopher Aristotle, pointing out that 'men create the gods after their own image', also indicated a vital link between human groups and the kinds of Divine Idea they develop.

Durkheim was seen as quite a radical for arguing that the complex religion of the Australian Aborigines served an important social purpose. Throughout the nineteenth century, missionaries reported that these original inhabitants of the land had no religion at all. Others saw it as inherently 'primitive' because it was based on the worship of plants and animals. Durkheim argued that Aboriginal society was given its structure and its unity by a belief that different clans were associated with different animals and spirits, which they worshipped (so-called 'totem worship') in exchange for success.

Many have followed Durkheim's ideas, stressing the power that religion exercises in groups. Dietary laws, for example, exist to identify the group and keep them separate from others. We see this in the laws of permitted and forbidden food (*kashrut*) in Judaism, or in the custom of Christians eating fish on Fridays. On the face of it the specifics of *kashrut* seem totally irrational. They certainly are not instituted for health reasons, as has been claimed by some observers. But it is very difficult for a Jew to keep *kashrut* strictly when eating with non-Jews, whether in non-Jewish homes or in restaurants: environments where, of course, much social activity takes place. Dress can also act as a unifying force: the frock coats and fur-trimmed hats of Hasidic Jews stress their unity with each other, the nature of the group, and their non-participation in the society around them.

Within a religion, different kinds of worship express the different values and goals of different sections of society. In the Pentecostalist churches favoured by London's large Afro-Caribbean community, rousing music and songs are used to set a mood of enthusiasm. Worshippers frequently and spontaneously exclaim out loud in agreement with the minister. Yet in Westminster Cathedral worshippers speak out loud only in unison and at pre-arranged times, uttering a fixed form of words at established points in the Catholic Mass. Try entering the cathedral during a service, or even when no service is being held, and shouting 'Glory be to God! Halleluia!' That is not to suggest that Afro-Caribbean society is without strict notions of authority, respect, right and wrong – many of which it shares with people who worship at Westminster. But the attitudes expressed in worship are a way of demonstrating and celebrating difference. One of the hallmarks of multi-culturalism is that people have a right to be proud of their own identity and to assert it. This tendency of religion to express differences, whether from other religions or from other groups within the same religion, reminds me of the old joke about the religious Jew who, in the middle of a long ocean journey, encounters a storm and is ship-wrecked on a desert island. On reaching his deserted, uninhabited island, one of his first actions is to build not one synagogue but two – one of them being the synagogue he refuses to worship in.

Religious practices are often the means through which human communities root their identity in something superior to humanity, giving it a divine seal of approval. In many aspects of human society, the basic idea *God/the ancestors has/have commanded us to do this* provides a kind of imprimatur that gives authority to our actions, justifies them and makes them meaningful. It's not unlike the 'By Royal Appointment' legend which companies display on their products to entice customers – the notion being that this can of beans or that shoe polish is superior because it is used in the Queen's household. In both cases, something mundane is lent an extra significance by its association with something superior to, above and beyond the mundane world. Beards are mundane – but when God commands you to wear a beard, it has become a sacred badge. Baked beans are mundane – but if the titular head of our society eats them, then they are linked to history, to the very idea of kingship itself. In terms of religion, this idea can be used for varying ends: from setting up soup kitchens and defying oppressive regimes to torture and mass execution.

Religious beliefs – our very ideas about God – can also be used to galvanize community spirit in times of stress. The Pharisees, the Jewish sect which became prominent after the Romans destroyed the Jerusalem Temple in 70 CE, sought satisfaction through daily life in a period of exile and ruin. They saw God not as a transcendent being beyond the world who could be approached only through the rituals of the now-ruined Temple, but as a presence which suffused everything, like the writing running through a stick of rock. He could be experienced, and worshipped, whenever Jews studied the Torah (scriptures) together, whenever they ate a meal, whenever they showed charity to their neighbours. One of their favourite synonyms for God was Shechinah, 'the Divine Presence', a word with the same root in Hebrew as *shachan*, meaning to dwell, abide or pitch one's tent. Some rabbis even argued that the destruction of the Temple by the Romans had been a creative act, freeing the Shechinah from the Temple and allowing it to inhabit the rest of the world. They spoke of it moving from one synagogue to another – so that, wherever in the world Jews formed a *minyan*, the quorum of ten adult males required to hold a prayer meeting, the Shechinah would arrive to look over them. Thus the Jews of the post-Temple period used a new version of the Divine Idea to retain their community spirit.

The power of ritual

All religions involve ritual. Rituals also serve to bind communities together, for the simple reason that acting them out requires people to work together. An English wedding is a good example, even though nowadays it can be conducted without any mention of God, in hotels, register offices and even while bungee-jumping. Organizing a traditional wedding needs a great deal of co-operation – and this is often carried out by the parents of the bride and the groom. Sorting out the practicalities of the flowers, the dresses, the venue, the menu and so on serves in a real way to unite two families – although it can sometimes do the opposite, and drive them permanently apart.

The American anthropologist Victor Turner, who spent two and a half years living with the Ndembu people of Zaire, stressed the importance of ritual in forming a co-operative social group. Ndembu society was riven by inbuilt tensions and instability. For example, Ndembu people trace their descent matrilineally, that is, through their mothers, yet marriage is virilocal – the couple go to live with the husband's parents. Rights to property and wealth are therefore acquired through one's mother, but residence is determined by the men. Hunting and the production of cassava root, male and female occupations respectively, are both equally important means of subsistence in Ndembu society, but they are seen as competing and opposed. Hunting is ritualized and seen as sacred, whereas cassava production is seen as mundane drudgery.[9] Divorce and residential mobility are common.

In Turner's view, religion and ritual served both to express and to resolve these tensions. He devoted considerable energy to analysing the *chihamba* ritual performed to cure one particular woman's leprosy. The pretext may well have been to cure the afflicted patient, whose name was Nyamukola. But the complex rituals also served to settle intra-group conflicts. The sheer organization of a ceremony to be attended by 400 people required various factions within the village to co-operate closely and put their differences aside. The rituals reduced hostility towards Nyamukola herself, who had been accused of witchcraft in the past. They also lent prestige to the host village, strengthening ties with neighbouring communities. Moreover, the *chihamba* ritual provided the dramatic setting within which the core values of Ndembu society were enacted.

Turner, who came from a theatrical family, was fascinated by the dramatic power of ritual. He noticed its tendency, in all societies, to assault the senses with its vivid images, colours, smells, sounds and experiences. Such ritual drama has two effects. It demonstrates quite clearly that we are not in real life, ordinary life, but in a space and time that are sacred. Second, it makes the participants more receptive to absorbing certain truths about their society, its values and its origins. This is not a new idea. Aristotle noted the power of the performance of a Greek tragedy to induce a change in the people in the audience. Centuries later, the Marxist playwright Bertolt

Brecht employed what he called his *Verfremdungseffekt* (alienation effect) to suspend the normal theatrical conventions of illusion, continually reminding the audience that they were watching 'only' a play and that the actors were deliberately not trying to mimic reality. Brecht argued that this deliberate bewilderment of the audience would make them more receptive to the ideology behind the play – in his case, communism. Turner felt that ritual employed the same deliberate sensory over-stimulation and bewilderment to make participants absorb the values of their society.[10]

In 1976 *The Bar Mitzvah Boy*, a hilarious, BAFTA-winning play by the late playwright Jack Rosenthal, detailed the agonies of Eliot Green, a twelve-year-old boy, facing his first important ritual within the Jewish community. His loving, protective family make elaborate preparation for Eliot's coming of age, it all gets too much, and Eliot does a runner. The bar mitzvah (which means 'son of the commandments') is what anthropologists call a rite of passage. Like a wedding, it marks the recognition by the community that a person has changed, and acquired a new status. In Judaism, once a boy is bar mitzvah, he is regarded as a man and carries a new, adult set of religious responsibilities.

To prepare for his bar mitzvah, even a boy who is not particularly knowledgeable is often required to learn quite a complex passage from the Pentateuch and usually some verses from the Prophets as well. These he recites in a full synagogue, usually during the main morning service, using a memorized musical notation which is around two thousand years old. Thus, in addition to the ordeal of being repeatedly kissed and told he is handsome by hordes of exuberant relatives, and the inevitable, sometimes cringe-making speech at his party, the main task involves – for some boys at least – a kind of willing suffering, a public feat of knowledge, memory and recital (at an age of sometimes painful shyness) which demonstrates his readiness to enter the community. In many Jewish communities, boys of similar age study together in preparation for this ceremony, a communal activity that itself has some social purpose. By the preparatory work as well as the event itself, a young man's identity as a Jew and a member of the congregation is thoroughly established.

The bar mitzvah is not, of course, especially painful; on the contrary, it is a joyous occasion, the protagonist usually enduring nothing worse than embarrassment. But in many other societies rites of passage involve much more obvious suffering on the part of the initiates: they may be starved for long periods of time; they may be subjected to extensive tattoos or scarring; they may be taken out into the jungle for a fairly perilous exercise in hunting and on their return forced to undergo painful circumcision without showing any sign of suffering. Just as I maintain links with people with whom I 'suffered' during my rites of passage as a young doctor, these ceremonies of endurance can help to forge close bonds between young men and women of the same age. And these bonds endure throughout life – ensuring extra levels of cohesion in society.

Searching for the 'God gene'

Religion, then, apparently has power to unite communities, and humans have a long-standing history of living in groups. The communal nature of religion certainly would have given groups of hunter–gatherers a stronger sense of togetherness. This produced a leaner, meaner survival machine, a group that was more likely to be able to defend a waterhole, or kill more antelope, or capture their opponents' daughters. The better the religion was at producing an organized and disciplined group, the more effective they would have been at staying alive, and hence at passing their genes on to the next generation. This is what we mean by 'natural selection': adaptations which help survival and reproduction get passed down through the genes. Taking into account the additional suggestion, from various studies of twins, that we may have an inherited disposition towards religious belief, is there any evidence that the Divine Idea might be carried in our genes?

While nobody has identified any gene for religion, there are certainly some candidate genes that may influence human personality and confer a tendency to religious feelings. Some of the genes likely to be involved are those which control levels of different chemicals called neurotransmitters in the brain. Dopamine is one neurotransmitter which we know plays a powerful role in our feelings of well-being; it may also be involved in the sense of peace that humans feel during some spiritual experiences. One particular gene involved in dopamine action – incidentally, by no means the only one that has been studied in this way – is the dopamine D4 receptor gene (DRD4). In some people, because of slight changes in spelling of the DNA sequences (a so-called polymorphism) making up this gene, the gene may be more biologically active, and this could be partly responsible for a religious bent.

Dean Hamer, an eminent molecular biologist at the US National Institutes of Health in Bethesda, argues that genes controlling some neurotransmitters are very likely to be involved in religious feeling. In his own studies, he first identified a group of people who are capable of 'self-transcendence'. These people had certain personality traits in common. One was the attribute of 'self-forgetfulness', the ability to lose oneself completely in whatever one is doing – weeding the garden, meditating, reading. Such people tend to focus less on themselves and more on things outside. Another trait was so-called 'transpersonal identification', which Hamer describes as a feeling of unity with things in the universe. A third trait is a tendency to mysticism or spiritual acceptance. Such people sometimes feel that mystical experiences have changed their lives; or they believe there is something beyond the universe that science cannot explain. Sometimes they believe in mysterious human attributes, such as extra-sensory perception.

Dr Hamer took DNA samples from several hundred individuals who showed these traits in a questionnaire-based study. He looked in particular at one particular area on chromosome 10, in the region of the gene VMAT2, Vesicular Monoamine Transporter 2, which codes for one of the proteins responsible for regulating the monoamine neurotransmitters serotonin, norepinephrine and dopamine. These neurotransmitters are very much involved in how we feel – at peace, satisfied or rewarded, content, anxious and so on. It turns out that there are at least two versions of this gene with very slightly different DNA spellings. According to Hamer, one in particular seems linked to self-transcendence. Hamer, who is an excellent communicator as well as a good scientist, rather boldly published some of his findings in a popular book rather than in the peer-reviewed scientific press. Needless to say, the appearance of this book, entitled *The God Gene*, was greeted with scorn by many scientists, many journalists and some religious people.[11] Barbara J. King, who was writing her own book on religion at the time and might be thought to have a slight conflict of interest, published what was, in my opinion, a particularly vicious article: 'Oh for space enough and time to debunk Hamer's assumptions! ... If there were a literary analog to the emperor with no clothes, surely it would be *The God Gene*. Years ago, Hamer announced a gene for homosexuality. That "discovery" has been not just challenged but repudiated. Maybe Hamer needs to find a gene for recognizing fiction masquerading as science.'

But it is easy to suggest a mechanism by which religious beliefs could help us to pass on our genes. Greater cohesion and stricter moral codes would tend to produce more co-operation, and more co-operation means that hunting and gathering are likely to bring in more food. In turn, full bellies mean greater strength and alertness, greater immunity against infection, and offspring who develop and become independent more swiftly. Members of the group would also be more likely to take care of each other, especially those who are sick or injured. Therefore – in the long run – a shared religion appears to be evolutionarily advantageous, and natural selection might favour those groups with stronger religious beliefs. It's been suggested that early Christianity, despite starting off as a tiny sect within Roman society, was successful partly because it had an inbuilt enthusiasm for making babies. The birth rate of Christians was higher than that of the pagans around them, so the religion tended to spread.

But Christianity gets passed on through the culture people grow up in, not through their DNA. And we do not necessarily *need* religion to live a successful communal existence. Jared Diamond, author of the Pulitzer Prize-winning *Guns, Germs and Steel*, points out that the existence of religion does not mean that it is the basis for every form of morality – or for all kinds of social cohesion.[12] He takes the example of the Sikari people of New Guinea, who have an elaborate collection of religious myths and creation stories, but whose morality does not depend on religion at all. When Sikari tribesmen talk about their genocidal attacks on neighbouring villages, they do not invoke gods or animal spirits, or any other divine or supernatural being. Nor is their behaviour within the group determined by moral precepts derived from religion. Instead, their moral code is entirely secular and depends on the family relationships among tribal members and their duties and responsibilities.

Similarly, although religion might be useful in developing a solid moral framework – and enforcing it – we can quite easily develop moral intuitions without relying on religion. Psychologist Eliot Turiel observed that even three- and four-year-olds could distinguish between moral rules (for example, not hitting someone) and conventional rules (such as not talking when the teacher is talking). Furthermore, they could understand that a moral breach, like hitting someone, was wrong whether you had been told not to do it or not, whereas a conventional breach, like talking in class, was wrong only if it had been expressly forbidden. They were also clearly able to distinguish between prudential rules (such as not leaving your notebook next to the fireplace) and moral rules.[13]

This would suggest that there is a sort of 'morality module' in the brain that is activated at an early age. Evidence from neuroscience would back this up, to a degree. In my last book, *The Human Mind*, I noted that certain brain areas become activated when we engage in co-operation with others, and that these areas are associated with feelings of pleasure and reward. It also seems that certain areas of the brain are brought into action in situations where we feel empathy and forgiveness.[14]

So religion does not seem to be produced by a specific part of our psychological make-up. Is it more likely, then, that religious ideas are something of an accidental by-product created by other parts of our basic blueprint, by processes deep in the unconscious mind that evolved to help us survive?

Pushing primordial buttons

> Mary's strength to resist was ebbing away; she was like a tiny meteor drawn into the orbit of some great planet ... Marcus was brushing her breasts with his fingertips, all the time shushing and stroking her like a groom reassuring a frightened foal ... Waves of guilt coincided with her orgasm; an extraordinary release of emotion washing over her like breakers across a tide barrier.[15]

The extract above was shortlisted for the Bad Sex Prize, an award presented yearly since 1993 by the *Literary Review*. You might think that authors would be annoyed to receive an award for bad writing, but it seems the opposite is the case. The winner of the prize that year, Aniruddha Bahal,[16] even flew in from Delhi to receive his award. It seems there is a good-natured acceptance, even by literary egos, that it's extremely difficult to write about sexual experience – and that our attempts to do so can often result in scenes of coruscating embarrassment.

I don't believe that's solely because we in the west suffer from a degree of prudishness concerning sex. It has more to do with the fact that sexual arousal is partly unconscious, and correspondingly impossible to render in words. Across the world, religious mystics, describing their moments of union with God, have also found it hard to explain themselves in terms that ordinary folk can understand. Often they are driven to use quite startling metaphors. St Teresa of Avila, a Spanish nun who died in 1582, described one of her mystical encounters thus:

> In his hands I saw a long golden spear and at the end of the iron tip I seemed to see a point of fire. With this he seemed to pierce my heart several times so that it penetrated to my entrails. When he drew it out, I thought he was drawing them out with it and he left me completely afire with a great love for God. The pain was so sharp that it made me utter several moans; and so excessive was

> the sweetness caused me by this intense pain that one can never
> wish to lose it, nor will one's soul be content with anything less …
> So sweet are the colloquies of love which pass between the soul
> and God that if anyone thinks I am lying I beseech God, in His good-
> ness, to give him the same experience.[17]

You might have noticed that the imagery employed by St Teresa has strong sexual resonances: she is pierced by a fiery spear, whose withdrawal leaves her on fire with love; she talks of moans, intensity, sweetness.

Dr Andrew Newberg, Director of Clinical Nuclear Medicine at the Hospital of the University of Pennsylvania, hit on the idea of measuring blood in the brains of meditating Tibetan Buddhists. Using a technique called SPECT (single photon emission computed tomography), he has shown that certain parts of the brain are activated during spiritual experiences.[18] SPECT is particularly convenient for such studies because, unlike the more conventional method of brain scanning called magnetic resonance imaging (MRI), it isn't noisy and therefore does not intrude so much during meditation. He believes that mystical ideas 'co-opt' certain pathways in the brain that were developed during evolution for other purposes.

Newberg suggests that this could come about because of the way evolution works. It's a blind process of very gradual changes, changes which may have started out for one reason, but end up producing quite different effects. For example, a bird cannot fly without a full set of wings, but those wings did not appear overnight; rather, they developed over thousands of generations during which those emerging nubs of tissue would not have aided flight at all. It's likely that, a very long time ago, birds began to develop wing-like appendages that performed some other useful purpose – such as cooling the body, or scaring off predators. If these worked well, they aided survival and so passed into the genetic blueprint for birds, evolving into larger structures that eventually enabled the first flying bird to take to the air.

The same might be true for the areas of the brain involved in religion. Newberg notes that the language of mystics in all cultures is very similar to the language of love and sex. Visionaries talk of bliss, exaltation, rapture, ecstasy, of a loss of self in the other: all terms we use when talking of love and sex. More than that, Newberg thinks that mystic experiences use the very same brain systems as sex. We might also note that religions throughout the world are rife with sexual and bodily symbolism – the *yoni* and the *lingam*, the vagina and phallus, are common emblems in Hindu worship. Even religions which seem rather aloof towards human bodily functions, like Christianity, employ a vivid and very basic sensory symbolism in their rituals, using taste, touch and smell to arouse participants and put them in a certain, ultimately receptive frame of mind. Orgasm itself is achieved through rhythmic movements and is experienced as a series of strong pulses through the body – intense arousal giving way to a release of tension and sensations of peace and calm. In religious worship, rhythmic chanting, repetitive music and formal, syncopated movements are used to induce similar states. Personally I have found that, even in periods of my life when religion has had little meaning for me, I have still felt strong sensations of both arousal and peace at hearing the call of the muezzin in some eastern city or while listening to the haunting Gregorian chant. I would baulk at saying there was any-thing sexual about these feelings, but, like sexual stimuli, they elicited a very bodily response, suggesting that some of my primordial buttons were being pushed.

Newberg argues that mystical experiences might be accidental by-products of the sex mechanism. To say that is not to devalue them. Our modern-day appreciation of art, for example, doubtless evolved from more ancient structures relating to vision in the brain that would have served our more mundane survival needs. Nor is it to rule out the possibility that there may indeed be some ultimate reality beyond our daily experience. The bird evolved the capacity to fly, and the fact of its being able to fly is undisputed. Newberg suggests that perhaps we humans, accidentally or otherwise, slowly evolved an ability to perceive a higher truth.

But not everyone agrees that religion would have aided our development, even if it did evolve naturally from the way we behaved. In his book *In Gods We Trust*, Scott Atran argues that religion was an evolutionary blooper. 'From an evolutionary standpoint, the reasons religion shouldn't exist are patent: religion is materially expensive and unrelentingly counterfactual and even counter-intuitive. Religious practice is costly in terms of material sacrifice (at least one's prayer time), emotional expenditure (inciting fears and hopes), and cognitive effort (maintaining both factual and counterintuitive networks of beliefs).'[19]

The effort involved in maintaining beliefs that are 'wrong' – and Atran is speaking from the point of view of a confirmed atheist – places a heavy burden on our daily life. When Cain and Abel made their burnt offerings, they were destroying perfectly good food. When Aztecs were slaughtering thousands of victims in order to nourish the sun god, they were killing people who could have played a useful role in society. Clearly, from a pure time-and-motion perspective, religion is somewhat inefficient (never mind the plight of the poor suckers who were sacrificed).

Atran suggests that we do have an inbuilt potential for religiosity, but that it came about as an unintended and unwelcome consequence of other psychological adaptations. An example of this would be the kind of mechanisms that evolved to allow us to transmit information – such as what kinds of foods are poisonous – between individuals and down the generations. Facial expressions and language would fall into this category. But elaborate body and verbal language that is effective in passing along useful information is just as useful for passing along totem stories or creation myths.

Another example Atran uses is our ability to detect the presence of prey by taking in auditory or visual clues: for example, we hear a rustle in the grass and conclude that a sabre-tooth tiger might be stalking us. This is transformed into the ability to believe that we are seeing or hearing evidence of the supernatural world; the rustling in the grass is taken as evidence of an invisible spirit. Again, an adaptation that emerges for some practical and useful purpose is reworked and used for something entirely different, something that evolution did not select for in the first place, and something inherently un-useful.

Pascal Boyer, Professor of Anthropology at Washington University, also feels that religious ideas seem to 'squat' upon systems that developed for other purposes. He bolsters his argument by pointing out that none of the other explanations for religion really hold water.[20]

For example, we considered above the idea that religion provides people with certainty and comfort – which promotes psychological and even physical well-being. But this 'religion-as-comfort' argument works only when we are considering a benign and just God or gods who reward us if we do what we are told. Many societies, however, believe in a world dominated by malicious spirits and demons, who can visit harm upon humans; though, of course, the 'existence' of such demons might indirectly promote well-being, because a belief that we can propitiate them by offerings and sacrifices or by prayer might confer on believers a feeling that they have some control over the supernatural world. Even within

Christianity, the 'fire and brimstone' preaching found in the pulpits of the Free Church of Scotland, with its continual references to damnation, or the guilt felt by many Catholics, hardly provide evidence of a comforting belief system. In contrast, the benign character of New Age religion teaches us that we can achieve great things if we tap into some power both inside and outside ourselves. Yet this is a wholly modern, western phenomenon – attractive in a milieu where we are not regularly assailed by famine or life-threatening disease. If religion provided comfort, we might expect to find comforting religions where comfort was most needed, such as in the Third World, or in Europe's medieval past. And we don't.

So what about 'religion as social glue' – as the force providing unity and morality in all human groups? One of the key problems with this is that human history is characterized by division, disloyalty, squabbling and faction-fighting – in other words, it seems that society is not a harmonious unit, so religion doesn't appear to be doing its job very well. On the other hand, primate and hominid groups on the savannah would have comprised perhaps 50 to 150 individuals, and it would have been much more likely for a binding culture to develop within such a small group. Rituals might, as Victor Turner argued, provide a means for expressing and resolving conflict within a group; but what is it about the brain that made us develop rituals (and a belief in the supernatural) to do it, and not something else?

The harshest critics of religion dismiss it as a mass illusion. They note that some people are prepared to believe in anything – from water turning into wine to Elvis's face being visible on the moon. They also note that most religious ideas are irrefutable: we can neither prove nor disprove the existence of God or ghosts. Furthermore, in most human societies, refuting an established idea is much more costly than going along with it. If everyone around you believes that the spirits of the ancestors are present everywhere, then it takes far more effort, and involves more risk, to reject than to accept the idea.

But these observations are unsatisfying. For a start, people do not believe just anything. Remember the examples I used earlier in this chapter: some propositions seemed inherently more 'religious' to us than others. That is because there is a limited storehouse of supernatural ideas from which humans construct their religious belief. As Boyer observes, 'the idea that there are invisible souls of dead people lurking around is a very common idea; the notion that people's organs change position during the night is very rare'. It is true that religions involve the acceptance of some extraordinary claims – the son of a virgin impregnated by a ghost, water turned into wine, people springing to life after they have died – but there are still some rigid criteria in the brain, selecting certain propositions and rejecting others. Boyer believes these criteria are dictated by our evolutionary inheritance.

An interesting experiment by psychologist Tom Ward indicated this when he asked people to draw and describe imaginary animals. The results were pretty wacky and wonderful, but Ward was struck by how most of the drawings stuck to certain principles about the way animals look. For instance, these creatures were symmetrical – if they had ten legs, then they nevertheless had five legs on each side.[21] Even when we let our wildest imaginations run free, they still run according to a basic set of rules.

But what does this have to do with religion? Quite a lot, because it seems ideas about the supernatural follow a similarly rigid pattern. There is a basic category of reality, and then, added to it, something we do not expect, a violation of that reality. Let us take, for example, Mary the mother of Jesus. 'Women who give birth to sons' is a perfectly real and natural category. But added to that is the belief that Mary was a virgin. This contradicts reality and 'turns the idea supernatural'. Myths and religions are replete with this basic pattern – people who can turn into animals, for example, or people who can fly, or people who are raised from the dead. Psychologists Frank Keil and Michael Kelly waded through a huge quantity of myths and folklore to work out the 'rules of the supernatural'. They found that there was a common theme of people turning into animals and vice versa, and that there were some strict principles behind this idea of species-change. For example, people turn into animals more often than they turn into plants, or birds, or insects. People and animals are only very rarely turned into inanimate objects. We accept that the ugly frog turns into a handsome prince – but it seems we would be less happy to accept the heroine kissing an oily carburettor! It's not true to assume that 'anything goes' where the supernatural is concerned: our brains seem able to imagine only those things that conform to certain rules.[22]

Because there are rules governing what we can imagine, Boyer argues, there is a limited catalogue of supernatural ideas, formed from a basic mixing up of the various categories perceived by the human mind. So, for example, people can be represented as having 'counter-intuitive' biology, as in the case of virgin births or ghosts who can walk through walls. Material objects can be represented as having human properties, such as volcanoes that demand to be fed, or statues that bleed or cry.

It seems that ideas like these are more striking and better remembered than ideas that simply replicate reality, or are just 'odd'. Boyer tested people's recall of three types of ideas – normal ideas, 'far-out' ideas, and ideas based on a mixing-up of categories: for example, 'a table with four legs', 'a table made of chocolate' and 'a table that felt sad when people left the room'. The last – based on mixing up an inanimate object with a human emotion – was the best recalled. His theory is that these ideas are supremely 'attention-grabbing', due to the structure of the human mind, and therefore more likely to endure from one generation to the next.

The Kwaio people of the Solomon Islands believe that, after death, people turn into spirits. These spirits, the *adalo*, have the power to visit good and bad fortunes on living humans – a very common idea in supernatural beliefs, as is the idea that the spirits' actions can be influenced by the generous donation of sacrifices. What is significant is that the *adalo* are thought of in very anthropomorphic terms: as possessing minds, as being able to communicate with mortals, as needing food. As human-like beings, with extra powers added on, the *adalo* are highly complex. Their complexity allows the Kwaio people to use them in a range of circumstances. For example, if a hungry *adalo* is found to be the cause of a child's illness, and the child does not recover, the Kwaio may assume that the *adalo* was not offered enough food, or something was wrong in the way the sacrifice was offered. So the belief remains intact, and the Kwaio still have a means of explaining the illness and a plan of action. A spirit that was less human – in that it did not require food, or did not care how much food was offered or how – would not provide a satisfying answer in this case.

If anyone is thinking this seems very removed from their kind of religion, then they should consider the work of Justin Barrett, whose research might be termed experimental theology. Barrett asked Christians to imagine various situations in which they might have to pray to God to save others from danger – for example, if a ship was sinking on the high seas, the subjects were offered the choice of deciding whether God would help the ship to stay afloat with a broken hull, give the passengers the physical strength to stay afloat in freezing waters, or influence the mind of a captain of a nearby ship to come and rescue them. The subjects overwhelmingly plumped for the last option. In other words, they had a counter-intuitive notion of God as being all-powerful – but they also considered Him more likely to change someone's mind – a very human strategy – than to intervene in physics or biology.[23]

This is somewhat reminiscent of the tale of the observant Jew, Hymie, whose village in a valley was flooded when a dam some few miles away burst in a terrible rainstorm. Everybody in the village was told by the police to evacuate their houses immediately and leave for the higher ground, from where transport out of danger would be provided. Hymie, though, decided he would stay put because, he told the police, 'God will save me'. Soon he was forced up to the first floor because of the rising water. A passing dinghy with some villagers in it sailed past his bedroom window. In spite of the villagers' desperate pleas, he called out: 'Don't worry – God will save me!' and announced his intention to stay where he was until the water receded. But the water continued to rise, and so Hymie climbed out onto his balcony, and finally onto the roof of his house. A man on a passing raft tried to persuade him to jump across. 'No need,' he shouted, 'God will save me!' Eventually, the water rose so high that the roof was covered; Hymie was swept off, and he drowned. On arrival at the pearly gates, he demanded to see God. When God granted him an audience, Hymie was full of bitter complaints. 'Oh God,' he said, 'I trusted you, I put my life in your hands, I had faith in you, and you let me drown. How can I believe in you and your beneficence?' God looked him up and down. 'You nebbisch,' he replied, 'I sent the police, a boat and a raft – and you just stubbornly sat there.'

God is acting in a very human fashion in that tale, and a glib response would be to say that we give human attributes to the things we can't understand to make them more easily understandable. But this is only half an answer. Anthropologist Stewart Guthrie believes that we anthropomorphize, not because humans are simple to understand, but because they are the most complex objects we know. Our brains are driven to try to extract as much information as possible from any given situation – that is, to create the maximum possible number of inferences.[24] Seeing a human face in the clouds is not inherently comforting – it may be frightening. So is the idea that the dead are all around us – as many a horror movie has exploited. Yet these comfortless ideas persist in the religious frameworks of many human groups, suggesting that they have a function. Maybe superimposing a 'human' element onto things we do not understand provides us with the richest possible way of thinking what to do. If that face in the clouds has a mind like me, maybe it will be pleased if I offer it food. Maybe, in return, it will give rain. If it's in the sky, it can see this whole area, like I can when I climb a tree – so maybe it can protect me from leopards.

Evolutionary psychologists believe that religious ideas – weeping statues, gods who can be pleased or angry – are just produced by the mind as one of the ways it makes sense of the world around us: effective in some situations, but deadly in others.

Fundamentalism: all in the mind?

The Fort of Masada, in Israel, has to be one of the most haunting sites in the world. It is situated atop an isolated rock at the edge of the Judean desert, nearly 500 metres above the Dead Sea. Formerly a residence of King Herod, and then a Roman garrison, Masada fell into the hands of Jewish rebels, called Zealots, under the leadership of Eleazar ben Yair. Here, in 72 CE, a pitched battle took place.

Seeking to destroy this outpost of resistance, the Roman governor Flavius Silva marched on Masada with – according to the ancient historian Josephus – the whole of the Tenth Legion and thousands of captured Jews who were forced to fight for the Romans. The Romans built a high wall around the base of the Masada rock, and laid siege for nine months – pitting 960 rebels against thousands of well-equipped soldiers. At last the Romans broke through the rebels' defences, only to come upon an eerie sight. Everyone inside the fortress was dead, except for two women and five children. These survivors gave themselves up, telling the Roman victors that Eleazar had persuaded everyone to commit suicide rather than surrender.

In the 1960s, the Israeli archaeologist Yigael Yadin excavated the Masada site and found much that backs up Josephus' account. Storerooms were found full of food, indicating that the mass suicide had not been dictated by hunger. A synagogue equipped with a large library of religious texts suggested that this was a community of the faithful, as much as one of revolutionaries. They seemed to have made a conscious choice not to surrender to the Romans and so, from a modern perspective, the dead of Masada seem much like early fundamentalists, prepared even to give up their own lives in pursuit of a cause. One wonders whether their thinking was perhaps not so very different from that of the early Christian martyrs, or, more recently, of some Muslim suicide bombers.

We evolved in order to survive and pass on our genes. So it seems impossible that any set of beliefs encouraging people to take their own lives could have an evolutionary origin. Nevertheless, perhaps the mental blueprint we inherited from our ancestors can help us to understand fundamentalism.

Some people choose to see fundamentalism as the natural extension of religious belief. If the Divine Idea can create strong bonds between certain people, and oppose them to others, and motivate them by making them feel they are carrying out the divine will, then it follows that they will be prepared to act in an extreme fashion. Others think fundamentalism is nothing to do with religion, but merely the way oppressed groups seek to overthrow their oppressors. Maybe both are missing the point. Even if fundamentalism were a natural consequence of religion, that wouldn't help us to explain why some people in certain times and places choose it, and others don't. And if it's just a motivating factor in power struggles, why do people choose such an extreme and often ineffective way of rebelling?

Perhaps, to explain this phenomenon better, we should return to our hunter–gatherer past, where co-operation and mutual reliance were of paramount importance. The group was the optimum survival unit, and everyone in it had a role and a purpose. Strong penalties existed for people who contravened the codes of co-operation and support. In ancient Israel, the penalty of *karet* meant abandonment – being thrown out of the group to survive alone: easy in modern urban societies, a swift route to death in the past. Throwing someone out of the group brought both a benefit and a cost, of course: you lost a cheat, but you might also lose a valuable pair of hands. If more and more people were thrown out, there might be serious consequences when it came to the next hunt. So the business of belonging and exclusion was not something that our ancestors treated lightly – and its importance came to be embedded in our mental make-up. Even if we are not conscious of it, this element of our psychology still affects the way we think and act.

Nowadays people have far more mobility in every sense. Women can choose when or whether to have babies. People can choose, in a broad sense, how to live – and it does not have to be within a moral tradition supervised by a church. Crucially, people can choose how to live without being ostracized. The fact that there are no penalties for opting out of the group means opting out is not only possible, but likely. To our primitive psychology, this spells danger.

Pascal Boyer uses the example of army platoons in wartime – close-knit units utterly dependent upon every man being prepared to obey orders and act with complete disregard for his own safety. People who demonstrated cowardice – that is, who opted out of the group – were severely punished. But, importantly, a lot of time and effort was spent in trying to root out potential cowards *before* battle – through painful ordeals and initiation rites. As I write this chapter, the British army is confronting challenging evidence about the treatment of new recruits, suggesting that this is a deeply ingrained tradition. Such activities are directed not so much at the possible cowards in the ranks as at the others, on whom they aim to imprint a certain message: if you do act in a cowardly way, this is what will happen to you. Defection carries a high price. Allowing people to opt out of a group with no cost is very dangerous for the group. So the more you put at risk in joining up, the higher you want the stakes to be for opting out.

Fundamentalism can be seen in this light – as a means of reminding people that opting out is very dangerous, and that if they do opt out they will end up dead. It's notable in this light that many fundamentalist movements are concerned with public displays of anger against people who are breaking the rules. The football stadium amputations in Afghanistan or mass protests outside abortion clinics in America suggest that fundamentalists are concerned with making a public example. The numbers of people actually harmed are tiny compared to the numbers of people intended to see the spectacle, which suggests that it is being carried out primarily to deter other people from defecting. Fundamentalism is far from being solely a matter of groups from different cultures attacking one another – the Taleban vigorously repressed their *own* people, as did Christian witch-hunters in Europe. What matters, it seems, is the behaviour of people within the group, not those outside it. Of course, the men who flew hijacked planes into the Twin Towers were motivated by the idea of *jihad*, a religious war in the name of Islam. But psychologists might argue that, in some unconscious way, the terrorists were also trying to send a message to Muslims.

Whether or not religious extremism arises from the structures of our psychology, it's clear that the Jews of Masada who committed suicide rather than surrender to the Romans were motivated by a faith that ran contrary to the demands of survival. This faith stressed the power and significance of a single, transcendent, unknowable God, which offended the Romans, with their pantheon of male and female deities characterized by very human faults and foibles. This struggle – between those following many gods and those following the One – is right at the heart of the story of the Divine Idea. As we shall see in the next chapter, it was by no means restricted to one place or time.

3

Finding the One God

Yazd is one of the oldest cities in the world. It also seems to be one of the most difficult to reach; it certainly requires a good deal of sweat and determination to get there. Until recent years, the bumpy, potholed road going south-east out of Tehran was not nearly so well surfaced as it is now. My first trip there, as an impoverished medical student driving alone in blistering heat over three hundred miles from what seemed like any civilization, was at times accompanied by a howling wind. Little tornados of spinning dust, 'dust devils', swirled across my view of the dessicated landscape; often the cloud of desert dust was so thick I had to slow my battered vehicle to walking pace to see the roadway at all. When the fusillade of sand and grit was not being whipped up around the windscreen of my Land-Rover or preventing me changing a punctured tyre, the stifling, shimmering air distorted the view of the empty road in a mirage, making false promise of a lake ahead. There are few villages once you are past Isfahan, and the bleak Iranian salt flats and featureless sandy deserts beyond are not exactly inviting. But always, in the distance to the south, the mountains rising 4,000 metres high provide a dramatic backdrop through the hazy afternoon light. In such a hot and arid country, it is a surprise that sometimes even in late spring they are snow-capped.

Just outside Yazd, only three or four miles from the city, there are two smaller hills, red-coloured and lifeless, completely devoid of vegetation. At their foot are one or two ancient empty buildings constructed of mud bricks, long since abandoned to the barren sand. The tops of these rocky hills are crowned with something more bizarre and rather more mysterious than the melting snow of the Iranian winter. Approached by a long, rocky, winding path, each summit provides a remote seat for a solitary, round-walled turret. These man-made structures most resemble large, windowless fortresses. They are the Dakhmas, or Towers of Silence, of the Zoroastrians. This is where, for two thousand years, the bodies of their dead, stripped of clothing and weighted down over slabs, have been laid out. These are the places where the bones of the corpses of the faithful have been picked clean of flesh by vultures.

Zoroaster: a Bronze Age democrat?

Some three thousand years before the birth of Christ, nomads emerged from the steppe-lands of southern Russia. One tribe separated from this migration and headed into Persia. They slowly crossed the same desert that I negotiated far more comfortably five thousand years later, and settled near the same mountains that I saw on my journey. Presumably water was easier to come by then, because these so-called Indo-Iranians established a somewhat more permanent existence in this area by constructing various settlements and trading with their neighbours. They inherited the ox-cart as a means of transport from nearby Mesopotamia and learned to extract metals such as copper and tin from the mountainous territory surrounding the plateau.

In an unsettled age the Indo-Iranians put their metallurgical skills to use in warfare, and their language developed numerous military terms, such as 'warrior' and 'charioteer'. The archaeological record suggests that they spent as much time feuding and raiding among their own settlements as they did fending off attacks from outsiders coming from further away. But at some point, a new creed emerged out of this turbulent age of bloodshed. It proclaimed equality, personal responsibility, and the dominion of one God.

Before Zoroastrianism, these Indo-Iranians usually worshipped a range of gods, employing an intricate system of animal and plant sacrifices superintended by specialized priests. They had been a wandering people living close to the elements, so they worshipped nature – water and fire, the sky, moon and wind. Being nomadic, they did not build elaborate temples: they simply carved a trench around a small rectangle of ground, in the centre of which the priest sat, with a fire for the performance of a sacrifice. As befitted a moving community, needing to travel light, they had no 'sacred objects'; after use, the tools of the sacrifice were ritually cleansed with water, and could then be touched and carried, packed away and transported by anyone. If this seems an odd point, consider how you might feel if you saw someone using a crucifix to prop a door open or stir some soup.

Like many warrior societies, these people developed a code of ethical conduct – they believed that truth and decency were basic laws, called *asha*, akin to the rising and setting of the sun and the passage of the seasons. Whenever it was necessary for a man to swear an oath, or forge a pact with another, they invoked specific gods to watch the proceedings. A system of elaborate ordeals existed, to be brought into play if deals were broken.

During the water ordeal, for example, the accused had to declare his honesty before the god Varuna before submerging himself completely in water. As he did so, an arrow was shot by an archer, and the tribe's fastest runner sent to fetch it back. If the submerged man was still alive by the time the runner returned, he was innocent. In some cases, where a man was considered to have broken an oath, molten copper was poured onto the breast of the accused: if he survived, this meant that the relevant god, Mitra, was not angry, because the person had not broken his word. A particularly significant member of their pantheon was Ahura Mazda, the Lord of Wisdom, who, as we shall see, came to have a powerful influence on the history of the region.

The society of this people was rigidly ordered. Women, children and everyone from the lower orders could expect little after death, except a shadowy existence in the underworld. They depended upon their surviving descendants and relatives for sustenance, obtained through sacrifice. But the warriors and priests expected to ascend to an eternal paradise, reached by a bridge heading upwards from the centre of the earth. Those who met the entry requirements would be reunited with their physical bodies within a year after death, to enjoy all the delights of the afterlife.

Zarathustra, or Zoroaster as the Greeks later called him, was one of the priests of this religion. Most of the little we know about him comes from the Gathas, a set of seventeen hymns believed to have been composed by him. These were transmitted orally to his followers and eventually written down. Debate rages about the true age of Zoroaster and the Gathas: one sect of Zoroastrians considers Zoroaster to have lived at least 6,000 years before Christ, while others argue for 600 BCE; scholars, meanwhile, have converged on a date around 1600–1500 BCE, on the basis of similarities between the Gathas and the Vedas, the texts of the Hindus, which were known to have been composed around this time. Whichever group is most nearly correct, Zoroastrianism is certainly one of the oldest continuing faiths.

Zoroaster referred to himself as a *zaotar*, or 'leader of the ceremonies', which means that he would have been ordained as a priest while still a teenager, probably at fifteen, which is the age when, according to Zoroastrian beliefs, a boy becomes a man. When he was thirty he was going about his priestly business, obtaining water for the spring festival, when he experienced a vision. A shining being led him into the presence of Ahura Mazda, the Lord of Wisdom. Throughout the rest of his life, Zoroaster experienced a number of encounters with this being, sometimes seeing him directly, sometimes hearing his words, sometimes merely sensing his presence. For Zoroaster, Ahura Mazda was the one uncreated God, eternal and the begetter of all other gods.

Zoroaster believed the world was a battleground between two forces, embodying good and evil. Ahura Mazda was pitted against another formidable power, Angra Mainyu, the force of all that was malign and ignorant. As Zoroaster developed his teaching, he reordered the earlier pantheon of gods, accepting six of them as lesser, but important, creations of Ahura Mazda, but divinities in their own right. He also rejected others, most notably Indra, the god of the warriors.

From a western point of view, Zoroaster's belief in a whole set of divinities does not look quite like monotheism.[1] Even the gods he rejected he considered existed; it was just that they were not important. But in lauding Ahura Mazda as the only God of importance, and declaring that others, long revered in his society, were evil, he was edging towards a new idea. Zoroaster had a clearly defined social purpose. His two forces of good and evil had, in his view, chosen what they were going to be at the outset of their existence; and all humans had a similar choice in front of them. This led to a radically new vision of the afterlife. Paradise, said Zoroaster, was open to all genders and classes; entry depended strictly upon the amount of good thoughts, words and deeds one had stored up in life. The soul, the *urvan* – another new concept – was literally weighed upon death. If the owner had stored up enough good, a beautiful maiden – a reflection of his or her own virtue – came to lead him along the mythical Chinvat Bridge and into everlasting bliss. If the soul was predominantly bad, the bridge narrowed to the width of a razor blade and the unhappy owner was cast down into the dark abyss of hell by a grotesque hag.

Zoroaster did not, as other religious rebels have done, say that the priesthood was unnecessary. He did not claim that sacrifices were worthless. He did not prophesy that society was corrupt and facing imminent destruction. As we shall see, these are common features of the journey towards monotheism, but they were not present in the outlook of this Bronze Age Persian. Even so, in stressing the importance of individual goodness, and warning the mighty that they could end up in hell, Zoroaster was taking a big risk. However, perhaps because his ideas seemed so entirely preposterous to the leading classes, he was ignored for a number of years and succeeded only in converting one of his cousins. Tradition has it that, disappointed by this indifference, Zoroaster moved to another area whose king, Vishtapa, proved sympathetic. The king's habit of going around the country proclaiming the new faith annoyed his neighbours, and he was swiftly attacked. But Vishtapa won this conflict, eventually allowing Zoroastrianism to establish dominion over a wide expanse of territory.

Like Judaism, Christianity and Islam, Zoroastrianism has its sacred texts. Collectively, they are known as the Avesta, a canon which also contains the Gathas, the teachings of the prophet Zoroaster. One tradition has it that when Alexander the Great conquered the area in the fourth century BCE, he carried the Avesta away with him, never to be seen again. In fact, the Avesta survived the Greek invasions, and to an extent helped the faith to stay alive while the Greeks systematically killed all the Zoroastrian priests. The Avesta contains not only the Zoroastrian view of creation, but also various laws, hymns and prayers. The purity of the elements – fire, water, earth – was important to Zoroaster: there were elaborate rituals designed to avoid defiling them, and complex rules to avoid humans being made unclean. Hence the origin of the Towers of Silence. Burial of the dead and cremation desecrated the earth by making it unclean; to Zoroastrians, the purest way of disposing of a corpse was to allow it to be eaten by dogs or birds on a remote hilltop where men would not stray. When Zoroastrians found themselves living in urban centres, sometimes with neighbours who were hostile to their practices, the forbidding Towers of Silence, hard to gain access to and hard to climb, prevented people from interfering with their burial traditions. In Zoroastrianism, death was viewed not as a natural event but as a sad calamity, the work of evil Angra Mainyu. As the Avesta says: 'O Maker of the material world, thou Holy one! Which is the place where the Earth feels sorest grief? Ahura Mazda answered: "It is the place wherein most corpses of dogs and of men lie buried." '

The Avesta sets out strict instructions on how to avoid the risk of impurity. So, for example, disposal of the dead could not be done simply by one person on his own:

Let no man alone by himself carry a corpse. If a man alone by himself carry a corpse, the corpse-demon rushes upon him, to defile him, from the nose of the dead, from the eye, from the tongue, from the jaws, from the sexual organs, from the hinder parts. This demon

falls upon him, [stains him] even to the end of the nails, and he is
unclean, thenceforth, for ever and ever.

If a death occurred in the countryside where no Tower of
Silence was available, the Avesta instructs the followers of
Ahura Mazda to

fasten the corpse, by the feet and by the hair, with brass, stones, or
clay, lest the corpse-eating dogs and the corpse-eating birds shall go
and carry the bones to the water and to the trees [thus defiling
nature].

The priests of the religion in which Zoroaster was
brought up had worn a woven cord as a sign of their
office. Zoroaster extended this to all members of the faith
– not abolishing priesthood, certainly, but symbolically
reminding everyone that they were each responsible for
their own salvation. To this day, Zoroastrians wear this
cord, the *kusti*, along with a white undershirt into which
a little pocket is sewn at the throat. This is to remind the
believer that he or she has a duty to store up 'spiritual
wealth' through good thoughts, words and deeds. In this
way, Zoroaster's new religion gave people tangible
symbols to hold onto, rather than merely abstract concepts,
and this would certainly have helped it to spread.

He also reorganized the traditional routine of praying
three times a day into a more extensive scheme of five
daily prayers, alongside a new calendar of seven ritual
feasts based around the seasons. A fixed prayer routine
was developed, which endures to this day. The believer
first washes his hands and face and then unties the sacred
cord. He then stands with the *kusti* in his hands, flicking
the ends to show his contempt for the evil Angra Mainyu,
with his eyes focused upon a sacred flame. These flames,
a remnant of the old fire sacrifice, have become a potent
symbol of the Zoroastrian faith. The 'fire temple' in Yazd

has at its centre a holy fire which has been burning continuously there for at least seventeen hundred years. It is tended constantly by a priest dressed in white, and whenever the fire is fed the priest wears a cotton mask to prevent the fire being contaminated by the breath of a human.

At one time – between 1500 BCE and the Islamic conquest of the area in the seventh century CE – Zoroastrianism was the most important religion in this region. When I visited Yazd again for the making of the BBC film, I was struck by the lonely ritual of the solitary priest in front of a glowing fire in an empty temple. How tenuous has this ancient religion now become in Iran. Life has been far from easy for Zoroastrians since the Islamic Revolution of 1979, and in these uncertain conditions many have chosen to emigrate or marry out of the faith. The fire temple seems more like a deserted tourist attraction than a place of worship. By law, the dead now have to be buried in the ground. When I climbed a hill to study a solitary Tower of Silence, I found it defaced, its walls hung with massive banners and posters left by zealous Shi'ite youths after their Ashura festival the week before.

Since the Islamic Revolution the Iranian authorities have tolerated other monotheistic religions to some extent, but Zoroastrianism seems only grudgingly accepted in modern Iran. Although Zoroastrianism is not perhaps truly monotheistic in the strictest sense, the faith Zoroaster established was certainly one of the first to proclaim an afterlife open to all people on the basis of personal goodness, rather than on that of wealth, status or ritual observances. This was a theme that later came to the fore in Judaism and Christianity and, later still, Islam.

So why do we call Zoroastrianism a monotheistic religion? Perhaps because so much of its history has been written from a western Judaeo-Christian viewpoint.

During the fifth century CE the Zoroastrian king in Persia, Yedzegerd II, forbade his Jewish subjects to recite the Shema (key scriptural passages forming part of the daily liturgy), which starts with the line: 'Hear O Israel, the Lord Our God, the Lord is One.' He wished to eradicate the belief in the Jewish God as Creator. To the Zoroastrian, the notion of one God being the god both of good and of evil was bizarre; to the Jew, only one God could be permitted. To ensure that the Shema would not be recited in synagogues at the beginning of the morning service, Yedzegerd stationed guards in synagogues at dawn – the time for its recital. To get round this restriction, the rabbis instructed that the whole prayer could be said at home, while to comply with the need to recite the prayer with an entire congregation, they hid the first verse only in a recitation of another part of the liturgy. This fulfilled the religious obligation, as in an emergency Jews are required only to recite that first verse. Fifteen hundred years later, that verse is still recited in the same place – tucked into a part of the liturgy to which it has no real intellectual connection.

Perhaps we call Zoroastrianism monotheistic because we can't help seeing the notion of a single God as inherently superior to polytheistic faith systems. Consequently, it seems to me, those coming from Abrahamic faiths – Judaism, Christianity, Islam – tend to project monotheism, unrealistically, onto various figures in the past, to prove that the idea has a truly ancient pedigree. This is probably the case with the Egyptian pharaoh Akhenaten, who lived in the fourteenth century BCE. Past generations have often souped up the legend of this myopic, irascible, iconoclastic monarch, crediting him with being the first monotheist, as well as a champion of homosexual rights and vegetarianism, among other worthy causes. In reality, the reforms that Akhenaten introduced to Egyptian religion were closely bound up with his desire to protect his status as king. But they nevertheless look very much like a step towards monotheism.

Akhenaten: a modern ancient Egyptian?

In 1798, Napoleon Bonaparte took a party of 139 renowned scholars with him on an expedition to Egypt, a mission conducted largely for the purposes of negotiating a swift route to India. The locals were extremely hostile, and this cohort of great minds was protected from bullets and stone-throwers by a sizeable platoon of soldiers. Nevertheless, they were able to visit a site called El Amarna, which contained the ruins of an ancient city where images of the sun were etched into ruined temples and tombs. Compared to the majesty of the pyramids, this collection of broken stones was rather unimpressive, so the French scholars, having rapidly logged the area and drawn up maps, moved on swiftly.

In the 1840s the English archaeologist Sir John Gardner Wilkinson paid closer attention to the site, and uncovered what seemed to be a pharaoh's tomb. He was particularly excited by the presence of a new form of painting, found nowhere else in Egypt. This novel art style depicted a highly stylized, almost feminine pharaoh, in scenes of unusual closeness and intimacy with his wife and children. In each case, the royal family was bathed in the rays of a disc-like sun god, and, curiously for any ancient Egyptian site, there seemed to be no other representations of the gods. The decipherment of hieroglyphics, at that time a new discipline, allowed scholars to conclude that El Amarna was the city built by Akhenaten, the father of King Tutankhamen, in the second half of the second millennium before Christ. Wilkinson's work was complemented by that of a German archaeologist, Richard

Lepsius, whose visit to the site was funded by the Prussian king. As the result of an intense period of research on the site, lasting just twelve days, Lepsius boldly stated that Akhenaten had 'abolished the whole religious system of the Egyptians and . . . [placed] in its stead the single worship of the Sun . . . He erased all the gods' names, with the exception of the Sun-God, from every monument that was visible.'[2]

Over time, a myth of Akhenaten developed, championed by British Egyptologists such as James Henry Breasted, in which this pharaoh was seen as the first modern, a bold, far-sighted revolutionary who attempted a thorough revision of Egyptian society – from new forms of art and architecture through to a religion based on the worship of one god. Excavators like Lepsius had noticed that, in later generations, the names of Akhenaten and his immediate successors had been deliberately scratched out and erased from inscriptions listing the royal lineages, suggesting an attempt to remove him, quite literally, from the historical record. This added lustre to the romantic image of Akhenaten as a misunderstood prophet of modernity, a genius who had the misfortune to be born 'before his time'.

The picture of Akhenaten as a pioneering religious reformer was somewhat challenged by a cache of clay tablets found in the 1880s. These, the so-called Amarna letters, had been dug up inadvertently by a local peasant woman searching for phosphate to fertilize her crops. Considering them worthless, she sold them on for a few pence. They eventually ended up in the hands of the keeper of Syrian and Egyptian antiquities at the British Museum, one Ernest Alfred Wallis Budge, who soon saw their true value. This handful of inscribed tablets was an archive of diplomatic correspondence, much of it between Akhenaten and his father, Amenophus III, and assorted neighbouring monarchs of the period. Their exchanges were dominated not by theological debates but by wheeling and dealing over women. For example, 'to Milkilu, the ruler of Gazru . . . send extremely beautiful female cup-bearers, in whom there is no defect'. Many of the letters suggest that the young Akhenaten had little respect for his neighbours. One rather pained missive from King Tushratta to the pharaoh includes the words: 'as to the tablet you sent me, why did you put your name over my name? And who now is the one who upsets the good relations between us . . . ? My brother, did you write to me with peace in mind?'

But for all this, the period covered by the Amarna letters seems to have been one of relative peace. Akhenaten's predecessors had begun this process, opting for a conciliatory foreign policy, and expanding the size and complexity of their civil service. His grandfather, Tuthmosis IV, had tried hard to save a discredited monarchy by depicting himself as receiving his power from Re, the sun god. The Aten, the physical disc of the sun, had previously been worshipped as the perceptible aspect of Re – but it now began to appear more like a god in its own right, a circle sprouting a pair of arms. When he came to power himself, as King Amenhotep IV, Akhenaten began to develop this solar disc into the focus of a major cult. He renamed himself in the fifth year of his reign, adopting the title by which we know him, meaning 'he who does the work of the Aten'.

His reforms were wide-ranging – and probably not popular. The images of other gods were smashed and vandalized in an orgy of destruction from which not even tiny personal objects such as rings and make-up boxes were exempted. The old temples, situated in dark, indoor and secluded locations, were abandoned in favour of open-air sites where the god himself, the sun, was visible. As part of this reorganization the priesthood was reduced in both size and scope, its responsibilities restricted to administering offerings at the temples. In former ages, priests had been powerful officials of the state.

Akhenaten consolidated his cult by building a new capital at the site of El Amarna. The symbolism of this is evident: situated in the very heart of the country, the new city was where the Aten and his agent on earth, the pharaoh, would reign supreme, unchallenged by priests or other gods. So the new cult seems to have been driven less by a religious imperative than by a desire to centralize power around the king. It is possible that this might have been prompted by an assassination attempt. One text, found on the edges of the El Amarna site, notes Akhenaten's intention to build a new city. Alongside the usual formulaic statements, and oaths of obsequious support from his courtiers, a curiously real-sounding lament is recorded, its repetitive style suggesting that it may have been a verbatim account of a speech by the pharaoh. In this, he expresses his shock and dismay at some recent but unnamed event. Scholars have suggested that he could only have given such importance to an attempt on his own life. The curious new art style, a mixture of intimacy and deliberate weirdness, suggests that the pharaoh was trying to make himself seem un-human, distant from his subjects, a god in his own right. This could have been a strategy adopted to reinforce his rule and deter any potential rebels from further attempted coups.

The cult promulgated by Akhenaten is not, of course, true monotheism – a better term would be *cathenotheism*, meaning the belief that one god is superior to all others, even if those others do still exist. But in seeking to destroy all references to the other gods, Akhenaten seems to have been trying to obliterate them from the public mind. And even if it was largely fuelled by political motives, his religion of the Sun Disc nevertheless had certain beliefs at its core. Akhenaten devised a slogan to sum up his religious movement: *ankh em maat*, which roughly means 'living according to the right order'. A similar modern soundbite might be 'back to basics' – the line coined by the British Conservative Party in the early 1990s – for his aim was to try to purify and simplify a life that had, in his view, strayed from righteousness and become corrupt.

The *ankh em maat* campaign was focused very much upon doing the right thing in the here and now: conforming to religious disciplines and the established order. It stressed that statues and paintings were not themselves gods worthy of worship, but just man-made representations. The afterlife – a very important concept in ancient Egyptian religion – was stripped of its relevance. The Inyotef Song, which some believe was composed by the pharaoh himself, includes the following lines, expressive of a sort of *carpe diem* ethos: 'Make holiday, do not weary of it! Lo, none is allowed to take his goods with him. Lo, none who departs comes back again.'

Akhenaten's religion seems to have had little to say on the subject of suffering or virtue.

> His Disc created the cosmos . . . and keeps it going, but he seems to show no compassion to his creatures. He provides them with life and sustenance, but in a rather perfunctory way. No text tells us that he hears the cry of the poor man, or succours the sick, or forgives the sinner. The reason for this is that a compassionate god didn't serve Akhenaten's purpose.[3]

So what was his purpose? Debate will undoubtedly continue – but one interesting theory is that Akhenaten's reforms were influenced not so much by political ambition as by personal illness. A number of archaeologists have argued that Akhenaten was a sick man; several have suggested that this sickness may have been genetic. One piece of supportive evidence is the high death rate at young ages of people born in his family line.

Further circumstantial information is that Akhenaten seems to have been kept hidden from public view during his father's reign, as sick members of royal families often are. The Canadian archaeologist Alwyn L. Burridge, of the University of Toronto, has boldly identified a particular genetic disorder from which Akhenaten might have suffered. He published an interesting paper in which he analysed the paintings of this pharaoh and concludes that these indicate he may well have suffered from Marfan's Syndrome.

Marfan's is a genetic disorder which affects about one in 5,000 people in the UK and USA. If it runs in the family, on average around half of any offspring will inherit the damaged gene that causes it. Quite often, inheritance of this gene is compatible with a reasonably normal life, but it can have manifestations of greater or lesser seriousness. Typically, people with Marfan's Syndrome are very tall, slender, and loose- or 'double'-jointed. Generally the sufferer may have arms, legs, and toes that are disproportionately long in relation to the rest of the body; some have arachnodactyly – spider-like long fingers. A person with Marfan's Syndrome often has a long, narrow face, and the roof of the mouth may be arched, causing the teeth to be crowded. Sometimes there are other abnormalities of the skeleton, with a breastbone that is either protruding or indented, curvature of the spine and flat feet. Many sufferers have poor vision or are even largely blind, as Marfan's frequently affects the lens of the eye, dislocating it from its attachments. Also, as if these unpleasant manifestations were not enough, changes in the membranes around the brain may cause chronic headaches. Weaknesses in other connective tissues means that many people with this complaint have severe stretch marks in their skin and may have an abdominal hernia. But the most serious and frequently fatal problem it can cause is heart disease; often the valves in the heart do not close properly, eventually weakening the heart muscle. A form of heart failure can develop, with excessive tiredness and breathlessness on slight exertion or even, occasionally, at rest. A linked problem is weakness of the big vessels taking blood from the heart; this can lead to their rupture, causing severe chest pain and sudden death.

When you look at the very odd pictures of Akhenaten which remain after nearly four thousand years, there seems a striking possibility that Professor Burridge is right. Presumably, the many depictions of Akhenaten represent reasonably accurately what he looked like in life. He is often shown as being taller than those around him, with a long, angular face, a misshapen head and curvature of the spine. Sometimes he apparently needs a stick for support. Quite intriguing evidence of a genetic disease of this sort is found in the skeleton of an unborn child of Akhenaten's; it has a full range of features suggestive of Marfan's. And in pictures of the pharaoh with his family, everyone is shown sitting or standing very close, often touching; just possibly, this may have been less to do with intimacy, and more to do with poor eyesight. Perhaps it is not too fanciful to imagine that Akhenaten focused his religious attention upon the sun because this was the only 'deity' he could clearly see, given his relatively poor or fading vision.[4]

Akhenaten is always likely to remain a fascinating puzzle. The whole story of this perplexing figure is cloaked in mystery. Some pictures of him naked show him without any genitalia, and a few people have suggested that he was not a man at all, but a woman. It has also been argued that the curious drawings of him, the like of which are not seen in any other period of Egyptian art, were specific to the Amarna period – an exaggeration of the human form to set this king aside from mere mortals.

It is not known where Akhenaten was buried. It is possible that his mummy was removed from El Amarna and taken to the Valley of the Kings, where many of the pharaohs are interred. A sarcophagus bearing his name was found there, but his mummy has never been identified. Until the mysteries of his burial are resolved we are unlikely to be very much wiser about him. It would be nice to think that, if a body were discovered, some DNA might be obtained from the mummified tissue. DNA has been collected from other mummies in the past, and sometimes, apparently, sequenced with partial success.[5] A few diseases, such as tuberculosis and malaria, where only a short segment of the DNA is needed for identification, appear to have been identified in some mummified tissue. But DNA decays quite rapidly in the kind of temperatures common in central Egypt; and though tissues would be kept relatively cool in the tombs, it still seems very unlikely that a genetic condition as complex as Marfan's would be easily detectable, even with the most sophisticated techniques at our disposal.

There is another curious twist to the Akhenaten story. Some people have argued he was Moses. His mother was Queen Tiy, a descendant of a Hebrew tribe; and he lived roughly at the right time, in the mid-fourteenth century BCE. Because Moses may have started life as a high official in the court of Amenhotep III, the story does have faint credibility. But although, like Akhenaten, Moses was a renegade, he led his people the Israelites out of Egypt to freedom and to the worship of one God – a God of a very different kind from any recognized by Akhenaten.

The goat that laughed

True monotheism really starts with Judaism, as we shall see in the next chapter. But even with the Israelites this shift did not happen very quickly. The Israelites' ideas about God underwent a huge transformation in the years between 1800 and 500 BCE. At this time a leap of under-standing was also taking place in India, but its direction was very different. Instead of starting to think of a single God who controlled the whole world, religious Indians embarked on a journey inwards.

There's a tale concerning the Buddha, who was asked by a group of disciples whether there was any point in sacrificing goats and sheep as offerings for recently departed relatives. The Buddha replied solemnly that there was never any value in taking a life. A skilful teacher, he told a little tale to illustrate his point.

A Hindu priest once ordered some of his students to take a goat down to the river, bathe it, brush it, put a garland around its neck and return it to him. While they were doing this, the goat began to laugh – and then, just as suddenly, began weeping. The students asked the goat why it was doing this. 'Repeat your question when we get back to your teacher,' replied the goat. The priest was equally astonished, and asked the goat himself why it laughed and wept. 'In past times, sir,' the goat replied, 'I was a priest like you. Like you, I sacrificed a goat for the dead. Because of that one goat, I have had my head cut off 499 times. That's why I laughed just now. Today is the last time my head will be cut off.'

The priest asked the goat why, if he was so happy, he cried. 'For you,' the goat answered. 'Because if you kill me, you may be doomed to lose your head 500 times. I cried out of pity.' 'In that case,' said the priest, 'I shan't kill you.' 'It won't help,' replied the goat. 'The power of my past evil deeds will ensure that I am killed again today.'

The priest ignored the goat at this point and untied it, commanding his students to follow it and make sure it came to no harm. The goat began to graze. It stretched out its neck to reach the leaves on a bush growing near the top of a large rock. A lightning bolt hit the rock, breaking off a sharp section of stone which flew through the air and neatly cut off the goat's head. A crowd of people gathered around the dead goat and began to talk excitedly about the amazing accident. They realized that taking life would only bring them misery and abandoned their sacrifices.

Like Zoroastrianism, Buddhism emerged from a religious tradition based on the sacrifice of animals. Zoroaster's religion came about because a group of nomads from southern Russia settled in Iran. Another group, calling themselves the Aryans, continued eastwards, and by the seventeenth century BCE they had arrived in the Indus valley. In later centuries, a quasi-historical myth grew up, popular with the Nazis, of the Aryans as a tribe of proto-Europeans, conquering the primitive east. This was nonsense: in the Indus valley, complex cities like Mohenjodaro and Harappa are evidence of a highly developed existing culture that existed some five thousand years ago – well before the time of Abraham.

Today, the site of the prehistoric city of Harappa lies just about 100 miles from Multan in Pakistan, almost halfway to Lahore along the fertile Indus valley. It was discovered by two tourists in 1830 and was visited a few years later by General Alexander Cunningham (1814–93), the first westerner to describe the site in any detail. Not far away, on the west bank of the river, is the city of Mohenjodaro. Unfortunately, the archaeological site of Harappa was badly damaged when the railway track between Lahore and Multan was laid and workmen ransacked the remains for loose ballast. But subsequent excavations indicate that Harappa was an important trade centre, and that its residents adorned themselves with copious amounts of jewellery and ornament, including precious stones some of which may have been obtained from traders en route from Mesopotamia.

These ancient cities of the Indus valley, where the

Hindu religion seems to have originated, were highly sophisticated. The drainage system was efficient, with drains and chutes ensuring hygienic conditions. Town planning was advanced: streets and buildings were laid out on a grid system, incorporating granaries, municipal buildings, workmen's quarters and communal baths. The impression of order and symmetry is everywhere, and building regulations were obviously strictly enforced: the substantial houses were built to a regular pattern, most of them two storeys high. Even the bricks that were used throughout the valley cities were of a uniform size.

The monotony and regularity of these cities suggests that theirs was a strictly governed, even authoritarian society. This civilization, which was established around 3000 BCE, was undoubtedly highly advanced; and yet we have no real idea how or what these people worshipped, beyond what we can glean from the Aryans who took over the area. Even though no idols have been found, and there is no evidence of pagan practice, there is no proof either that these people had a high priest or a monarch – or worshipped one god. There is no building which could be considered a dedicated temple. These people had writing, but so far nobody has been able to decipher what they wrote. It was a civilization which died leaving a most enigmatic puzzle in ancient oriental history.

We do not know why these ancient Indus valley civilizations died out. Maybe there was a natural disaster; maybe the arrival of the Aryans from the west had something to do with it. The 'death' was not total, of course. The texts of these Aryans, combined into a book called the Rig Veda, express great contempt for the indigenous people, describing them as *dasyas*, the untouchable outcasts: black-skinned, noseless, worshippers of the phallus. None the less, the Sanskrit language developed by these settlers contains many words with non-Indo-European roots, suggesting contact and assimilation between the two populations.

The Rig Veda is a vivid source of information about the people who were later to develop Hinduism. As in Iran, their society was rigidly structured into classes or *varnas*, dominated by the brahmins (the priests) and the kshatriyas (the warriors). These people enjoyed music and dancing and playing with dice; they intoxicated themselves with drinks called *soma* and *sura*. As we can see from the tale of the goat above, their religion centred on sacrifices to the gods, controlled by the brahmins. The blessings they sought to secure through sacrifice were very similar to the concerns of most societies of the time: health, long life, sons, cattle, wealth. They addressed these concerns to a vast pantheon of gods, dominated by Indra, a warrior god, made known in the thunderstorm and the lightning bolt; Varuna, the god of law and order; and Agni, the god of fire. The Vedic pantheon was very male – reflecting a society where priests served a warrior elite.

But a change began to occur in Vedic religion. We see this in the ideas expressed in texts called the Upanishads and Aranyakas, which were added onto the Vedas at some time between the eighth and fourth centuries BCE and thus earned the collective title of Vedanta, or the end of the Vedas. The message of the Vedanta texts was that the many gods were only an *expression* of a greater, underlying power. This power, called Brahman, was not unknown to the readers of the Vedas – just as YHWH was already familiar to the Israelites before He became the One God. Brahman was regarded as the power or essence that was addressed during the sacrifice – indeed, the name for the priests, brahmins, reflects this belief. But the Upanishads said that the individual gods were simply reflections of this underlying essence.

The Upanishads – themselves anonymous texts, giving no clue as to who composed them – stressed that this essence, Brahman, was unknowable. But they went much further than that. It could not be addressed as Thou. It did not speak to humankind. It could not meet us. Our actions neither please nor offend it. Instead, our fortunes, bad and good, were interpreted in a new way, through the workings of a natural law called karma, of which Brahman was the source.

Karma is the idea that bad and good actions have, respectively, bad and good consequences for the people who commit them. Whereas other peoples, in other times, have interpreted misfortune as the punishment of God or gods or ancestor-spirits, karma is an impersonal force in the universe, like the law of gravity.

Many of us subscribe to a limited belief in karma – there is a general view that people who act in a bad way will meet with a 'sticky end' and that, if we are decent towards others, this is how we will be treated ourselves – even if we see this less as the working of a cosmic law, and rather simply as a chain of cause and effect. I find my own theories of karma shaken by the fact that many of the world's wickedest personalities have lived to a ripe old age without enduring so much as a bad night's sleep. Conversely, you can stub your toe while nobly taking a cup of tea to your wife in bed. But the Upanishads had an answer to this apparent contradiction – bound up with belief in the reincarnation of the soul. If we do evil, they said, then the evil might return to afflict us not in this life, but in the next one – when we might be reborn into a lower caste of society, or as a reviled animal or insect. Similarly, a life of goodness might bring its reward not instantly, but in a later existence, depending upon how much stored-up 'bad karma' we had to work through. The tale of the goat at the start of this section refers to this idea.

It might seem that the Upanishads were taking religion out of the hands of the gods, and substituting a law of the universe in their place. So they would have – except that few humans would have tolerated it. Instead, then, the

Upanishads stressed that Brahman was also the force that pervaded and sustained us, as it did the whole universe. Whereas the early Vedas upheld an intricate system of sacrifice, the Upanishads advanced the spiritual practice of *yoga* (which literally means 'yoking', and stems from the same Indo-European root-word). Yoga, through disciplines of posture, breath control, fasting, exercise and meditation, sought to discipline the senses, detaching them from the external world and turning them in upon the mind, and the reality that lay behind the mind itself.

As we shall see, this is a different religious path from the one suggested by the prophets in ancient Israel, but both they and the writers of the Upanishads – along with Zoroaster – were expressing dissatisfaction with the cult of sacrifice, and stressing the importance of individual action. The Israelite prophets wanted social justice and an end to 'foreign' religious practices. The Upanishads urged people to journey into their minds and discover the presence of Brahman that lies within each of us, and is known as the *atman*, which we might loosely translate as 'soul'.

Another important concept was that of moksha or liberation. This essentially is the way to break free of the continued cycle of reincarnation. If moksha is not achieved during one's life, then one will be reborn in a form reflecting what one deserves. Moksha can be attained by action (karma), knowledge (jnana) or devotion (bhakti). Devotion is the usual way to achieve liberation, mostly by pujas – the acts of prayer and sacrifice usually made in the temple service.

In early Vedic religion the most important figure was the brahmin priest, but in the Upanishads it was the yogi – the mystic who divorces himself from the world to pursue the inner journey into the mind and spirit. One such individual was a wealthy prince in Nepal called Siddhartha Gautama, who in about 530 BCE, at the age of twenty-nine, left his wife, Yashodhara, gave up his claim to his father's throne and his luxurious life, and went in search of enlightenment. Just as the leap towards monotheism created conditions whereby Judaism, Christianity and Islam could emerge from ancient Israelite religion, so the revolution of the Upanishads enabled Buddhism to emerge as a rival force to Hinduism.

The young Gautama, despite his affluent upbringing, was deeply sensitive to suffering, and it was his mission to discover if it could be avoided, or if all humankind was doomed to suffer. The mission is described in a collection of texts written in the Pali language in Sri Lanka during the first century CE. They relate that four sights in particular had motivated Gautama's journey – an old man, a sick person, a corpse and a mendicant beggar. So Gautama fasted until he became skeletal. He endured countless yogic mortifications of the flesh. He sat at the feet of various Hindu gurus or teachers; none could help. Eventually, however, seated under a fig-tree, he experienced a joyful epiphany. At this point a demon called Mara appeared, tempting Gautama to stay where he was and enjoy his own bliss; there was no point trying to tell others – they would not understand. But then two figures from the traditional Hindu pantheon appeared and begged Gautama, who had now become Buddha, 'the Enlightened One', to share his insights with others. For the next forty-five years he tramped the earth, visiting the great new cities of the region – Benares, Uruvela, Rajagriha – even, according to some legends, travelling as far as Sri Lanka. He died in his eighties of a digestive complaint. His last words were: 'Impermanent are all created things; strive on with awareness.' He did not go on to be reincarnated. Through his insights, he had overcome the cycle of *samsara*, the constant repetition of death, rebirth, suffering and further death.

Gautama's insights were interesting because, echoing

Zoroaster and the prophets' insistence on personal piety, they suggested that the individual is reponsible for his or her own salvation. Buddhism is sometimes referred to as an 'atheistic religion', having no God or gods. This is a judgement that can only be pronounced by people who haven't seen much of it in action, or read many Buddhist texts. Buddhism does certainly admit the existence of gods, but argues that they are not much use to humankind because they, like us, are bound to a perpetual cycle of reincarnation and suffering, the realm of *samsara*. But far from being pessimistic, Buddhism offers practical hope to suffering gods and men alike, through the Four Noble Truths:

> All existence is suffering.
> The root of suffering is craving.
> By stopping craving, we can cease suffering.
> The way to cease craving is to follow the
>   Eightfold Path.

The above are translations of the Buddha's Four Noble Truths. They are not unique. We find very similar ideas expressed in Greek stoicism, which emerged in Cyprus in the second century BCE. They emerge from a basic observation of existence – that everything is impermanent, and yet we desire permanence. If we eat a meal, then in time, the pleasant sensation of fullness gives way to hunger again. We cling to life, yet know that inevitably we will die. As a scientist, I struggle for years to make a breakthrough, that single goal becoming the thing I most desire. Eventually I achieve it, only for my feelings of satisfaction to fade away over time, and for a new and difficult quest to preoccupy me. As a consumer I save up for the latest labour-saving device and buy it, at great cost, only to enjoy a few hours of pleasure in it before the television advertisements alert me to some new 'must-have' item.

This experience, repeated again and again at so many levels in our lives, can induce a profound feeling of disaffection and the question 'Is it all worth it?' – which was exactly what motivated Gautama to go on his quest. And the conclusion he reached was neither earth-shattering nor unique, but it was a very clever bit of thinking. If we can stop desiring things, then we will stop suffering when these things pass away, or we can't have them.

The Eightfold Path was an all-encompassing programme of action and belief for would-be Buddhas – including ethical behaviour towards others and yogic practice. The stress was very clearly on individual action, rather than on higher powers, or on the powers invested in priests. Across the eastern world today, images of the Buddha are venerated and presented with offerings, as they have been for centuries. But the Buddha himself strongly cautioned against any veneration of him, or indeed of his teachings. He declared that his words were meant to be like a raft, a necessary tool to get people from one state to another and then to be discarded. Although Gautama's disciples founded the institution of the *sangha*, the Buddhist monastic community, Buddhism was a route that could be taken individually.

As the tale at the beginning of this chapter suggests, the success of Buddhism owed much to the character of Gautama, who made his truths accessible to the people through little stories and anecdotes. Jesus Christ is a similar figure in the story of religion – not just in his storytelling, but as an enlightened person, unable to keep the contents of his enlightenment to himself. But before we explore just how dangerous that was, we need to consider the first truly monotheistic religion: Judaism.

4

The World's Greatest Book

A few scholars have advanced the possibility that the cult of Akhenaten coincided roughly with the time when Moses lived. Some have even gone further and suggested that the prevalent cult of Akhenaten may have been influential in helping to forge the monotheistic beliefs of the Israelites – and, as we have seen in chapter 3, some have gone further still, suggesting that Akhenaten and Moses were the same person. But in ancient Israel itself, the idea of one God was not associated primarily with one individual. It was advanced by a class of religious rebels, the prophets, and was very much shaped by Israel's often turbulent relations with its neighbours. Thanks to the remarkably rich source material of the Old Testament, this is an idea whose journey we are truly able to trace from its earliest roots.

Although the Bible is one of the best-selling and most widely used books on the planet, the great majority of people are remarkably ignorant about what lies between its covers. Nearly all of us can lay our hands on a copy. Sue Lawley gives it as an automatic present to her *Desert Island Discs* castaways. Even hotel bedrooms supply their lodgers with this consolation for being alone and far from home – though sadly, I am told, the authorities in some of our wonderfully regulated National Health Service hospitals are removing all the Gideon bibles from where they may be most needed, on the grounds that they are an infection risk. Could this be to prevent the leprosy of Leviticus? Yet even many well-informed religious people have a very sketchy knowledge of the Old Testament; I know to my own shame, even though I am pretty conversant with classical Hebrew and Aramaic – the original languages in which the Old Testament or Tenakh was written – how much of the Bible seems completely novel or unknown to me when I read or reread it.

The story of a people with no word for 'history'

The Hebrew name for the Old Testament, Tenakh or TeNaKh, is an acronym for the three main volumes of the scriptures – 'T' for 'Torah', the five books of Moses; 'N' for Nevi'im ('the Prophets'), including the books of Joshua, Judges, Samuel and Kings, along with those of the three major prophets, Isaiah, Jeremiah and Ezekiel, and the twelve others, such as Amos and Hosea; and 'K' for Ketuvim, 'the Writings' (the vowels are inserted simply to help pronunciation). The five books of the Torah, which is the oldest section, contain the story of Creation, the history of the patriarchs – Abraham, Isaac, Jacob – and the canon of law, the most important part of which is the Ten Commandments. Nevi'im, the Prophets, contains twenty-one books in all and is partly narrative, recording the early history of the Jews after they entered the Promised Land, and partly the prophetic, poetic and literary creations of the great orators of the times. The third section, Ketuvim or Writings, is almost certainly the part of the Old Testament which was most recently written down. It is a varied collection of thirteen books. These include liturgical poetry – for example, Psalms and Lamentations; secular love poetry – the Song of Songs; history – for example, Ruth, Esther, Chronicles, Ezra; philosophy – Proverbs and Ecclesiastes; and a curious blend of history and prophecy in the book of Daniel – which, incidentally, like bits of the book of Ezra, is partly written in Aramaic as well as Hebrew. Job, one of the most enigmatic books in Ketuvim, is thought by some scholars to have been written in Aramaic and then translated into Hebrew.

In order to understand what I consider to be the origins of true monotheism, it is necessary to examine its greatest literary work in some detail. The Old Testament, unlike the sacred books of other religions, such as the Avesta of Zoroastrianism, the Vedas of Hinduism or the Qur'an of Islam, has a narrative structure. The Jews are the only people alive anywhere today who have such a historical record. It is somewhat ironic that in classical Hebrew there is no word for 'history'.

While some of the stories in the Bible stand alone, as diversions from the main history of the Jewish people, the Bible describes the growth of the monotheistic faith of the Jews sequentially. The idea of monotheism occurs very early, along with the central tenet of God-given morality. According to rabbinical authority, God presents Noah with seven universal laws which apply to every human on the planet and which are considered vital to human society.[1] 'Who ever sheddeth man's blood, by man shall his blood be shed; for in the image of God made He man.' Human life is sacred because humans are made in the image of the one God.

Setting aside some parts of the five books of the Torah from which modern-day Jews derive their complex system of religious laws, the Hebrew Bible can be read as a history, from beginning to end. This is particularly evident if you read it in its Christian form, because Christians have placed the writings of the prophets, which contain hardly any narrative, at the end. The Tenakh is the account of the development, the maturing, of a people, and it is the document which describes their struggle – a wrestling match between a people and their one God. It is essentially the story of how a nation coming from a pagan environment gradually accepts that there is just one deity. Let me, rather arrogantly, attempt to sum the Bible up in a single paragraph.

God takes dust and from it creates man in His own image. Although He gives him and his mate Paradise to live in, they disappoint Him and God throws them out, forcing them and their descendants to wander. Generations later, God sends a massive flood to wipe out the sins of humankind, but He saves one honest man, Noah, and his immediate family. God promises Abraham, a descendant of Noah, that he will make his offspring a mighty nation. Among his descendants are the three patriarchs, Abraham, Isaac and Jacob. Each of them is a believer and each is vividly depicted in turn, in a remarkable story of the life of a desert family, with all their virtues and their flaws. Joseph, Jacob's favourite son, is sold to a wandering tribe by his jealous brothers and is taken to Egypt where, going from rags to riches, he eventually becomes the pharaoh's most powerful minister. During a subsequent famine, Joseph's family are reunited and settle in Egypt. A new pharaoh enslaves their descendants physically; mentally they are enslaved by a pagan society. God appears to Moses and recruits him to lead the rescue of the Israelites from bondage in Egypt and into freedom. They cross the Red Sea into the wilderness, but the leadership of Moses is repeatedly tested. Moses ascends the mountain in Sinai, leaving his brother Aaron the priest in charge of the Israelites below. Moses is to receive the Ten Commandments directly from God. These are the universal laws that the Israelites are required to follow and to give to the world. But, as happens repeatedly, this recalcitrant people loses faith while Moses is absent up the mountain and builds a golden calf. Moses returns; God pardons the Israelites, demanding their adherence to a strict code of laws. He rewards them by feeding them in the desert and helps them prevail over all the enemy tribes that try to destroy them in battle, but when successful the Israelites continue to be intermittently rebellious. After forty years of wandering, they enter the Promised Land, a 'land flowing with milk and honey', under a new leader, Joshua, carrying the Ark of the Covenant containing the Ten Commandments. The twelve tribes of the Israelites settle in different parts of this land, Canaan, a new loose-knit nation based on the notion of justice and on the moral laws God gave them. After a period of being led by judges, during which time the Israelites repeatedly make war against the neighbouring nations but also often worship their gods, the Israelites adopt the notion of appointing a king to help them in their conflict with the Philistines. Under their last judge, Samuel (eleventh century BCE), they are transformed from a loose confederation of tribes into a monarchy. The charismatic warrior King Saul is anointed. After his suicide in a lost battle against the Philistines, his estranged son-in-law, David, another great warrior, secures the safety of the state, Judea. David's son, Solomon, builds the first Temple (tenth century BCE) in Jerusalem, the capital city. Ten of the northern tribes break away to form the Kingdom of Israel, while David's descendants remain in Judea. The people of the kingdom repeatedly continue to only pay lip-service to the God who has protected them, often preferring the nature cult of Baal and other pagan practices, and committing injustices against one another. Prophets warn the sinful nation of its impending doom. The Kingdom of Israel is invaded by the Assyrians in 700 BCE and its people are killed. Eventually, in 587 BCE, the Temple is destroyed and the people of Judea, who have also become sinful, are carried off into captivity by the Babylonians. After their return from exile in Babylon some fifty or so years later, a second Temple is built and the walls around Jerusalem are restored.

This is where the narrative of the Old Testament finishes. It does not tell the full story of the repeated invasions of the Holy Land thereafter, nor does it give a full account of the tyranny of the Persians and, later, the Greeks. Nor does it describe the subsequent uneasy relationship with Rome (though some idea of this is gained from the Christian New Testament); nor does it tell about the final destruction of the Second Temple in 70 CE. Much of this can be learnt from the books of the Apocrypha and from the written evidence in some of the Dead Sea Scrolls. The Roman destruction of that Temple, rebuilt by Herod – a kind of Jewish Quisling – is based on solid, verifiable historical evidence, for example, the writings of the first-century CE historian Josephus and others. But the overall point of this long story is that the ruination of their homeland and the repeated exiles they experience eventually result in the Israelites arriving at a fresh understanding of God. While a few Jews (as they are now known) and their descendants remain in Palestine, most of them become wanderers, sustained by their belief in a single God. Their tortuous journey has taught them the truth of monotheism.

Orthodox Jews regard the oldest part of the Tenakh, the Torah, as the word of God, and therefore literally true. But this by no means implies that Jews accept the text slavishly. I recently visited Arizona and Kentucky, meeting Christian fundamentalists who sincerely believe that every word of the book of Genesis, particularly the opening eleven chapters, must be taken entirely literally. They argue that scientists have repeatedly lied about evolution, claiming that Adam and the dinosaurs were on the surface of the earth at the same time, six thousand years ago. Even more oddly, they believe with all seriousness that the Grand Canyon in Arizona was formed by the waters receding after Noah's flood. They admit that they are reading the Bible in an English translation, which has in turn been taken from another translation of the Greek edition of the original Hebrew, but all this seems irrelevant to their deeply held belief. To them, 'sacred' means 'immutable'.

Jews take quite a different view. They agree that every word of the Hebrew Torah is holy. In one sense, they go further, maintaining that not even a single letter may be changed in any way. If a scribe writing a scroll of the Torah makes a mistake, it must be corrected. If mistakes cannot be corrected or if the wording cannot be read properly, then that text has to be treated like a human body and reverentially buried. But none of this prevents the Orthodox Jew from trying to interpret the text – indeed, it is incumbent upon him to turn it and turn it again, to examine nuances, to reflect on various shades of meaning. This is why so many commentaries have been written by Jews on the Torah – there is almost certainly no book in the world, certainly including Shakespeare's works, which has been more carefully scrutinized, dissected, cross-referenced and argued over.

According to Jewish tradition, revelation has given us the raw material of the Torah – the word of God. But the rational faculty of the human mind is essential in seeking to understand its material and its insights. The Torah stands or falls on reason, and rational methods are required to attain truth. When King David (in Psalm 119) says 'I have chosen the way of faith', this does not imply the way that leads to or ends in faith; it means a ceaseless and continual search to acquire a fuller understanding of the Torah. The man of faith trusts God's wisdom, and because man is made in God's image, he also trusts man's reason as far as it will go. This is essentially why Judaism is committed to enquiry and intellectual exploration, and why probing the universe and uncovering its mysteries – what we now would call science – is a key activity for the Jew.

This constant state of taking nothing for granted, of always approaching 'given' knowledge in the spirit of restless enquiry, is exemplified by a rather nice story. It was told by Nachum Rabinovitch, who was once principal of Jews' College in London, an important centre for higher education. In the yeshiva – the Jewish religious university where many young adult Orthodox men go to learn after finishing their childhood education – argument over texts can get pretty heated. It is often said that when two scholars get together in such a place, they are like warriors in combat. The secret ambition of every young scholar is to find a *kashya* – some real or apparent slip in a chain of reasoning. Not surprisingly, there is a hierarchy of *kashyot*. To find a fallacy in the teacher's discourse is a matter of joy; to find a *kashya* in a recently published book of importance even more satisfying. But to uncover a fresh *kashya* on an older, more widely studied tome is a real accomplishment. As Rabinovitch says, to find a logical contradiction in, for example, a text of Maimonides, the great philosopher whose works have been pored over for a thousand years by the most acute minds in Jewry, would be regarded as a signal achievement by any leading modern scholar. And higher still in the aristocracy of *kashyot* would be a valid criticism of a text in the Talmud – the great compilation of the commentary, discussions and arguments of the great rabbis around Babylon in the third century CE. So Rabinovitch recounts how a student contemporary of his once burst into the assembly hall where all the students were studying hard. He was wildly elated and stammering with delight: 'I have found a *kashya* on the very Torah itself!' Older students just smiled knowingly, but none the less all congratulated him on the attempt. No text is regarded as immune to critical analysis, but none of his fellow students entertained the idea for a moment that simply because he had found such a *kashya*, he would as a result suspend his commitment to the Torah.

Jews – an argumentative people

As we have seen, the Tenakh constantly shows the Jewish people arguing with God. From Abraham, through the time of Moses, up to the end of the Old Testament with Job, Jews indeed indulge in a divine wrestling match. This tradition of argument seems to be a feature peculiar to Judaism. There seems to be no other religion where God, and His justice, are held up to scrutiny and examination in the same way. The tradition of argument continues well after the conclusion of the Old Testament. Jews take the written law as only part of the rules of their observance. The detailed regulation of day-to-day life is laid down in the Mishnah and the Talmud – the oral tradition as opposed to the written tradition.

Judaism depends on there being two sets of legal codes. The Torah is the written, unchangeable law, in Jewish belief directly God-given through Moses. But the written law needs authoritative interpretation – and this is where the oral law comes in. As the great philosopher and poet, Rabbi Judah HaLevi (1075–1141) says rather succinctly in *Kuzari*, 'that which is plain in the Torah is obscure'.[2] At many points the five books of Moses lack clarity and definition. The important prohibition on working on the Sabbath in the Ten Commandments does not specify remotely what constitutes 'work'. The notion of an 'eye for an eye, a tooth for a tooth, hand for hand, foot for foot' in Exodus is inexplicable simply in terms of blinding or maiming the culprit who causes such an injury. As the Talmud points out, this cannot be taken literally. For example, where a one-eyed man takes out someone else's eye, that attacker's eye is clearly worth a great deal more than the one eye the victim has lost. Nor, for example, in the Torah are any laws of marriage laid down; and divorce is mentioned only, as it were, in passing. Indeed, the tradition of there needing to be an oral law is itself laid down in the Torah – that is to say, the Torah states that interpretation is needed: 'if there be a matter too hard for thee, . . . thou shalt turn unto a judge that shall be in those days' (Deuteronomy 17: 8–9).

The distillation of the oral law is found in the Talmud. The Talmud is massive and is divided into a set of six volumes called the Mishnah and another set called the Gemara. The Mishnah is the compilation of the discourse of the rabbis concerning Jewish law, redacted by Rabbi Judah the Prince in about 200 CE. The Gemara, which was compiled by the rabbis over the next three hundred years or so after the Mishnah, and comprises about two and a half million words in numerous volumes, consists of detailed commentary on the Mishnah. To make things even more complicated, there are two quite different versions of the Gemara, one set of volumes written in Jerusalem, the other set emanating from the academies in and around Babylon; occasionally, the versions of the Mishnah from the two centres conflict with each other. Remarkably, although the Torah is recognized as divine, the discussions and rules laid out in the Talmud are in some sense more important in day-to-day life – for these formulations dictate how a Jew should behave, think, observe. There are in turn, various commentaries written on the Mishnah and Talmud.

So the oral tradition is rather like an onion, with layer

after layer of more commentary. At every stage, the rabbinical tradition is one of dissection and argument. Indeed, the Gemara frequently reads much like a conversation – often unstructured, sometimes illustrated with jokes or stories, often quite confrontational and invariably delineating differences of opinion among different great scholars of the past. The Talmud reflects the Jewish character; Jews are a very argumentative lot.

Britain's Chief Rabbi, Dr Sir Jonathan Sacks, tells a nice story. Yeshiva University – the Orthodox Jewish university in the United States – decided to form a rowing eight, complete with diminutive cox. They had an extensive fixture list, but though they trained vigorously, ate kosher steak and large amounts of carbohydrate before each regatta, and prayed regularly, they invariably came last in every race in which they competed. Eventually, a member of the crew surreptitiously took the plane to Boston to watch the Harvard team practising on the Charles River. He came back to New York, highly excited. 'I've got their secret – it's incredible!' he said. 'Only *one* of them gives the instructions . . .'

Much has been written about the authorship of the Bible, particularly the five books of the Torah. Although the Orthodox Jew regards the Torah as divine, this has not prevented Jews from engaging, vigorously and at length, in biblical criticism. In the 1830s, one Rabbi Abraham Geiger from Frankfurt, an outstanding orator renowned for his poetic use of German, denied the divine origin of the Torah. He also ridiculed the dietary laws and advocated abolition of circumcision – thus out-pauling Paul himself. He described Jewish laws and observances as empty utterances without any moral relevance in modern times. But Geiger was a Reform rabbi, and almost certainly hoped to assimilate into his non-Jewish surroundings, where anti-Semitic attitudes and restrictions were all too prevalent. Such views as his are, needless to say, regarded as heretical by the Orthodox community. The Reform movement of which he was part, and the Age of Enlightenment from which it sprouted, was the origin of one of the great schisms in the Jewish community.

It has to be said, however reluctantly, that some of the 'higher' criticism of the Old Testament has anti-Semitic roots. Gerhard Kittel, a German born in 1888, is an infamous example. He was a Christian theologian who became a mouthpiece for the most vicious Nazi anti-Semitism. Possibly he, more than anybody else, has the dubious distinction of making the extermination of the Jews appear theologically respectable. One of his several books set out to use science to demonstrate that the Jews are a criminal race with inherited degenerate traits. Above all, Kittel campaigned vigorously against the Jewish elements in Christianity. It seems reprehensible that biblical texts translated and published by Kittel, who was convicted as a war criminal at Nuremberg, are apparently still used in some seminaries.

Kittel, together with another Nazi theologian, Emmanuel Hirsch of Göttingen, tried to peddle a new theology, namely that the Old Testament, far from being divine in its origin, only had vague sentimental interest for Christians. Further, they claimed that Jesus was not a Jew, but rather was an Aryan of 'pure' race. Another 'scholar' of this period, Franz Delitzsch, held the Jews responsible for Germany's catastrophic defeat in 1918, receiving many compliments from the Kaiser for his lecture on 'Babel und Bibel'. He argued that Judaism – as seen in the Old Testament – condoned ritual murder, and that modern Jews used their biblical text to promote this practice.

Jews believe that the words of the Torah are the very words that God wrote – or at least, the words that Moses wrote as they were dictated to him on Mount Sinai. But denial of divine authorship of the Torah goes back a long way. As early as 250 years after Christ, Father Origen argued that Moses was not the writer of the five books. Jean Astruc, a French physician who made a detailed study of the Torah,[3] tried to prove that the bulk of these books were derived from two separate sources, called E and J – so-called because God is referred to as either Elohim or Jahveh in different parts of the text. More recently, other critics have added other presumed sources – P for the priestly code of writings in the Torah, and D for the writings in Deuteronomy. Sometimes these four sources, it is claimed, are intertwined with each other, sometimes they apparently contradict each other, and sometimes they appear to run parallel to each other. To this somewhat tangled hypothesis, another character is often added – R, the redactor – who is supposed to have been the collator and proofreader of the Bible. Clearly, though, given the many contradictions that remain, he did an editorial job outstanding for its inadequacy. More recently, other critics have ascribed part of the scriptures to other literary sources, known in scholarly shorthand as C, K, S, Pg, P1, and P2. As the Jewish scholar Emanuel Feldman has written, 'the Scriptures were starting to look more like algebra than a Bible'.

All this may sound very dismissive – a Jew denigrating Christian scholarship where it concerns the Old Testament. This is very far from the truth. Many of the best critics and analysts of the Old Testament have undoubtedly been Christian scholars, with a detailed knowledge of Hebrew and Aramaic and great respect for the integrity of the texts. They flourished in many of our universities in the eighteenth and nineteenth centuries, leaving an important body of commentaries, concordances and dictionaries. One of my most prized possessions, in my quite large library of Judaica, is an outstanding commentary on what I regard as the most difficult book in the Bible – the book of Job. It is inspired by Dr Hermann Bernard of Cambridge University, who was born in Germany and died in 1857 in England. The commentary is edited and published by another non-Jew, his faithful pupil Frank Chance, BA, who studied with him in Cambridge. Chance draws attention to Bernard's insistence that a superficial knowledge of Hebrew is quite insufficient, indeed very likely to be misleading, in attempting to understand the more difficult biblical texts. Bernard emphasized that many years of exposure to classical Hebrew, a good training in the rabbinical Hebrew of the Talmud, and a sound understanding of Aramaic as well were essential preconditions for any really accurate translation and meaningful commentary. Alas, such scholarship nowadays is much less common, and even many Christian theologians do not always have a really sound working knowledge of the Hebrew of the Bible.

Abraham's God – the only God?

How much of the biblical account really happened? The story of Creation aside, we know that, at the end of the third millennium BCE, society in the Middle East was disrupted by a series of invasions from the east. These interlopers caused great upheaval in Egypt and Palestine – we can see from the archaeological record that whole towns were destroyed and abandoned. These invaders, coming out of present-day Iraq and heading towards the Mediterranean, spoke a West Semitic language. A series of Mesopotamian tablets refers to them as Hapiru or Habiru – in other words, Hebrew. Egyptian sources call them Abiru. But these were not desert nomads, for whom other terms already existed. Instead, these Habiru were rather more difficult to classify and because of that, just as modern-day traveller groups do in Britain, they attracted hostility and suspicion. Sometimes they stayed in one place, sometimes they went on the move. Sometimes they worked as mercenaries, or servants, or merchants. Each group of Habiru had its own leader, who wisely attached himself to the rulers of whatever region he found himself in. By comparing archaeology to the biblical record, we gain a fascinating picture of Abraham and his kind in history.

This history commences with the journey of Terah, who, the Bible tells us, left Ur of the Chaldees along with his son Abraham and the rest of the family. The archaeological record shows that until about 1900 BCE (very approximately the time when Abraham may have lived) Ur was a comfortable, thriving metropolis, whose inhabitants had security and a good standard of living. It is not clear why this family decided, as the Bible says, to leave Ur, though the record suggests that at this time the city was in economic decline. The Akkadian dynasty there had lost an important battle against the Amorites, and things were probably a good deal less stable as a result. Genesis tells us that Abraham and Terah set off for Haran, a city prospering from its position on a major trade route, 1,200 kilometres to the north-west (just inside present-day Turkey, north of the Syrian border). *Haran* in Hebrew and Akkadian means 'road' or 'route', and in the ancient Hittite language *harvana* has the same origins – giving rise to our modern word 'caravan', possibly through a Persian word, *karwan*. The trek to Haran would have been no mean journey – but it was certainly possible, as is shown by excavations of numerous clay tablets from this region. And Haran was taking over from Ur as the focus of the Moon Cult at around this time – which might have been at least part of the motive for Abraham's father in undertaking the journey.

After Terah had died in Haran, the story of Abraham

begins in earnest. It starts with a command from God: 'Get thee out of thy country, and from thy kindred, and from thy father's house, unto a land that I will show thee' (Genesis 12: 1). Abraham has a revelation that a multitude of different gods – worshipped by his ancestors, and still revered by his relatives, friends and neighbours – count for nothing, and that in fact there is just one God, referred to in the Hebrew text as YHWH (often referred to as 'Jahveh' by non-Jewish sources).[4]

So Abraham, in response to his new belief and to escape his pagan environment, starts travelling once more. This time he travels even further west, to Canaan – a substantial journey of some 700 kilometres. We cannot know exactly who this Abraham, the first of the patriarchs, was. The archaeological record cannot tell us. But the Bible gives an extraordinary amount of detail about him and his environment; and it also tells us a great deal about what he was like as a man. He was almost certainly one of the Habiru tribal sheikhs who spoke the Western Semitic language. We know that he was a tent-dweller, though not a nomad. But until he bought the land in Hebron as a place for his wife's burial, he had never owned any land anywhere. His family in Mesopotamia were unlikely to have been farmers, for farmers would have not have willingly left the region of Ur, where they would have had a stake in the harvest. Perhaps Terah, his father, was a merchant – possibly he even traded in the idols of the Moon Cult that were worshipped in Ur and Haran. We certainly know that soon after Abraham left Haran for Canaan he became a man of considerable power and influence, as might befit a tribal chief. Genesis chapter 14 tells us that he was able to muster a substantial force – 'he led forth his 318 trained servants, born in his house' – to pursue the captors of his nephew, Lot. Earlier, in Genesis 12, we see him having a sufficient status to deal with the Egyptian authorities, when he was escaping a serious famine in Canaan.

Most importantly, as with all the leading players among the forebears of the Israelites, the Bible does not seek to whitewash Abraham in any way. He, and his immediate descendants, are portrayed warts and all. Thus his relationships with the Egyptians, whose protection he seeks during the Canaan famine, are characterized by unease and deceit. Knowing that there is a possibility that they may lust after his wife Sarah, a notable beauty, and kill him to procure her, he tells Sarah to pretend to be his sister – a ruse that would not necessarily save Sarah; Abraham seems primarily concerned to save his own skin. Moreover, he subsequently repeats this failed 'trick' in his dealings with Abimelech when journeying through Gerar.

The Bible presents a novel view of Abraham's God. He intervenes directly in history, changing the ethnic patchwork of the Near East by ordering Abraham to move on. But this is not the dawn of strict monotheism. Nowhere does the book of Genesis tell us that this YHWH is the only God. In fact, there is considerable disagreement on the matter. An episode which gives us considerable insight into this question revolves around the position of Eliezer in Abraham's establishment.

When Abraham complains in sadness about his lack of an heir, he says: 'Oh Lord God [YHWH – Elohim] what wilt Thou give me seeing that I go barren; he that shall be possessor of my house shall be Eliezer of Damascus?' Eliezer, although a servant, is effectively Abraham's adopted son. Archaeology provides corroborating evidence for this arrangement. A huge number of tablets from the Nuzu site, excavated in north-west Mesopotamia in the 1920s and 1930s, have now been translated. Nuzu is part of the area from which the people of Haran came, and not so far from Damascus from where Eliezer presumably originated. Many of these thousands of cuneiform texts concentrate on family relationships. A common topic is that of adoption. Couples without children would adopt an heir like Eliezer on the understanding that the adopted person would relinquish any privileges of inheritance if a natural child were born subsequently. And when Abraham finally does have a child, Isaac, it is to Eliezer he turns for help, sending this most trusted member of his large household on confidential family business to find a suitable bride for his son.

But while Eliezer may have been Abraham's favourite and his adopted son, he prays in a different way from Abraham, and may even have a different god. When he is sent by Abraham on his highly important mission to find a possible bride for Isaac, Abraham's son, he prays: 'Oh Lord, the God of my master Abraham, send me, I pray thee, good speed this day and show kindness to my master.' Once his mission is accomplished – Rebecca is found – he prays again: 'Blessed be the Lord, the God of my master Abraham who hath not forsaken His mercy and His truth towards my master.' Even though Eliezer is a member of Abraham's family, the repeated wording of the Hebrew text with its reference to 'the God of my master Abraham' implies that Eliezer may worship a different God or gods, but on this occasion specifically seeks YHWH's help.

Incidentally, the tablets found at Nuzu give other evidence broadly supporting the biblical account. Faced with her husband Abraham's long-standing pain at not having his bloodline continued because she is sterile, Sarah, his beautiful and deeply beloved wife, offers her handmaiden, Hagar, to her husband. This very early example of surrogacy was an attempt by Sarah not only to assuage her husband's grief, but to gain status. If Hagar has a child on her behalf, Sarah says, 'Perhaps I will be builded up through her.' The Nuzu tablets indicate that this kind of surrogacy agreement was a widely accepted custom in this part of Mesopotamia. When a wife was infertile, a favoured handmaiden might be presented by the wife to her husband, so that he could have children. And some of the tablets imply that the wife would have 'authority' over any child that was born.

Those are the bald facts; but the story in Genesis does not hide the potential for human disaster when such arrangements are contemplated. Sarah and Abraham do not come out of this episode with their reputations unstained. When Hagar conceives, she mocks Sarah. After Hagar gives birth to Ishmael (Genesis 21: 8–21), Sarah becomes hugely jealous and has her thrown out of the household into the desert. Without adequate shelter and provisions, Hagar and her baby are clearly likely to die. And Abraham does nothing to prevent Sarah. He simply refuses to take any responsibility: 'Behold,' he says, 'the maid is in thy hand; do to her what is good in thine eyes.' His only intervention is simply to give Hagar a bottle of water and a loaf of bread to take with her on her journey into the desert. It is at this point that God intervenes. He sends an angel (the first time an angel is mentioned in the Bible) who persuades Hagar to return to Abraham and tells her that her son Ishmael will eventually become the father of a great people.

At this point God assumes another name. He appears to Abraham once more and introduces himself as El Shaddai: 'God Almighty' or 'God the Powerful'. An alternative translation might be 'God the Dispenser of Benefits'. Yet, while the name may have changed, there is no doubt that this is the same God as before. He reiterates his promise of many children for Abraham, making a bargain with him: 'Walk before me and be whole-hearted.' That can be taken to mean that Abraham has to be whole-hearted about this God – to follow only him. And at this point God draws up a covenant between Himself and Abraham. He asks for this promise, and, as a sign of commitment to the covenant between them, instructs Abraham to ensure that all male members of his household, and all their descendants, are circumcised.

This one God is different from other gods who have gone before. He accepts that Abraham is not wholly good – no man is – but does not criticize so much as exhort him to better things. Furthermore, he is capable of bargaining with man, of holding a rational exchange – indeed, uniquely, this is a God who is prepared to allow men actually to argue with him. When God threatens to destroy Sodom, the wicked city in which Abraham's estranged nephew, Lot, lives, Abraham is forthright. If just fifty righteous men live there, would it be morally right to destroy it? he challenges. And the debate is bold, blunt, scathing, almost sarcastic. 'That be far from Thee to do after this manner to slay the righteous with the wicked, that so the righteous be as the wicked; that be far from Thee; shall not the Judge of all the earth do justly?' When God accepts that if there are fifty good men in the city He will spare it, Abraham continues to bargain unashamedly: 'Possibly there may be five short of the fifty; wilt Thou destroy the city for lack of five?' And God agrees; if there are forty-five good men he will spare the city. Abraham continues to reduce God's position almost to the point of absurdity. What about forty men? Or thirty? Twenty? Or just ten? And there Abraham stops – even though the implication is that there may be even fewer righteous inhabitants, perhaps just one.

But Abraham's single God is all-powerful, all-knowing. It is therefore unthinkable that he could not precisely know the moral demography of Sodom's population. So, although this passage seems to be a test of God, it is surely in a sense rather a test of Abraham. In Abraham, God is provoking man to show himself as a man, with a sense of justice. This theme sets the tone for much of the Hebrews' monotheistic religion. Thereafter, the concept of justice will recur time and again as a critical feature of their belief and of the observance of Judaism.

Abraham's debate with God initiates another critically important feature of this religion, too. Man argues with God; like Jacob, he wrestles with God. This theme continues right through the Old Testament, up to the last person in the Tenakh to whom God appears. Job is tested by God. Is he truly righteous? Would he still worship you, Satan asks God, if you took away all the good things in his life and afflicted him? And when Job is brought low, he does not submit meekly. He argues with God. He asks God to justify his actions. It is almost as if, rather than Job being on trial, it is God himself, and his innate justice, that is being tested.

Moses, too, does not immediately accept God and what he says automatically. When confronted in the desert by the burning bush, Moses hides his face. He is too frightened to confront God; yet he still refuses to obey him without discussion. Who am I, he asks, that I can go to the pharaoh and ask him to let the children of Israel out of Egypt? And then, extraordinarily, Moses asks God his name; without a name he will be powerless to explain to the Israelites that he, and they, really are supported by an all-powerful God. 'Behold, when I come unto the children of Israel and shall say unto them, The God of your fathers hath sent me unto you; and they shall say unto me, What is his name. What shall I say unto them?' God replies enigmatically but definitively: 'I AM THAT I AM.'[5] This is a declaration of uniqueness and sufficiency – of the unity and spirituality of the divine nature, in complete distinction from the polytheist worship of the Egyptians, who practise idolatry and pray to human, animal and celestial forms. The Hebrew expression 'I am', EHYH, has the same root (*hayah*) as the name given to this God, YHWH.

For the greater part of the journey out of Egypt, the journey from slavery to nationhood, Israel does not whole-heartedly accept this idea of one God. Israel's idea of God still seems closer to that of the Egyptian pharaoh Akhenaten – a superior god among many. As soon as Moses, who has embraced the notion of a single omnipotent God, disappears up Mount Sinai, the people lose confidence and call on Aaron, Moses' brother, to help them make a golden calf. Monotheism, an extraordinarily revolutionary idea which marks a watershed in human history, is an unfamiliar notion. It took time to grow, in specific historical circumstances. The great Jewish commentator Maimonides suggested that the reason why Moses led the Israelites through the wilderness for forty years was not so much to get the Israelites out of Egypt as to get Egypt out of the Israelites.[6] They were wedded to paganism and needed time, and tribulation, to help them accept this notion of a new, solitary God.

When YHWH concludes the covenant with the Israelites in the desert, prior to their re-entry into the land of Canaan, it is on the basis that they reject other gods and worship him alone. The commandment 'Thou shalt have no other gods before Me' reflects the idea that there certainly *were* other gods who could be worshipped. To this day, Jews recite the prayer known as the Shema, which begins: 'Hear O Israel, the Lord Our God, the Lord is One.' But perhaps when that verse was first read to the Israelites, the phrase *Adonai ehad*, 'the Lord is one', would not have implied a monotheistic notion – only that *our* Lord was the only one *we* were to worship.

All this tells us something about the Israelites, possibly about all humans. Not all ideas about God are popular. Some are very difficult to assimilate into daily life. Most views of human history are built around an assumption that we adopt only those beliefs which assist our survival and reproduction – a thread that runs right through the book of Genesis. Some anthropologists tell us that every ritual serves a purpose in bonding communities together.

Marxists tell us that the Divine Idea is a way for one group to dominate another. Psychologists tell us that religious ideas merely 'squat' on top of other unconscious mental systems developed for survival. In this light, religion seems inherently 'sensible'. But when we return to the accounts of people who actually lived in these belief systems, we see a different picture. The Divine Idea can cause pain and misery. It can impose demands that run contrary to a peaceful existence. And yet it endures.

Jephthah's vow: a Greek tragedy in Israel?

There is a poignant tale in the book of Judges illustrating the Israelites' ambivalent attitude to monotheism. It tells the story of Jephthah, the illegitimate son of Gilead, conceived after his father consorted with a prostitute. The rabbis see the story as a cautionary tale, condemning Jephthah as a man who takes a rash vow and ends by doing something deeply sinful.

Early in the story, Jephthah is ejected from his father's home by his elder brothers because they don't want to share their inheritance with an illegitimate half-brother. The exiled Jephthah falls into bad company ('vain men'), joining a group of bandits. At this time the Israelites have settled in Israel, but the south of the country is still surrounded by hostile foes – the Philistines to the west and the Ammonites in the east – and Jephthah and his new cronies seem to end up serving as mercenaries when the Ammonites stage a hostile attack. Because of his fearless performances on the battlefield, the Israelite leaders invite this untamed outcast to lead their army. Jephthah agrees; but, having longed for respectability from his earliest days – 'Did you not hate me, and expel me from my father's house, and why are you come to me now when ye are in distress?' – only on condition that they make him their leader in peacetime as well.

Jephthah proceeds to engage in long diplomatic negotiations with the Ammonites. He recounts the history of the settlement of the Israelites since they crossed over into the Promised Land, and argues that they have a right to be there on legal grounds. While he acknowledges that God helped this conquest at times, the tone of the negotiations is that of a diplomat rather than a committed monotheist. Jephthah then makes a solemn oath to the God of Israel, asking for success in war and pledging in return that, if he gets home safely, he will sacrifice as a burnt offering 'whatsoever first cometh forth from the doors of my house to meet me'. With God on his side, Jephthah defeats the Ammonites and seizes twenty cities 'with a very great slaughter'. On his victorious arrival home, the bargain he believes he has struck with God takes a darkly ironic turn. For Jephthah, who has long wanted a settled existence, has no other family than his beloved, beautiful daughter – and it is she who first rushes out joyfully to him, 'with timbrels and dances'. He tells his daughter of the oath he has sworn and that he cannot go back on his word to God; and his daughter accepts the contract her father has struck. She asks only for two months' grace, so that she and her companions might 'go up and down upon the mountain and bewail my virginity' (Judges 11: 37).

The story of the flawed Jephthah is a double tragedy. Not only does Jephthah, having finally found stability in family happiness through his only child, find himself bound to destroy her; but the vow itself is an indication of how little the Israelites understood God, for human sacrifice was a heathen practice.

In Israel, this painful passage towards monotheism continued for centuries. The Israelites were hardly ever loyal to the covenant they had forged with God. When King Solomon built a temple in Jerusalem, it was dedicated to

YHWH, but nevertheless structured just like a Canaanite temple. Solomon himself had many pagan wives, who continued their own religious practices undeterred. The worship of YHWH was almost swamped in the early ninth century BCE, when King Ahab's pagan wife Jezebel imported great numbers of priests of Baal to the northern Kingdom of Israel. This move was challenged by the wild prophet called Elijah, a man who was fed by ravens and stalked the earth in a hairy coat. When he challenged the priests of Baal to a rain-making duel and won, Elijah was not content to let them abandon their mistaken faith. Instead, he had them all put to death on the spot. Paganism, with its many gods, is a rather elastic kind of religion – there's always room for another god or two. But the emerging monotheism of Israel was very different: it demanded repression and denial of the validity of other beliefs. This exclusivity has to some extent characterized much of the history of the three faiths that sprang from it – Judaism, Christianity and Islam. But Judaism, unlike Christianity and Islam, soon stopped seeking to convert people of other faiths, preferring the view that 'the righteous of all nations shall inherit the world to come'.

But not even the thunder and lightning brought down by prophets like Elijah were sufficient to bring the idolatrous people of Israel into line for long. The Kings Manasseh and Amon, his son (687–42 and 642–40 BCE), actively encouraged the worship of Canaanite gods alongside YHWH. King Manasseh even put up an effigy of a Canaanite goddess in the Temple at Jerusalem. The idolatry of the Israelites had never been worse since the time of Moses. At this point in Jewish history, there occurred the extraordinary succession of an eight-year-old boy to the throne. As soon as he was able, King Josiah, 'who did what was right in the eyes of the Lord' (II Kings 22: 2), decided to put a stop to the religious and moral depravity into which the kingdom had fallen. He started by giving commands that all the idols be ground into powder, and the groves where sodomites and prostitutes lurked be destroyed. When he was twenty-six years old, he began making extensive refurbishments to the Temple in Jerusalem. During the building work, carried out under the supervision of Hilkiah, the High Priest, some work-men – their foremen, the Levites, are named in the book of Chronicles – found an ancient manuscript in the Temple foundations. The Bible recounts how Hilkiah handed the scroll to Shaphan the scribe, who, after read-ing it, brought it to the King. It seems that the Temple priests were claiming that this scroll was the fifth and last book of the Torah – Deuteronomy – a sensational find, if true. Shaphan read it aloud to the King, who, on hearing it, is reported to have immediately 'rent his clothes'.

Presumably the reason why Josiah reacted so violently was because the scroll (if indeed it was the lost book of Deuteronomy) contained the most terrible curses: this, it said, is what God promises to do to the people if the children of Israel do not adhere to the Law. Deuteronomy, the fifth book of the Torah, is substantially different from the four preceding books. It is essentially a sermon, a discourse seemingly written by Moses, often in the first person. It contains little history, beyond a retrospective summary of the travels that Moses made through the wilderness of Sinai. It combines exhortations by Moses to keep the Law, and emphasizes the ceremonial rules con-cerning such matters as the centralization of worship, the prohibition of idolatry, tithes, clean and unclean foods, and the solemn festivals. Criminal law is also covered extensively, with statements about homicide, property and false testimony. There are sections on family rights and moral conduct in war, as well as on education, the supremacy of justice, democratic institutions and the dignity of labour. Key to the book is a reiteration of the Ten Commandments and the opening statement of the Shema, central to Jewish monotheism: 'Hear O Israel, the Lord Our God, the Lord is One'. Although the author of the book seems to be Moses, there is some difficulty about the last eight sentences of the book, where his death and burial are described. Some rabbis writing in the Talmud suggest that these sentences should be ascribed to Joshua, who took over the leadership. Others suggest that they were dictated by God to Moses before his death.

Even today, the section of the book containing the curses does not make easy reading. If the people do not follow the Law, every imaginable horror – drought, heat, floods, boils, pestilence, famine, sterility, early death, death of children, madness, blindness, destruction of civilized society and exile – will be visited upon them. The language is exuberant and overpowering, but is interrupted periodically with promises of good times if the people are righteous. It is easy to imagine the profound impact of such a proclamation on a young king like Josiah, altruistic and God-fearing, and his horrified reaction as the scribe Shaphan read the scroll aloud to him. And because the curses are national – rather than individual – Josiah must have been acutely aware that he, as the national leader, must take responsibility for changing things. So Josiah's first act was to take the scroll back into the Temple. There he called an assembly of the people, and personally read it aloud to them.

Some critics have argued that the finding of this scroll was extraordinarily serendipitous – so much so that it may well have been a hoax. They claim that it seems extremely unlikely that a scroll of this importance could just have been lost in the Temple, that the scroll contains certain anachronisms (such as the reference to the conquest of Canaan, which had not taken place by the time of Moses), and that it does not seem to have been written in an homogeneous manner. On this level, one explanation could be that the scroll was an ideal way for the priesthood – through the King – to reassert its religious authority.

Deuteronomy has been described as a 'mission-statement for xenophobia'. Moses tells the Israelites that when they arrive in Canaan they are to have nothing to do with the inhabitants. There must be no intermarriage, no social relations of any kind. Moreover, they are to seek out and destroy all signs of the Canaanites' religion among the Israelites: smash their temples, set fire to their gods. This, it has to be added, is not because YHWH is the only god, but because YHWH is the only god who should matter to the Israelites. So the accusation of xenophobia is not entirely fair. Deuteronomy makes it clear that it was only the Jews who should not be idolaters. God created other nations, too; but as long as other nations did not indulge in immorality or crime, their practices were to be tolerated.

This is why I do not believe that this book was just a clever 'hoax'. It surely was not devised merely to combat the practice of fusing YHWH-worship with elements of Canaanite belief – so-called 'syncretism' – as has been claimed. But, whatever its provenance, it is quite plausible that the scroll had been hidden in the Temple by an earlier generation of priests, possibly at the time of Hezekiah or Manasseh. And there is also another possibility: namely, that the scroll that was dug out was not Deuteronomy at all, but some other writings specifically composed to help the King reform the country and realign its religious practice.

The publication of the scroll did signal something of a national resurgence. But this was only temporary. Still Israel refused to recognize YHWH as its sole god. In time, a new wave of religious ferment, spearheaded by prophets, began to claim that this was very dangerous. Political developments – mostly the rise of powerful aggressive nation-states on the fringes of Israel's territory – were interpreted as signs that YHWH was angry with

Israel. It was this movement that gave birth to true monotheism.

Monotheism: why bother?

Before we look at what the prophets had to say, it is worth asking a question or two. Why did Israelite religion view itself as so opposed to Canaanite religion? What was so wrong with Baal worship that its proponents had to be put to the sword? And why did it remain so attractive to the Israelites, in spite of so many dire warnings from men and punishments from God?

The Israelites had a unique approach to time and history. In the belief systems of their Near Eastern neighbours time was cyclical, renewing itself like the natural world in periodic festivals, such as the Akitu, the Babylonian New Year. Try as it might, the Israelite religion never quite divested itself of these traces: its legends, as we have seen, share much in common with other local stories, the names of its sacred places retain the linguistic traces of an earlier tradition, its rituals and festivals retain elements of a shared belief.

Yet at the head of the Israelites' religion was a God very unlike the gods of their neighbours. This was a God who interfered, made events happen, yet who remained above and beyond the world. His actions could be seen in the world; but, after Moses, He appeared as one who was unknowable and unseeable. When Moses asked God to explain who He was, God replied with the elliptical utterance quoted above. Ancient Hebrew employs a form of words like this to express uncertainty and mystery. For example, when we see a phrase in ancient Hebrew that means literally 'they went wherever they went', this means 'they went somewhere', 'we don't know where they went'. God's words to Moses seem less like an explanation and more a rebuke. He was expressing His essential beyond-ness, His separateness from the world, the inability of men to know Him.

Across time and space, ideas about God differ most crucially over this aspect. Is the god transcendent – above and beyond the world; or is he immanent – present within the world, in nature, in a child's smile or a satisfying meal? Every major world religion contains elements of both views within it, sometimes happily co-existing, sometimes at war, sometimes alternating, one replacing the other. Ancient Israelite religion, at a certain point, came under the sway of the transcendental view.

What might be the reasons for this? One possible, speculative answer is that the 'immanent' view is best suited to a relatively peaceful, settled existence. God is all around us, in the cyclical patterns of the year. But for a people on the move, not quite nomads, not quite settlers, permanently clashing with their hosts, alternately enslaved and escaping, this God-idea might not have been so helpful. In a world of such uncertainty, it is better, perhaps, to put one's faith in a God who stands well beyond the world.

Humans seem to have as strong an instinct not to belong to certain groups as to belong to others. Human history is a story of one group fighting with another. However large the group, we feel a distinct drive to set its limits somewhere, and express its separateness from other groups. We can see this now in the interminable debates surrounding the European currency. Although Britain is undoubtedly a 'part' of Europe in a myriad cultural and political ways, many British people fear being subsumed into a pan-European state, equating this with the loss of their identity. In every country and culture, a variety of methods exist to stress our unity with people who are like us, and our difference from people who are not: language, dress, eating and drinking, attitudes to sex and of course, beliefs about God.

The ancient Israelites were beleaguered. Constant movement from place to place engendered considerable insecurity. Did they stay and adopt the customs of the land they were in, or did they fight to retain their own ways, and thus remain conspicuous and an object of suspicion? This is part of the dilemma experienced by many immigrants in Britain today. Some of them respond to the crisis by reaffirming their cultural identity in the most uncompromising terms possible. We can see Islamic integration in this light, as well as, for example, the world-rejecting ethos of Rastafarianism, whose dreadlocked adherents describe white, western civilization as 'Babylon', meaning corrupt and barbaric.[7] So we should not be surprised if the ancient Israelites expressed their insecurities in a similar way. They developed a Divine Idea that was right for them in being utterly unlike the Divine Idea of their neighbours, the Canaanites. By worshipping this God above all others, and by maintaining that their unique cultural traditions (the prohibition on eating pork, for example) were the commands of this God, the Israelites had a powerful way of retaining their own identity. Certain Israelite laws certainly seem to have existed specifically to differentiate the Israelites from their neighbours. Modern-day Jews refuse to eat meat and dairy products together, and this principle of *kashrut*, the dietary laws, stems from a biblical prohibition on 'seething a kid in its mother's milk'. This was a Canaanite dish. Tattooing and scarification, common practices in neighbouring regions, were similarly outlawed for the Israelites.

But if the Israelites had such a strong desire to retain their identity, why do we find so much evidence of people turning to Baal worship? Perhaps because the Canaanite religion was simply more attractive; it certainly seems more hedonistic. The immanent Divine Idea is much easier to grasp than some transcendent 'Other' who cannot be seen or even accurately spoken of. We see something similar in more modern Catholic Christianity, where the personal aspect of the divine is revered in the form of the Virgin Mary, seen as the intercessor between us and the transcendent God Almighty. Buddhism supposedly rejects the notion of God altogether, yet throughout south-east Asia people make statues of the Buddha and give offerings to them. We desire a strongly 'material' element to our beliefs. An absent Otherness offers no satisfaction to the bulk of people; we want to see our gods, and approach them in meaningful rituals. If people look to God for explanation and consolation and help, it follows that they need the God in question to be accessible. Israelite religion may have helped retain Israelite identity, but it was a difficult strategy to maintain, requiring intermittent bouts of repression to prevent people from turning to other beliefs. It fell to a later generation of religious men to arrive at a more satisfying solution.

Of prophets and priests

The Christian prophet Simon Kimbangu was born in western Zaire in 1887. Converted to Baptism by missionaries in 1915, he worked for several years as a catechist, a religious teacher who prepared candidates for baptism. After a vision in which God commanded him to preach, he healed a young woman in his native village of Nkamba by praying and the laying on of hands. He cured blindness, and on one occasion is said even to have raised a child from the dead. Reports of further miraculous healings followed, and thousands of people flocked to Nkamba every day. Kimbangu – whose name means 'he who has been called to speak' – never baptized new converts, but sent them to the nearest missionary stations. But the missionaries who had colonized the country were divided about his following. The most hostile reaction came from the Roman Catholic Church, which saw him as a troublemaker.

As Kimbangu's popularity grew, and he began to prophesy that Africans would be freed by God from the Belgian colonial government, he attracted unfavourable attention. People were refusing to work; hospitals emptied themselves of patients. Mass rallies gathered to hear Kimbangu speak. The fear of popular rebellion led to the installation of machine-gun posts in the capital's streets. There was also an element of envy in the attitude of the European missionaries. For decades, their African subjects had treated them and their faith with suspicion, believing that the white men kept the true secrets of the gospels to themselves. Kimbangu opened Christianity to Africans, claiming that the God of the Christians was identical to Nzambi, the local deity, and declaring himself an *ngunzi* – a traditional prophet. He had a remarkable attitude to nature, respecting what he saw as the primate cousins of human beings. Kimbangu did not permit his followers to kill other primates. The reasons he gave included the notion that we share a common ancestry with other primates, so that to kill a monkey is tantamount to murder. Kimbangu pointed to the way in which, when left to fend for ourselves in the forest, we mimic the activities of our primate cousins. In seeking to survive we are forced to forage as they do; indeed, to become as they are. Kimbangu taught that it is part of God's plan to unify humans and animals. One day we will lie down with the lion in peace, and this meant not treating nature with arrogance. He was totally opposed to the consumption of bush-meat but also hostile to many other local customs, declaring that God was displeased with the practice of polygamy, with erotic dancing, witchcraft and drums. To the shock of the missionaries, his followers accepted these prohibitions without a murmur.

For once, Protestant and Catholic missionaries co-operated. Acting together, they persuaded the Belgians to arrest and imprison the prophet. An initial attempt to kidnap him at a rally backfired, but after this Kimbangu handed himself over to the authorities. He was flogged and condemned to death, but the sentence was commuted into life imprisonment by the King of Belgium. His twelve 'apostles' and thousands of followers were prohibited from acknowledging him in any way. Kimbangu festered in gaol for thirty years before his death in October 1951. But it is said that he has some 12 million followers still believing in his teachings.

Simon Kimbangu's story tells us something about the nature of prophecy. Wherever people claim to be able to see into the future, and to receive messages from God, others will want them to keep quiet. Prophecy itself can be a force for change and upheaval, a thorn in the side of the powerful.

Take thou away from me the noise of thy songs, for I will not hear
  the melody of thy viols.
But let judgement rain down as waters
And righteousness as a mighty stream. (Amos 5: 23–4)

So spoke a simple shepherd from the hills of Tekoah close to Jerusalem, more than 750 years before Christ. Amos believed that YHWH was angry with a number of cities, such as Damascus, Gaza and Tyre, and that their inhabitants were about to be punished. Whole nations would fall – Syria, Philistia, Edom – but the Kingdom of Israel to the north was singled out for particular opprobrium: 'For crime after crime of Israel I shall grant them no reprieve.' YHWH had, in the past, been a powerful ally, assuring the Israelites of victory on the battlefield; as a result, the Israelites may have viewed themselves as invincible. But that had led to complacency. A number of things were now at fault in the nation, and Amos was about to remind them in the most uncompromising terms. Nor was it just the Kingdom of Israel that stood in peril; Judea, where the Temple stood, was under threat. God, said Amos, was not interested in sacrifices and grand pageantry in the Temple. He wanted strict obedience to the Law, honest, equal dealings between men, and an end to any hint of Canaanite worship. The consequences of not listening to Amos's message would be dire: destruction, death and captivity. 'Lo, the days shall come upon you that he will take you away with hooks, and your posterity with fish-hooks' (Amos 4: 2).

Unsurprisingly, the deliverer of this unwelcome message was treated with hostility. When Amos spoke of the destruction of the house of King Jeroboam, Amaziah, the High Priest of Bethel, reported him to the royal authorities. He ordered him to leave Bethel, 'for it is the king's chapel', and go and prophesy to the south in Judah and Jerusalem. 'Go and eat bread,' he commanded (Amos 7: 12), dismissing Amos as a mere Temple prophet, who issued comfortable, optimistic oracles in return for food. Hosea, another prophet and a contemporary of Amos, was treated in similar fashion.

Amos was just a shepherd and a gatherer of sycamore fruit; but he had a message to deliver, and nothing would stop him. We don't know what became of this brave herdsman, whose conviction was so powerful that he was prepared to walk into the King's sanctuary and prophesy the downfall of the King himself. He may well have been put to death – Amaziah did, at least, give him a chance. But, like so many religious men and women over the years, Amos felt he could not be silent.

There is a poignant, indeed, horrifying, bas-relief in the British Museum in London. It comes from Sennacherib's palace at Nineveh. Just thirty or so years after Amos was prophesying, the Assyrian King Sennacherib invaded

Judah. He laid siege to the city of Lachish, the second largest in the country after Jerusalem, and in 701 BCE razed this well-fortified, seemingly impregnable place to the ground. While the siege was in progress, Sennacherib watched in comfort from his tents just 200 yards away from the city walls. Assyrian tablets describe Sennacherib's siege and conquest of the city, and what happened to the people that survived. The bas-relief shows in detail how the siege engines were ponderously wheeled up to the ramparts, and how numerous archers inundated the skies above. But also it shows the men and women, as well as many children, being taken away from Judea – some impaled on spears and carried away on hooks.

Fortunately for Jerusalem, just a few miles away, King Hezekiah learned his lesson from the fall of Lachish. When Sennacherib was hammering at the doors of the capital city, the King listened to Isaiah. He repented, donned sackcloth and ashes, and prayed. And one night in the middle of the siege 5,180 of Sennacherib's troops succumbed to a plague. By the morning they were dead. Sennacherib returned forthwith to Nineveh.

It is alleged that when he first had his vision of God, Simon Kimbangu ran away, seeking to escape his destiny. Throughout history, prophets are often reluctant to undertake their troublesome task of proclaiming what they know will be an unwelcome message. Isaiah experienced his call to prophecy most painfully, as a burning coal placed on his lips, and Jeremiah baulked at the task before him, declaring: 'Behold I cannot speak, for I am a child.'

Another characteristic of prophets is that they sometimes enter states of trance, or ecstasy, in which they experience hallucinations – such as the terrifying visions of Ezekiel and Isaiah. Their behaviour in general is characterized by a sort of 'otherness' – they may wear strange clothes, don magic iron horns, or perform symbolic acts. As well as being intrigued by them, people quite often classify them as 'mad' – as they did the prophet Hosea. In fact, hallucinations and a sense of 'mission' are quite common traits of schizophrenia. Ezekiel was ordered by God to lie on his left side for 390 days and on his right side for 40 days, each day symbolizing a year of Israel and Judah's wickedness respectively. He made bread from dung and was ordered to eat it. But then, under siege conditions, people are not too concerned about the wholesome nature of their diet. He shaved off sections of his hair and beard, alternately burning them and scattering them in the wind to symbolize the forthcoming fate of his nation. Men like Ezekiel must have produced fear and disgust in people, as well as awe.

Unwelcome as it generally is, prophecy is a long-established part of religious practice. But at the same time, their behaviour as outlined above makes them very unlike those other religious 'officers', priests. Priests derive their authority not from personal qualities or being individually chosen by God, but purely from their having *studied* how to be priests. Their job is exercised, not through having visions or seeing into the future or doing magical acts, but through the regular enactment of formal rituals. Prophets speak whenever the mood takes them; priests act in accordance to a fixed calendar. And to my mind prophet and priest have different roles. Priests and rabbis are concerned with religious observance, with the detail of the laws, with the conduct of the liturgy and sometimes the rituals. Prophets are there, at least in part, to speak out about moral issues. Perhaps one of the deficiencies of modern Orthodox Judaism is that comparatively few religious leaders feel able or inspired to give moral leadership about the complex issues which face us. This was a duty that in former times was incumbent on the prophet.

The German sociologist Max Weber (1864–1920) was

among the first to notice these two types of religious 'officer', and to observe that there always seemed to be a tension between them. Later scholars have developed this idea further. Ioan M. Lewis, in his work *Ecstatic Religion*,[8] argued that prophecy is a means of revolt against a stiffly hierarchical society, of which the 'priest' type is the bulwark. Others, such as the cognitive anthropologist Pascal Boyer, think the tension is mainly economic. In many religions (though admittedly not Judaism), trained 'guilds' of priests protected their interests like any other commercial group. They were often the only ones who could read and write, and interpret sacred texts. But inspirational, prophetic pronouncement required no books – just the charisma of an individual. For that reason, its spontaneity contrasted starkly with the staid, regular practice offered by the priestly guilds, who offered, by necessity, a set, uniform religious 'product'. In other words, you never know what a prophet is going to do, but a Catholic Mass is the same no matter what priest celebrates it.[9] The latter may offer a comforting stability – rather like tuning in to *The Archers* of an evening. But the former will always be more entrancing because of its unpredictability and exuberance.

The prophet-type frequently tends to get on pretty badly with the priest-type. So if any one message can be said to unify the outpourings of the various Israelite prophets, written over several centuries, it is hostility to the practices of the Temple, where YHWH was worshipped through animal sacrifices. Amos summed it up when he declared, in the name of God, 'I hate, I despise your feast days, and I will not smell in your solemn assemblies. Though ye offer me burnt offerings and your meat offerings, I will not accept them: neither will I regard the peace offerings of your fat beasts' (Amos 5: 21–2).

For Amos and the other prophets, the worship of ancient Israel had become empty and formulaic – mere lip-service. The people 'fulfilled' the covenant merely by offering sacrifices at the Temple. This is a familiar message, one we hear now from many a pulpit: the rabbis and the vicars of the modern suburbs echo the sentiments of the ancient prophets when they tell us that God isn't ultimately interested in seeing us coming to synagogue or church once a week in our best clothes. What He demands is something both simpler and more complex – an inner goodness, expressed in the way we behave. Unfortunately, where religious leadership still frequently falls down is on the really big and important issues of the day. Only too frequently – unlike the prophets – priests, vicars and rabbis do not seem to want to be controversial.

The prophets in ancient Israel came from widely divergent backgrounds. Isaiah was a member of the aristocracy, Amos a simple shepherd, Jeremiah even appears to have been working as a priest. But they had very similar notions of what it was that God wanted: an end to the superficial pomp and ceremony of the cult and a deliberate turning towards social justice. The oppression of the poor by the rich, the stealing of land, debauchery, drunkenness, vacuous displays of wealth – all are condemned in favour of a simple piety.

But if the prophets were broadcasting a message of tolerance, they were not extending this to people practising other religions when they were sinful. Indeed, their whole attack on the sacrifice religion of the Temple seems to have been based on the idea that it contained too much that was foreign, and specifically Canaanite. Amos points out in chapter 5, verse 26: 'Ye have borne the tabernacle of your Moloch and Chiun [Canaanite gods], your images, the star of your God which ye made to yourselves.' Later, in chapter 8, verse 13, he promises, referring to various holy places of the Canaanite religion: 'They that swear by the sin of Samaria, and say thy God, O Dan, liveth and the manner of Beersheba liveth, even they shall fall and never rise up again.'

The book of Hosea can be seen, moreover, as an attack against Canaanite worship in its entirety. Hosea is a mystifying figure who, on the command of YHWH, married a prostitute because the whole land of Israel 'hath committed great whoredom, departing from the Lord'. The phrase used for 'prostitute' implies that his wife, Gomer, was not so much a street-walker as a woman who performed rituals within the Baal cult, which involved copulation. Hosea went on to give his children rather unfashionable names, calling his daughter Lo-Ruhamah (Unloved) and his younger son Lo-Ammi (Not My People). Hosea focused on symbolizing through his own life the fracture between YHWH and His offspring, Israel.

As we read on, it becomes clear that actually Gomer turned to the Baal cult only after her marriage to Hosea, suggesting that the whole story of his being commanded to marry a prostitute may be a gloss applied later. The book of Hosea reads like the lament of a jilted husband, but a husband who sees parallels between his situation and that of YHWH and Israel. The people of the covenant were running after foreign gods – a symbolic kind of infidelity. His lamentations, however, end on a note of hope: he buys his wife back from her new master and takes her back home, because he still loves her. His actions are a metaphor for YHWH being prepared to give Israel a second chance.

In other respects the prophets' messages are far from uniform. Some show great insight into the political situation of the day, with Israel facing hostile incursion from a variety of powerful neighbouring nations. The prophets seem, perhaps, less like prescient seers and more like shrewd analysts of unfolding history. They all agree that Israel is facing disaster as a result of the twin facts of its idolatrous relationship with 'foreign gods' and a general decline in morality.

The historical background to these prophecies is important. Israel was under considerable pressure in the years between 800 and 500 BCE. When Isaiah had his dramatic epiphany in the Temple, the Kingdom of Judah was under the sway of King Ahaz, who encouraged full-blown Canaanite worship. Meanwhile, in the northern kingdom of Israel, anarchy reigned, while the Assyrians began a hungry plan of expansion. This culminated in an attack of 722 BCE by King Sargon II of Assyria, who conquered the northern kingdom and deported the population, resulting in the loss of ten of the original twelve tribes of Israel. And, as we have seen, in 701 his successor Sennacherib invaded Judah, laid waste its cities, and tortured or deported some 2,000 people. In 587 Nebuchadnezzar set his sights on Jerusalem and deported the entire population to Babylon. In 539 the Persian king Cyrus conquered Babylon and set the exiles free – but only a proportion of them returned to the Promised Land. From then onwards, the people of the covenant were to be a people in exile. These changes, in turn, radically altered their ideas about God.

The hostile political situation of the ancient Near East probably provided another reason why Israelite religion was so concerned with rooting out and destroying elements of Canaanite practice. The enemy outside – Assyria, Babylon – became the enemy within: Baal worship. Fears about something uncontrollable – a mighty superpower – became expressed towards something that was controllable: religious practice. Israelite religion was on the verge of making another leap. And it was the experience of invasion and exile that brought it about. In fact, we can see the roots of this belief in the time before Israel was divided and destroyed by its voracious neighbours. In a political climate where a return to exile was looking likely, the prophet Isaiah delivered the words 'Holy, holy, holy is the Lord of Hosts, the whole earth is filled with His Glory' (Isaiah 6: 3). The reference to the 'whole earth' is important. It expresses the idea that YHWH was no longer a God confined to the Temple at Jerusalem. This was an all-powerful God whose actions were to be felt everywhere, throughout the world. This is further progress towards the acceptance of the idea of one God. Isaiah's predecessors saw YHWH as wholly partisan, protecting Israel, punishing other nations when they threatened the Chosen People. But in the vision of the later prophets, God was seen as on the offensive against Israel *and* its neighbours.

If I had been a contemporary of Amos or Isaiah, perhaps I'd have wanted to ask: What about *their* gods? If YHWH can use Assyria like a chess piece, then surely He is more powerful than the gods of the Assyrians? Suddenly, YHWH may seem less like a god who has power only for Israel and more like the mightiest god around. There is no longer any need to tell me not to worship other gods, because it only makes sense to worship the most powerful one. Indeed, if other gods are less powerful, then they are less god-like altogether. The *only* god who really deserves the name is YHWH. The prophetic view of history is edging closer and closer to monotheism.

God on the move

Few ancient Israelites, perhaps, would have engaged in this form of speculation. But the question became more palpable when thousands were forced into exile. For centuries, the Israelites' religion had been centred on a specific place – the Temple at Jerusalem, the territory of Israel itself. But what happens when these places are destroyed and invaded, when the people are sent away or taken to live elsewhere?

Clearly, some people must have felt that YHWH had deserted them altogether and given up their beliefs. We see this in the lament of Psalm 137: 3, 'By the waters of Babylon, we sat down and wept when we remembered Zion.' The lament offers a very human and poignant insight into the experience of exile. The Babylonians mockingly ask the Israelites to 'Sing us one of the songs of Zion.' This verse has resonances with modern times, when the Nazis forced Jewish musicians to sing and play as their homes were destroyed and their communities emptied. The ancient Israelites respond: 'How shall we sing the Lord's song, in a foreign land?' They console themselves, not by singing, but by thinking of the day when Babylon will be destroyed, and when its young will have their brains dashed out against stones. It is a bitter hymn that echoes the anger and hopelessness that Jews and other exiles must have felt over the centuries.

But, as a counter to this mood of despair, we need to turn to the second half of the book of Isaiah. Nearly all modern commentators agree that the second half of the book, from chapter 40 onwards, is written by at least one other person, someone who has experienced exile and captivity. Clearly he comes from a much later age, and these chapters depict a time when the gods of the Babylonians will be carted away, mere man-made objects without power. Crucially, the writer reports YHWH saying: 'I am the first and I am the last; and beside me there is no God' (Isaiah 44: 6). Here, as in the Shema written in the Torah so much earlier, we find overt reference to monotheism.

This theme, developed throughout the words of the second part of Isaiah, is directly related to the experience of exile. The religion of the Israelites could quite clearly have perished in Babylon. But instead it became a force for unity and renewal – thanks to a reworking of the Divine Idea. God was no longer bound to a specific place, the Temple, where the Israelites could not now be with him. He was everywhere. And because He was everywhere, He was the only true God. This Isaiah takes the radical notion a step further, envisioning a time when not only the Israelites, but also the peoples of Assyria and Egypt, would recognize YHWH as the only God, and He would bless them accordingly.

5

The God of Change

Around twenty centuries ago, an obscure man by the
name of Yehoshua caused a stir in Roman-occupied
Palestine. With a prophet's typical instinct for controversy, this man denounced the decline of the Jewish people
into vice and proclaimed the imminent destruction of the
Temple in Jerusalem. He did so, moreover, during one of
the most important holidays, the Feast of Tabernacles.
People came from all over the country for the festival and
the narrow streets of the city were jammed with cohorts
of the faithful. Word of Yehoshua's claims reached the
Sanhedrin, the Jewish court, and it swiftly had him
arrested. This was an act of political expediency. The
Romans viewed Palestine as a troublesome province with
a hair-trigger where public disorder was concerned. And
the Jewish memory still bore traces of the cataclysmic
invasion – albeit five hundred years before – by
Nebuchadnezzar's Babylonian troops, who destroyed the
Temple and carried off large numbers of the population
into exile. The Sanhedrin felt it needed to be especially
careful because public holidays were occasions when
the Romans were particularly vigilant: an uproar in the

Temple could spark a revolt, rioting and bloody reprisals from troops who were hot, bored, far from home and resentful of the turbulent locals. The Pharisees needed to ensure that peace was preserved at all costs.

The court offered Yehoshua a chance to repent, hoping to quash the matter before it reached Roman ears. Yehoshua refused and was flogged – but even this failed to silence him. Increasingly anxious, and well aware that there was an ever-present current of official hostility towards the Temple authorities, the Sanhedrin handed the troublesome preacher over to the Roman governor Albinus. This was yet more pragmatism on their part. If they were seen to be co-operating in this way, there was less likelihood of the Romans over-reacting to any potential disturbances, and one death was better than thousands.

Finding him undeterred in his convictions, the Roman governor had the prophet Yehoshua flayed to the bone – a gruesome but common spectacle, which none the less must have turned the stomachs of those witnessing it. Any man in this state, his skin in shreds, his body weakened from blood loss, his senses in turmoil from the blows of the lash and the jeers of the crowd, would be forgiven for changing his mind, or losing it altogether. But Yehoshua remained steadfast. The governor was baffled and had him flung into gaol, where this pitiful wreck of scar tissue continued to declare the doom of the Temple from his cell. Observing that he seemed to have no followers among the population, Albinus declared that he was a harmless lunatic and had him freed.

Yehoshua's story might seem rather familiar. Indeed, it has parallels to an event, or string of events, that had taken place some sixty-two years earlier. Another voice decried the Temple practices during a public holiday – this time overturning the tables of the money-changers within the Temple complex. Another man called Yehoshua was arrested by anxious Jewish authorities and handed over to the Romans. Another body was flayed in a futile bid to secure a recantation. The crucial difference was that this Yehoshua, or Jesus as the Greeks came to call him, had followers. In addition, he hailed from Galilee, a zone to the north beyond Judea, outside the Romans' control and known to be a source of insurrection. Worse than that, he said that he was the Messiah foretold in the Jewish religion, a messenger from God, sent to usher in a new age.

All these things made this Jesus more dangerous, and he was put to death by crucifixion. This was not, as the gospels tell us, the result of angry Jewish authorities clamouring for his death to a fair-minded Roman administration, but rather the responsibility of Pontius Pilate, a notorious hanging judge with no sympathy for upstarts. For all his apparent opposition to the Temple, Jesus was sufficiently well connected to have allies within the Sanhedrin; he was arrayed in relatively smart burial garments and given the grave of an affluent person. But, three days after his death, his body was reported missing. Several of his followers reported having met with him. People believed that he had been reborn – like the ancient gods of the Near Eastern myths, with which most were still very familiar.

Two tales of troublemakers, shunted between the indigenous authorities and the ruling occupiers. But Yehoshua, son of Ananias, the freed lunatic, disappears from the history books after his release from captivity. The crucified Jesus, son of Joseph, did not merely enter history with his crucifixion, but changed its shape irrevocably.

Waiting for the Messiah

A view of ancient Palestine is easily coloured by the Monty Python version of events depicted in *Life of Brian*.

It is not difficult to imagine rumours spreading like wild-fire and the Roman centurions standing back baffled as the local population ecstatically embraced the notion that this crucified man had been the true Messiah.

But religions are not born like that. And Christianity took four centuries to emerge into anything recognizably Christian. It began as a wholly Jewish movement, not suppressed by the authorities. Jesus was just one of several men who had claimed to be the Messiah – which, since Judaism expected a Messiah, was not in itself so controversial.

Belief in a Messiah was a relatively recent development in Judaism, probably emerging from the conditions of invasion and exile. One key text occurs in the book of Daniel, written either during the Babylonian captivity or, more probably, just about 200 years before Jesus' birth. Daniel is hardly a thumping good read. Its language is difficult and much of the writing is hard to understand even in the best English translation. Unlike most early Hebrew literature, it is frequently eschatological – that is to say, it contains notions of an apocalypse, of the end of the known world. Much of it is written in Aramaic, which is obscure to the modern reader, and is about curious visions. But towards the end, in fairly simple Hebrew, Daniel foretells the downfall of those who oppressed the Jews, and seems to predict the dawn of a new age when a Jewish Superman will establish a worldwide dominion of righteous people. Daniel has a revelation while at prayer:

> Know therefore and understand, that from the going forth of the command to restore and build Jerusalem, until Messiah the Prince, there shall be seven weeks and sixty-two weeks; The street shall be built again, and the wall,' even in troublesome times. And after the sixty-two weeks, Messiah shall be cut off, but not for Himself . . . (Daniel 9: 25–6)

Years later, Daniel, now an old man in exile, experiences another vision when in a remote spot by a river. He has not washed or eaten, or drunk wine, for three weeks. These are obviously good circumstances in which to be in a receptive frame of mind. His vision speaks about the resurrection of people after death: 'at that time thy people shall be delivered, everyone that shall be found written in a book. And many that sleep in the dust of the earth shall wake up, some to everlasting life, and some to shame and everlasting contempt' (Daniel 12: 1–2).

Apocalyptic ideas involving a day of judgement, an afterlife, heaven and hell, and resurrection may sometimes be born out of extreme social conditions – as, maybe, in modern times suicide bombers have been enticed into their acts of destruction by promises of their salvation. Whether this be the case or not in current times, throughout history and across the world, from the Plains Indians of North America to the island-tribes of Melanesia, peoples experiencing deprivation or oppression have sought comfort in visions of 'the end of the world' and the new era that will replace it. We call these ideas 'millenarian', and they crop up among all kinds of religious people, from the Jehovah's Witnesses to some Hasidic Jews. Perhaps conditions in Judea at the time were ideal for apocalyptic ferment.

It is not immediately obvious, though, that there was great unrest or deprivation, or indeed any particular increase in messianic fervour, at the time of Jesus' birth, or just afterwards. Indeed, Palestine had been experiencing a period of considerable stability. Apart from punitive Roman taxation – the *fiscus Judaicus* – the Jewish community was enjoying considerable affluence from profitable trading. However, this uneasy stability was brought about under the rule of one the most bizarre and most flamboyant characters ever to appear in Jewish history.

Herod sets the scene

Herod's Jewish origins were dubious to say the least; one account suggests that his father was an Edomite slave in Ashkelon. Certainly his mother was not Jewish and therefore, by rabbinical law, he was not Jewish; some sources argue that he converted to Judaism. Herod was probably a bit of a chameleon and most sympathetic to the secular Greek way of life. He first made his mark in 47 BCE, when, supported by his father Antipater, he ruthlessly suppressed a Jewish rebellion in Galilee. Once victorious, he illegally executed a number of King Hezekiah's followers without trial and then established himself as governor of that region. His regime was characterized by corruption and bribery, but eventually it seemed that justice had caught up with him, when he was dragged to Jerusalem to stand trial in front of the Sanhedrin (the rabbinical court). However, he arranged to be accompanied to the capital city by some of his well-armed brigands, and soon managed to intimidate the court. While the Sanhedrin was pondering his sentence (the death penalty was probably being considered), he escaped to Syria with the help of his ruffians.

Once in Syria, Herod soon ingratiated himself with the Roman local government. When news arrived of Julius Caesar's assassination in Rome, Herod espoused the cause of Cassius, one of Caesar's assassins. When it suited him, he switched sides and, with support and considerable military help from Mark Antony, eventually took control of Jerusalem after a protracted siege. He came to power as king in 37 BCE, and almost his first act was to suppress the Sanhedrin, the body that had attempted to try him for murder ten years earlier, putting to death forty-six of its members in an act of revenge.

Herod was king only by the grace of Rome, which was content to let him govern under its own ultimate control and prepared to be tolerant for the sake of reasonable calm and a degree of stability in a troublesome province. Herod did a deal to obtain exemption for Jews from observance of the Roman state religion – an unusual dispensation, because around the Empire religion and state went hand in hand, the Romans tending to deify their emperors. As part of this deal, he was able to impose heavy taxation on Judean citizens, using the income thus acquired to increase his power. Perhaps this is why tax collectors figure so negatively in the gospels.

Once king, Herod exercised virtually total legal and religious control over land and people. A true oriental potentate, he set about strengthening his position by marrying a member of the Jewish ruling family – a Hasmonean, Mariamne. He then embarked on a series of dynastic murders which included most of her family, and eventually, in a fit of wild rage, Mariamne herself. He appointed Mariamne's seventeen-year-old brother Aristobulus as High Priest, and then, when Aristobulus proved too popular with the people, engineered his death by drowning in a swimming pool in Jericho.

Like many despots before and since, Herod embarked on a massive building programme during his reign. Some of his buildings have been preserved to this day. One palace he built was the fortress to the north-west of the Temple Mount in Jerusalem; this was almost certainly the building where Jesus was subsequently tried by Pontius Pilate. The famous fortress at Herodium, and the fortress/palace high in the mountains at Masada, were others whose ruins are well preserved. He built the port city at Caesarea (which he named in honour of Caesar Augustus), which had the most advanced harbour on the Mediterranean coast and extended Judean trade. He also built the huge castellated edifice which still stands over the Tombs of the Patriarchs in Hebron. But by far the most ambitious of all his projects was the rebuilding of the

Temple in Jerusalem. There is little if any evidence that this was driven by his own religious feelings; rather, the rebuilt Temple represented a monument to his power and affluence and probably an attempt to gain popularity among his Jewish subjects. This was the ultimate sop to the masses, who had been far from well disposed towards their profligate king.

The rebuilt Temple was no ordinary municipal project. It took around 10,000 workmen and 1,000 priests almost ten years just to build the walls around the Temple Mount, of which only the Western Wall (the so-called Wailing Wall) still stands. The area of the Temple Mount was massive – far bigger than it had been under Solomon, or after the rebuilding in the sixth century BCE – around seven times the size of the Millennium Stadium in Cardiff. Perhaps as many as five million Jews lived in the Roman Empire at this time, and this large area would accommodate the huge influx of those who came on pilgrimages to Jerusalem three times a year to worship during the Foot Festivals – Pesach (Passover), Shavuot (Pentecost) and Sukkot (Tabernacles). The Temple itself was lavishly constructed alternately of white and blue marble, and the Holy of Holies extravagantly decorated with gold. According to Josephus, the contemporary Jewish historian, the building was too dazzling to look at in full sunlight.

Though a tyrant with despotic powers, Herod was not without redeeming qualities. In addition to being a brave soldier, he was undoubtedly a skilled administrator and a talented diplomat, and was clearly capable of being charming when needed; but he was equally capable of being highly devious. He was an opportunist who had few ties of loyalty or friendship and fewer morals. He was wise enough to be fairly assiduous in trying not to offend religiously observant Jews: thus, for example, to avoid any accusation of idolatry, he did not put any human image (not even his own) on his coinage. But his beautiful Temple was no house of God along the lines of that of Solomon or Josiah. To curry favour with Rome, he erected a massive Roman eagle at the entrance. This caused outrage and eventually a number of young men attacked the edifice at night, smashing the eagle to pieces. Herod sought out these insurgents and had them dragged to his palace, where he ensured that they faced the horrible fate of being burned alive.

After Mariamne's death, Herod entered no fewer than ten more marriages, and had numerous children in and out of wedlock. In 17 BCE his two eldest sons, both Mariamne's children, returned from Rome, where he had sent them to be educated. For years after their return, Herod's court was a place of intrigue and slander, for both young men were deeply bitter at their mother's murder and repeatedly challenged their father's authority, while Herod had chosen his illegitimate son, Antipater, as his heir. Eventually Herod had Mariamne's sons arrested, tortured, brought to trial and executed. Subsequently, Herod – ever the paranoiac – believed that even his chosen Antipater was plotting against him. He too, was tried and sentence of death passed. It is said that the Emperor Augustus, hearing this, exclaimed, 'It is better to be Herod's pig than his son!' Ironically, as the sentence of death was being carried out, Herod himself lay dying in his palace in Jericho.

There is little doubt about Herod's legacy at the time of his death in 4 BCE. His tyrannical manipulation of the mechanisms of the state, his dissolution of most of the pious religious hierarchy and his encouragement of the ever-growing Hellenistic influence among the Jewish upper classes were morally and nationally corrosive. Herod's misgovernment had created a resurgence of Jewish nationalistic feelings. The Temple was now run more as a business and a tourist attraction than a centre of religious observance; its hierarchy was degenerate, controlled in large part by the Sadducees, the ultra-conservative religious group associated with the well-heeled upper class who collaborated with the Romans to keep power. The God-fearing and observant Pharisees, and the Zealots, an extreme religious minority, were marginalized. Judea was now a kingless kingdom, its moral values in tatters, its observant religious leadership and its responsible secular government long since destroyed, and the country held in thrall to an aggressive military occupying power. The threatened remnant of religious Judaism was fighting a spiritual battle against Greek and Roman paganism.

The province of Judea was rumbling, a smouldering volcano, with Jerusalem the place most likely to erupt – and the place where Rome was most prepared to reinforce its authority. The Jews needed a new king, a saviour – a David. This was the Jerusalem in which Jesus of Nazareth was tried, sentenced, and executed on the Cross within four decades of Herod's death.

Early Christianity: Paul, Jesus and God – and Rome

Some years after his death, the movement around Jesus split into two. One branch was centred on members of Jesus' family in Jerusalem, and one on Saul of Tarsus, later known as Paul, whose teachings found an enthusiastic reception among the urban populations of Jews who had emigrated to Greece.

Paul's own vision of what had occurred in Jesus' life, and what his teaching meant, led him to develop a broader idea of the message. He expressed this message in a series of letters to communities of early Christians in the various cities of the Greek Empire – the Thessalonians, the Corinthians, the Philippians. These mission statements by Paul form a substantial part of the New Testament, those books of the Bible concerned with the life of Jesus and the early foundations of the Christian religion.

In Paul's letters Jesus was turned into a divine figure, because this would have been more appealing to his Greek audience, who had no pre-existing ideas about what a Messiah was. The Greeks were conditioned by their beliefs to turn anyone they admired into a god. In another book of the New Testament, the Acts of the Apostles, we read that Paul and another early Christian missionary, Barnabas, were mobbed by crowds believing them to be incarnations of the gods Hermes and Zeus. Paul never actually declared Jesus to be a god, but over the years the nature of his writings changed, as he spoke of Jesus in increasingly divine terms and referred often to his resurrection.

When religious movements are centred on a charismatic leader, grave problems can occur if the leader dies without naming a successor or having any offspring. This was one of the reasons for the Sunni/Shia split in Islam, with both groups claiming to be the true successors of the prophet Muhammad. Generally, where there is no clear line of succession, the religious movement has to rethink some of its beliefs and activities. This, effectively, seems to be what happened in the years after Jesus' death.

As Paul proclaimed that Jesus was a god, he also made his message more universal. This was no longer a Jewish Messiah who was going to make everything good for the Jews; on the contrary, Paul said, Jesus had intended all men, all nations to be part of a new covenant between God and humanity, sealed with his sacrifice. What this covenant offered was not protection, as offered under the old treaty of the Israelites with their YHWH, but eternal life.

This attractive message appealed to Jews and pagans

alike. After the Romans under Vespasian, and then Titus, brutally crushed a revolt in Palestine in 66 CE (it is said that during the Jewish Wars the Romans killed around one million Jews) and four years later destroyed the Temple for the last time, Judaism no longer seemed a wise or popular faith to profess. In contrast to the earlier years of the first century, when Judaism had found converts among the pagans (including many Romans, who were known as 'Godfearers'), the faith now became more inward-looking; devotees of Christ were expelled from its synagogues because of their refusal to accept the Torah, the Jewish Law.

Christianity appealed particularly to women. The gospels, accounts of Jesus' life written in the first hundred years after his crucifixion, contained many references to Jesus' dealings with women, whom he seemed to treat as equals, even if his disciples were an exclusively male team. For example, we are told in the Gospel of Mark that, after his death, three women came to the tomb where Jesus was buried with the intention of anointing his body. He had been buried in a cave which was sealed by a rock. When the rock was rolled aside, the women found that the body was gone. Later, Jesus appeared to one of the women, Mary Magdalene, confirming that he had risen from the dead. The fact that this account survived in the gospels, instead of being doctored to make the principal witnesses male, suggests a radically different attitude to gender from that which prevailed in mainstream Judaism.

But it would be wrong to depict the evolution of this cult into the bedrock of western culture as a seamless process. Its early years were characterized by back-biting and faction-fighting. Saul/Paul, disillusioned with the Greek-influenced outlook of his peers in Tarsus, had become a Pharisee. Jesus himself was one. This was the sect that believed in the sanctification of daily life through adherence to the oral, rabbinical tradition of Jewish Law as well as what was written in the Torah (the Sadducees believing only in the written law, as laid down in the five books of Moses). A deep spiritual crisis, culminating in his vision on his way to Damascus, led him to reject Pharisaism and even Judaism. This, in turn, meant that he no longer had any common ground with the strictly Jewish Jesus-cult in Jerusalem. His letters to the new Jesus-worshipping communities around the Mediterranean suggest a great deal of confusion as to what Jesus had said in his teachings, and how his followers were to live.

It would also be wrong to suggest that the Roman Empire had a *laissez-faire* attitude to every cult that sprang up. Many historians have implied this over the centuries, suggesting that it was the Jews, not the Romans, who were intolerant towards the new faith. Although pantheism, by its nature, nearly always has room for another god or two, there were a number of reasons why Rome was deeply suspicious of Christianity. When the Christians made it clear that they were no longer Jews who belonged to the synagogue, the Romans viewed them askance. As Karen Armstrong notes in her *History of God*, Roman thinking was essentially conservative. Deliberate breaks with the past were seen not as creative, but as destructive and dangerous. Christianity, moreover, preached love and peace, and frequently non-violence too: this was a subversive religion which, if it spread, could undermine the very infrastructure of a militarized colonizing power. Aggressive, unfeeling soldiers who would not dream of turning the other cheek were likely to be much more effective than soft-minded individuals who wanted to spread friendship, compassion and harmony.

Roman society was much more accepting of rather militaristic and masculine belief systems – like the peculiar cult of Mithras, which appealed to soldiers and the ruling administrator class, made a virtue of strength and military might, and probably attracted quite a bloodthirsty group of followers.[2] Almost certainly this is why temples to Mithras were built by soldiers in the far corners of the Empire, and are still being dug up by archaeologists today. Also, the various colourful cults practised throughout the Roman Empire were for the most part highly practical affairs. People went to their gods to ask them for various things – luck, money, victory, a baby, a lost bracelet. The sense that they could do this, as their forefathers had done, and be heard, gave them a limited feeling of power and stability conducive to contentment with, or at least acceptance of, Roman rule. But this 'religion of consolation' was threatened when Christianity proposed that there was only one God, and that this one God mostly cared about inner humility, to say the least a somewhat alien notion to an occupying power.

The changes that Paul effected in the religion seem to have been directly influenced by his own character and experiences. A parable related in the Gospel of St Luke exerted a powerful hold on Paul's imagination, and contributed to Christianity's separation from the other faiths of the era. Jesus tells how two men, a Pharisee and a tax collector, went to the Temple to pray. The Pharisee thanks God that he is not like everyone else – the extortionists, the unjust and the adulterous. He recites a lengthy list of his worthy deeds: he fasts twice a week, he gives a tenth of whatever he earns. The tax collector, on the other hand, dares not even lift his eyes or his face to pray. He merely beats his breast and asks God to be merciful to him, as a sinner. And Jesus says that the tax collector, that most historically loathed of all professions, was the more just man. 'For everyone that exalteth himself shall be abased; and he that humbleth himself shall be exalted' (Luke 18: 14).

It would be easy to see this parable as a warning against self-righteousness and boasting, as well as a standard attack on the Pharisees, the 'bad guys' of the gospels. That's certainly how Luke himself – or whoever edited the book called the Gospel of Luke – interpreted it. But the spiritually torn Paul found deeper meaning in this simple homily. The Pharisee, being without sin, has no means of approaching God. The tax collector, the sinner, is the just man because for him the test of goodness is not virtue, but a childlike dependence upon God's mercy.

For Paul, this parable was the cornerstone of his belief in grace – an outpouring of divine love onto the whole of humankind, and not the Jews alone. But that, in itself, could not account for sin. How was it that God could simply pardon sin? That would imply that sin didn't matter. But sin did matter – the story of Creation in the book of Genesis told that humans were created with the ability to live for ever, until Adam and Eve sinned against God. Because of their 'original sin', all their descendants were doomed to die. But in Paul's personal theology – which was to shape Christianity in a very public manner – Jesus had balanced the books with his death. Thereafter, men could be saved from this ultimate price by absolute faith in the mercy of God, like that shown by the humble tax collector in the parable.

It's not hard to see that for people of a certain mind-set this was a disturbing idea. In every age, in every religion, there are those who prefer to consider themselves close to God through ritual observance alone. John Betjeman's 1940 poem 'In Westminster Abbey' sharply parodies this type of lip-service to religion:

> Although dear Lord I am a sinner
> I have done no major crime
> Now I'll come to evening service
> Whensoever I have the time
> So Lord reserve for me a crown
> And do not let my shares go down.[3]

But it is very unjust to see the Pharisees in the same light as Betjeman's deluded Chelsea worshipper. They were men of piety who followed the rabbinical oral tradition as well as the letter of the Torah. They stressed the values of justice and decency towards all their fellow men, whatever their nationality or religion. None the less, Paul's ideas, developed from this parable, constituted a threat to the whole world of observance, ritual and right action dictated by the Torah. The daily observance of the Pharisees and the ritual sacrifices of the pagans both had the advantage of giving the worshipper confidence in himself; both were ways of maintaining a degree of certainty in an uncertain world. But a childlike acceptance of one's own sinfulness, absolute abandonment of oneself into the hands of God – these were tricky concepts. Empires and Temples are not built, nor are wars waged, by humble men, but Christianity seemed to be stressing humility as the only virtue that guaranteed salvation. One of the most memorable lines attributed to Jesus was 'Blessed are the meek, for they shall inherit the earth' – a message that recurs throughout his teaching, perhaps was the very core of it.

So it is not surprising that many Romans viewed Christianity with a measure of suspicion, amusement and disdain, choosing to throw believers to lions for entertainment rather than adopt their tricky path of surrender to God's mercy: it wasn't just the Jews who felt the heavy hand of Rome. The fact that, in its first two centuries, Christianity was more popular among slaves and members of the lower classes only heightened the contempt of Roman rulers. Still more irritating was the tendency of Christians to offer prayers for the Emperor, but not sacrifices. The Romans would have seen this as deeply disrespectful. Accordingly, Paul was executed (tradition suggests in 62 CE by beheading), and in 64 CE the Emperor Nero began a sustained campaign of torture and hostility against the Christians. Not every Christian was prepared to stand up and be counted in the atmosphere that prevailed for the next two hundred years, and there was considerable hostility between those who had avoided or escaped martyrdom and those who were prepared to be put to the sword.

Nevertheless, the cult of Jesus continued to gain followers in the Roman Empire, particularly among the civil service and the higher ranks of the army. Certain feminist theologians have noted that women played a role in this process, with the persuasive wives of these men finding an appeal in Christianity's forgiving, loving ethos, as well as its stress on grace as a gift available to all, regardless of gender or race. It must also be noted that, however it was perceived by the Roman authorities, Christianity (like Judaism before it) was not a subversive cult. In Paul's letters to his companion, Timothy, Bishop of Ephesus, godliness is equated with conformity, and the author urges Christian slaves to submit to the rule of their masters.

Constantine, the first Christian Emperor

It was the divisions and weakness within the Roman Empire itself which led to Christianity's eventual triumph. Diocletian, realizing that the Empire was too big to rule alone, appointed a co-emperor. In 305 CE the two emperors, Diocletian in the West and Maximian in the East, abdicated, to be succeeded by their respective deputy emperors, Galerius and Constantius. Years of instability resulted in an impasse between Constantine, Constantius' son and the hero of the Roman army, and Maxentius, the son of Maximian, who had control over Italy and North Africa. Constantine, who had a Christian mother and Christian advisers among his retinue, went south from Gaul in 312 CE to attack his rival at the Milvian Bridge, north of Rome. On the eve of the battle, tradition reports, he had an important vision, which he later related to his biographer Eusebius of Caesarea: 'He said that about midday when the sun was beginning to decline he saw with his own eyes the trophy of a cross of light in the heavens, above the sun and bearing the inscription Conquer By This Signal [*hoc signe victor eris*]. At this sight he himself was struck with amazement and his whole army also.'

The next day, he attacked Maxentius' army 9 miles outside Rome. The luck of the day went with Constantine. Even though Maxentius' army was substantially bigger, his troops panicked and, caught in a bottleneck, attempted to run across the stone Milvian Bridge, which collapsed under their weight. Maxentius drowned under the weight of his armour and the soldiers left stranded on the north bank were all slaughtered or taken prisoner. Constantine was completely victorious. Nevertheless, another twelve years of battles lay ahead before he could finally secure the eastern half of the Empire and eventually become sole ruler.

Constantine did not formally convert to Christianity until his death in 337 CE, but he adopted a policy of tolerance towards it, and continued to rely on Christian advisers. He was particularly influenced by his biographer Eusebius, a historian who cleverly brought together views of divine kingship from the Greek-speaking world and Christian ideology. Constantine became a prototype European ruler, viewing his authority as coming from God, and his responsibility as guiding his subjects towards living as Christ required. For Constantine, peace and stability were bound up with the Christian rule of Christian leaders.

Such was Constantine's view of the religious importance of the Emperor that he became a guiding force in the historic Council of Nicaea in 325 CE. This meeting was convened to establish common ground between the various offshoots of Christian teaching and practice. Early

Christianity was in turmoil because Christians living in the east (Syria, Egypt and Persia, for example) leaned towards the beliefs of other monotheistic religions that were widespread where they lived. These included Mithraism, Zoroastrianism and possibly Judaism too. Christians living in Europe favoured a plurality to the Godhead – the Trinity – and this had some intellectual connection with pagan belief. Matters were brought to a head by Arius of Alexandria, who heretically denied the full divinity of Christ.

Arius and his followers thought that God, the Father of Jesus Christ, was unique, indivisible, incommunicable and transcendent. He was the singular divine being. Consequently, logic required that the Son was in some way different. He could not be of the Father's essence (otherwise that essence would be divisible or communicable or in some way not unique or simple). Christ had come into existence only by the Father's action, and while the description in the Bible of his having been begotten (or made or conceived) implied a special relationship between the Father and Son, Christ could not be seen to be on an equal footing with the Father. The Arians maintained of the Son that 'there was when he was not'; and that, having been created by God at some point in time, he was not immutable and was subject to moral change.

No records from the Council survive, but we do have the eye-witness account of Eusebius of Caesarea. His version of events is almost certainly somewhat coloured by his own sympathy with the Arian idea – indeed, he was himself in danger of being found heretical. The Council opened in the imperial summer palace at Nicaea, in Turkey, in May 325. Possibly over three hundred bishops attended, most of them from the east – Egypt, Persia, Palestine, Syria and Asia Minor – where this issue was a burning question. Some European bishops were also present, but the Pope himself, Sylvester

I, was unable to attend owing to age and infirmity.

Eusebius' memories of the moment when all the bishops assembled is poignant. Some came limping in, others were blind, maimed, or crippled, because so many of them had been tortured or persecuted by the old Roman world. Then the secular Christian master of this new Roman world entered, robed in scarlet and gold, and, before taking his place at the throne, bade them all be seated. There are moments in history when one would give a considerable amount to have been a fly on the wall, and this surely is one of them.

What followed seems to have been extraordinary. Although Arius is said to have had twenty-two supporters among the bishops, his position was condemned by the great majority at the Council as heresy. It is said that the disapproval of so many was so loud and vigorous that most of the 'Arian' bishops were intimidated. They rapidly withdrew their endorsement and Eusebius, who had himself been provisionally excommunicated earlier in the year and was anxious to clear his name, read a lengthy statement of faith. A new creed was drawn up:

> We believe in one God the Father Almighty, Maker of all things visible and invisible; and in one Lord Jesus Christ, the only begotten of the Father, that is, of the substance [*ousia*] of the Father, God of God, light of light, true God of true God, begotten not made, of the same substance [*homoousios*] with the Father, through whom all things were made both in heaven and on earth . . . Those who say: 'There was a time when He was not, and He was not before He was begotten;' and that 'He was made out of nothing;' or who maintain that 'He is of another hypostasis or another substance,' or that 'the Son of God is created, or mutable, or subject to change,' the Catholic Church anathematizes.

Even at this stage, eighteen bishops held out in favour of the Arian doctrine.

Constantine himself was probably quite untroubled about these points of theology. His main interest, as a statesman concerned primarily with stability, was to ensure that the church sorted itself out, and to this end he was determined to force unity upon it. He threatened exile to anyone who would not sign. After further discussion, all signed except two Libyan bishops and Arius himself. These three were exiled.

As it happens, this was by no means the end of the matter. The argument continued long after the Council concluded. Constantine himself wavered on the issue later in life – but, perhaps because he was a soldier and a general, he was a practical man and a pragmatist, not academically interested in the niceties of theology. Whatever his private views, however, this tough, charismatic leader married a persistent Middle Eastern cult to the might of the Roman Empire, and thereafter Christianity would flourish. In time, Europe would come to share his vision of divinely appointed kings. The cross on which Jesus had been crucified, the symbol he saw at the Battle of Milvian Bridge, would come to be the world's most widely recognized symbol. The characters of the story presented in the gospels – the Pharisee and the tax collector, Pontius Pilate, Judas – would enter the consciousness of nations from Iceland to Poland, throughout continents from Africa to Asia. Hundreds of millions would call themselves Christians, their beliefs generating important moral values, great works of art, some of the world's greatest architecture, acts of great charity – and sometimes, too, the most extreme bigotry and cruelty.

This truncated, potted history of Christianity shows how the Divine Idea can bring about enormous, permanent changes in the structure of the world. A brief glance at any history book will tell you how Christian faith altered the shape of national borders, imposed certain patterns of morality and ceremonial life and informed whole cultures, leaving its mark in artefact and custom from the cross-shape on a king's crown through to eating fish on Fridays. But, thought about another way, this seismic shift could be seen as brought about by no more than a chain of coincidences. If Pontius Pilate had been a different sort of man, like his successor, Albinus, Jesus might never have been put to death. If he had lived, as his prophecies failed to come to fruition his followers might have drifted away, or become caught up in various internecine struggles. Jesus himself might have undergone a crisis of faith as Paul did, and reconsidered his outlook from an older, wiser perspective. Without Paul's personality influencing the way he saw Jesus' teachings, Christianity might have remained limited to the Jews. Without Constantine, it might still have remained a low-key cult of slaves, soldiers' wives and urban artisans. Many Christians see this wondrous chain of slender chance as evidence of the inherent truth of their faith, but one doesn't have to believe in Christianity to agree that its career has been remarkable. It attests to the power of ideas, regardless of whether you accept God's role in creating those ideas.

But what were those ideas, and who was the man who first spoke them to his friends and followers? As far as Judaism is concerned, Jesus was an observant Jew who claimed to be the Messiah, like many did before and after, but got it wrong: the things he seemed to prophesy – the dawn of an age of peace, the establishment of the rule of God on earth – have not happened.

How Jewish was Jesus?

When the Roman general Vespasian laid siege to Jerusalem in 66 CE, some of the hard-line opponents of the Romans inside the beleaguered city wanted to engage his armies in battle. These militants, the *biryonim*, tried very aggressive tactics in a bid to force their fellow Jews to fight. In spite of the threat of famine that loomed in the surrounded city, they even audaciously set light to Jerusalem's stores of grain and oil. The ruling Pharisees, however, were opposed to the idea of an armed struggle. They thought their only chance was to make peace with the Romans. According to Jewish tradition, the great Pharisee sage Rabbi Yochanan ben Zakkai had a nephew who was a leader of the *biryonim*, and tried his best to reason with him. This nephew, Abbah Sikrah, could see the sense of his uncle's arguments, but said there was nothing he could do: he claimed to have lost control over the rest of the *biryonim*, who were committed to a bloody war. The story is told in the Talmud (Gittin 56).

Rabbi Yochanan and Abbah Sikrah came up with a plan to smuggle the sage out of the city so that he could make contact with the Romans. The rabbi was to pretend he was seriously ill, and Abbah Sikrah would spread the word so that everyone would enquire after him. Then Rabbi Yochanan was to place something really evil-smelling close to himself. When everybody thought he was dead, his disciples were to hide under the bed and lightly hold the bier down, 'because everybody knows that a living being is lighter than a corpse'. Then his body could be taken out of the city and buried. As it turned out this was a highly risky strategy, because it happened that when they reached the door of the bedroom they met some men who wanted to put a lance through the corpse.[4] Abbah Sikra saved the day by admonishing them 'Shall the Romans say [in disdain], they have pierced their master?' Having escaped through a gate of Jerusalem, Rabbi Yochanan rose from his filthy grave in his smelly clothes and made his way to see the Roman general. When he was ushered into Vespasian's presence, the rabbi bowed and said, 'Peace to you, O King, peace to you.'

Vespasian, unsurprisingly given his visitor's aura, wasn't terribly happy at the sight before him. He said he intended to kill the rabbi on two counts: 'One because I am not a king and you call me king, and again, if I am a king, why did you not come to me before now?' The rabbi replied that in truth he was a king and that the prophets Isaiah and Jeremiah had predicted that Jerusalem would fall into the hands of a king. 'And the reason I haven't come earlier is because the *biryonim* among us wouldn't let me go.'

At that very moment, a messenger arrived from Rome. He carried the news that the Emperor was dead, and Vespasian was to be the new Roman ruler. Just before this messenger arrived, Vespasian had been in the process of putting one boot on. Now, when he tried to put on the other, it wouldn't fit. Nor could he get the first boot off. The rabbi drily observed: 'Good tidings make the bone fat.' Not surprisingly (such stories always end like this), Vespasian decided not to kill the rabbi; instead, he would grant any request the old man made. Realizing that to ask the Romans to leave Jerusalem might be stretching his luck, Rabbi Yochanan instead asked for various favours for the Pharisees: protection for the scholars in the city of Yavneh, a doctor for a sick rabbi, and privileges for the descendants of the great sage Rabbi Hillel.

Does Rabbi Yochanan's request seem unambitious, even rather surprising? This story, rather, shows the canny wisdom and pragmatism of the Pharisees. Saving the city was impossible – indeed, it was utterly destroyed and the Temple burnt by Titus, Vespasian's son, not long afterwards. But the rabbis managed to save Jewish law and, above all, Jewish education. Yavneh became by far the most important centre of Jewish learning for hundreds of years, and it is only the teaching and tradition from there that ensured the survival of Judaism. The New Testament gives the impression that the Pharisees were fanatics, obsessed with every jot and tittle of the Torah. This is a caricature. The tales about them suggest that they were balanced and reasonable, and had a great sense of humour – a particular humour which Jews still display today. In practice, they showed a healthy scepticism towards extremism or potentially dangerous beliefs. This included the concept of a Messiah, concerning which Rabbi Yochanan archly commented: 'If you should happen to be holding a sapling in your hand when they tell you the Messiah has arrived, first plant the sapling and then go out and greet the Messiah.'

While belief in the Messiah is included as one of Moses Maimonides' thirteen principles of the Jewish faith,[5] this is only one strand of the religion. Historically it tended to rise to the fore, as I have noted earlier, in times of stress. Judaism, with its complex system of *mitzvot* (commandments) governing every aspect of daily existence, has always been based far more strongly on action in this world rather than expectation of another. Joseph Telushkin, in his work *Jewish Literacy*,[6] notes that Jews have good reasons to fear messianic ideas, because whenever such ideas appear, the consequences are disastrous for them. In chapter 7 we will see how a seventeenth-century 'Messiah' called Shabbetai Zevi created chaos and confusion and ultimate disillusionment for Jewish communities throughout Europe. Telushkin also refers to a plot uncovered in Israel in 1984, concocted by a group of Jewish madmen who intended to blow up the Muslim shrine at the Dome of the Rock in Jerusalem and thus, by triggering a major war, to clear the way for a new Temple and hasten the arrival of the Messiah. One shudders to think of the global consequences if they had got away with it. Although all Orthodox Jews still mention the Messiah in their morning prayers, the genuine expectation that a Messiah will actually come in anything like the near future is largely limited to a few sects.

Ideas of Jesus as the Messiah have also been pretty disastrous for Jews over the centuries – but not because of anything Jesus himself said or did. He was an observant Jew, who was circumcised at eight days old and died, according to some accounts, with the words of the Psalms on his lips. Although he later came to be regarded as the only Son of God, thanks to Paul's mythologizing and popularizing of the cult, Jesus never claimed this explicitly. He talks to God as his 'Father', but this was not an uncommon means of address. Indeed, Jews still view God as 'Father' to this day. He refers to himself as *bar Nasha*, the 'son of Man', seeming to stress his humanity and frailty. Moreover, even though others call him the 'son of God' in the gospels, this may not be intended in a literal sense, any more than saying someone is a 'son of Albion' refers to a belief that they emerged from the physical soil of England, or saying someone has 'blue blood' really means their blood is coloured blue. It seems the term was employed by his followers to show that Jesus was the genuine Messiah, no more.

Throughout his ministry, Jesus teaches precepts and ideas that are authentically Jewish and have their origin in the Torah. When, for example, Jesus heals the woman who has been crippled for eighteen years on the Sabbath (Luke 13: 10–17), saying to her after placing his hands on her, 'You are rid of your trouble,' in spite of the protests reported in the gospel text he is doing nothing which contravenes the Sabbath laws. Indeed, it is incumbent upon the Jew to save life wherever possible, and the religious Jew should break the Sabbath to do so. Here Jesus is employed in an act of *pikuach nefesh* (saving life) by effectively preventing prolonged suffering. When Jesus says, 'Do unto others as you would have them do unto you,' he is echoing a famous tale told about Rabbi Hillel. It was said that a Gentile approached Hillel and promised he would convert to Judaism if the rabbi could sum up the whole of the Law while he stood on one leg. Rabbi Hillel promptly said, 'Do not do unto others as you would not have done unto you. That is the whole of the Torah and the rest is explanation: go and learn it.'

Jesus would not have been viewed askance in the world in which he operated. The province of Galilee to the north of Judea, fertile and autonomous from Roman rule, teemed with wandering holy men who performed miracles – healing the sick, casting out devils, controlling the weather: these men were not unlike Jewish shamans, and many of the acts performed by Jesus suggest that he was part of this tradition. It is related of the great sage Rabban Gamaliel himself how he sent two of his disciples to see another rabbi, a healer called Chanina ben Dosa, because Gamaliel's son seemed on the verge of death due to a fever. Ben Dosa withdrew to an upper room and prayed. When the disciples returned to their master, they found the boy cured.[7]

The gospel accounts are also at pains to point out that Jesus' activities were entirely consistent with the writings of the Jewish prophets. Time and again when Jesus does or says something, the phrase 'this happened to fulfil . . .' is employed to demonstrate that Jesus was the fulfilment of a long-expected hope in Jewish tradition. Jesus himself declares that he has not come to abolish the Law but to fulfil it. Even the Gospel of John, which seems to have been written later than the other three gospels, Mark, Matthew and Luke, and at a time when Christians had an increasingly pressing need to dissociate themselves from the troublesome Jews to escape Roman persecution, contains frequent references to the prophecy of Zechariah.

The whole story surrounding Jesus' birth may possibly have been created in order to fulfil the prophecy of Micah, namely that the Messiah would come from Bethlehem: 'Bethlehem . . . from you someone will emerge for Me to be ruler over Israel; and his origins will be from early times, from days of old.' It's a story with which many of us are familiar. The Gospel of Luke tells us that Joseph and the heavily pregnant Mary had to travel to Bethlehem because the Romans had imposed a census, and everyone was required to return to their place of birth to register. Due to overcrowding, the infant Jesus spent his first night in a manger, a container for animal feed. But the Romans – assiduous bureaucrats – left no records of a census at that time, nor is there any reason why it would have required people to return to their place of birth: it was just a head count prior to a poll tax.[8]

The stated timing of the events surrounding Jesus' death, on the eve of and the day of the Passover festival, may also have been doctored with a specific aim in mind. Pesach, the Passover, is the most ancient of the festivals, when Jews commemorate their release from slavery in Egypt. Central to the Passover festival is the *seder* – a communal meal at which unleavened bread and bitter herbs are eaten. During Temple times, a lamb was sacrificed and eaten to celebrate the meal the Israelites ate on leaving Egypt. In later years, Christians saw Jesus as 'the lamb of God' and his death as a sacrifice which replaced the old obligations of the Jewish faith with a new covenant between God and the whole of humanity. Therefore, the placing of Jesus' death at Passover in the gospel accounts could possibly have been a deliberate device to support this interpretation. Other details in the account clash with this timing: for example, we are told that Jesus' followers drew their swords when he was arrested in the Garden of Gethsemane, but this is unlikely as observant Jews would almost certainly not have carried any weapons during the festival – unless they seriously thought their lives were in danger, which, of course, may have been the case. It is fair to say, at least, that we don't actually know for certain when Jesus really died.

But for all its continuity with Judaism, the Jesus cult, as

presented in the gospels, also seems to be fairly hostile towards the faith from which it sprang. We are told that Jesus is continually at odds with the Pharisees, scathing about their obsession with ritual, accusing them of hypocrisy in the strongest terms. Accordingly, the story is that the Pharisees are the ones who agitate for his death. By contrast, the Romans are treated as fair-minded and calm. The judge who tries him, Pontius Pilate, is reluctant to crucify Jesus. Because it is the Jewish Feast of Passover, he gives him a last chance by asking the mob which of two prisoners they want to spare, Jesus or Barabbas. The mob insist on Jesus' being crucified.

Some contemporary Christians and the popular media, for example, Mel Gibson's execrable film *Passion of the Christ*, present Pontius Pilate as a spineless individual, who cannot or will not enforce his authority as governor against the bullying Jews who demand Jesus' execution; merely an incidental player, reluctantly allowing the crucifixion to proceed. A. N. Wilson has pointed out in his *Jesus* that there is no record on any occasion of a Roman official granting amnesty to prisoners because of a local holiday. Barabbas was not a thief, as some Bible accounts tell us, but possibly a terrorist, since the Gospel of Mark records that he had committed a murder during a recent uprising. If Barabbas was indeed such an agitator, it seems very unlikely that the hard-nosed governor, Pontius Pilate, would set him free so readily. Later on in the narrative, as Jesus dies on the cross, a Roman centurion offers him a drink and accepts that this pathetic tortured person is the son of God. This seems like political storytelling. The Roman in question represents ordinary Romans, and they are made to seem empathetic perhaps because the early Christians wanted to avoid persecution.

Perhaps it is also significant that the gospels tell us Jesus was betrayed by one of his disciples, Judas Iscariot. This could be in large part invention. Judas, we are told, steals away from the house where Jesus and the rest of the disciples are having their final meal together to cut a deal with the chief priest, betraying Jesus in return for thirty pieces of silver. Later, after Jesus' arrest, Judas hangs himself. Yet how anyone would know that Judas received thirty pieces of silver, since he went off alone, is unclear. Nor is it clear why he betrays Jesus, or even how he does it. Later, in the Garden of Gethsemane, he points Jesus out to the arresting authorities by kissing him. But it is not obvious how the authorities know he is there – one might surmise that if they have the intelligence to follow Jesus, they would hardly need him to be pointed out to them in an act which would point the finger firmly at their collaborator.

The most interesting part of the whole story concerning Judas is his name. Iscariot seems to be derived from the word *sicarii*, the name given to a band of militant insurgents, mentioned in the Talmud, who concealed little daggers, called *sicae*, under their cloaks. They specialized in creating mayhem in crowds by stabbing people who were friendly to the Romans. Jesus also included among his disciples Simon, a Zealot – the Zealots being a group of particularly violent patriots who eventually committed mass suicide when under siege at the fortress of Masada.[9] In depicting Judas as the betrayer of the son of God, the early Christians may have been seeking to distance themselves from any suspicion of being just more Jewish terrorists.

These lacunae and anomalies suggest that there may be in the gospels an element of spin, applied for various political purposes. There is no doubt that Jesus had much to protest about. The Jewish King Herod had been deeply corrupt, and many of the people he put in power were equally bad. Much of what we know about the national situation in Judea strongly suggests that this was an opportune time for a truly righteous leader like Jesus who could lay some claim to the title King of the Jews. But the gospels go beyond a mere accusation of moral decline and lack of sound national leadership. Jesus' birth is moved to Bethlehem to stress continuity with Judaism; the Jews are reported as utterly vindictive, while the Romans (some of them, at least) are depicted as sympathetic; and the *sicarii* are delineated as traitors, possibly because of the need for Christians in later years to escape persecution. Unfortunately, these unsubstantiated additions to the gospels, together with the uncomfortable assertions about the Pharisees, have provided a theological justification for two millennia of persecution directed at the Jews. And a great paradox lies at the heart of this persecution. If, as Paul attests, Jesus' death was necessary to expunge the sins of humankind, surely it follows that the people accused of killing him, the Jews, have done the world a favour? None the less, the idea that 'the Jews killed Jesus', along with the mere similarity of the name 'Judas' to the words 'Jew' and 'Judaism', has informed attitudes to the Jewish people in many of the countries where they made their homes. To the shame of many Catholics, the Vatican did not expunge the idea that the Jews murdered Christ from its teachings until the 1960s, long after numerous persecutions in Russia and eastern Europe in the nineteenth and twentieth centuries, and the Nazi Holocaust, had pointed to the dangers of such a warped belief.

There is an undoubted historical problem about the identity of Jesus. We know that he existed: one independent source of evidence for his life is the account by the contemporary historian Josephus. He had a family – his brothers and his mother seem to have been involved in keeping the cult alive in the years after his death. We do not know whether he was married, but it would have been unlikely for a young man of this era not to have had a wife. He had a wide circle of friends, including both wealthy members of the Sanhedrin, such as Joseph of Arimathea, and terrorists like Simon the Zealot. That he was also remarkably gifted as well as being driven by high moral values, like so many prophets before him, is without doubt.

But there are enough inconsistencies in the detail of the gospels to render the real person of Jesus difficult to discern. He possessed the humility to wash the feet of his disciples after a long and tiring trek to Jerusalem. Yet on another occasion, when he was busy performing to the crowds, he refused to see his own mother. Indeed, he often comes across as haughty in his dealings with Mary. Luke's Gospel tells us that twelve-year-old Jesus disappears when visiting Jerusalem for the Passover feast. When his distraught mother finds him discoursing in the Temple with the rabbis, his response is seemingly far from docile or penitent. 'How is it that ye sought me? Wist ye not that I must be about my father's business?' How are we to arrive – can we arrive – at a true picture of this enigmatic man?

In 1999 the Church of England attempted to attract younger worshippers by depicting Jesus on posters as Che Guevara, the communist revolutionary, along with the slogan 'Meek. Mild. As If.' Certainly, Jesus' concern for the poor and Christians' belief in the universal brotherhood of men do have much in common with later socialist ideas. But it is both glib and reductive to depict Jesus as some has-been communist revolutionary; he was so much greater.

Much has been made of Jesus' revolutionary character, a persona that stems in part from one of the most vivid episodes before his capture and execution. The Gospel of Mark tells that Jesus made a number of visits to the Temple in the last week of his life. The complex teemed with visitors, priests and merchants; the air was electric with the clucking and squealing of frightened animals on sale for sacrifice, money-changers calling out their rates of exchange. This sight angered Jesus so much that he violently overturned the tables of the money-changers. He declared that the Temple was intended to be a house of prayer for all nations, and that the money-changers had made it a 'den of thieves'. Later, as he left, Jesus declared that the Temple would be pulled down and rebuilt in three days.

This prophecy has created a great deal of confusion. Was Jesus speaking merely figuratively? Mark hastily adds the words 'but he spake of a temple of his body', wishing to suggest that Jesus was prophesying the foundation of the Christian church. This looks very like a gloss added with hindsight: at the time it was written, the Temple had been destroyed and the Christians were now practising the Eucharist, symbolically eating the body and blood of Christ in their worship. It made sense to say that this was what Jesus had understood from the outset.

John (2: 19), on the other hand, gives Jesus a different line: 'Destroy this Temple, and in three days I will raise it up!' It does not seem here that Jesus is arguing against animal sacrifices – in which, as a Jew, he would have participated. Nor is he trying to undermine the Temple itself; indeed, he revisits it repeatedly later in John's Gospel, and the scriptures tell us that his followers were continually in the Temple after his death. The overturning of the tables seems more than just a protest; it is reminiscent of the actions of the ancient prophets, who performed striking gestures to make a point.

To discover what Jesus' point was, it is necessary to leave the turbulent scene in the Temple and consider some of his other utterances. In the Gospel of Matthew, Jesus delivers a lengthy and wide-ranging sermon from a mountain-top, apparently a clear echo of the time Moses received the Law on Mount Sinai. One of the most commonly quoted utterances from this 'Sermon on the

Mount' concerns adultery: 'Ye have heard that it was said to them of old time, "Thou shalt not commit adultery." But I say unto you, that whosoever looketh upon a woman to lust after her hath already committed adultery in his heart.'

As someone who has devoted much of his career to the study of human behaviour, particularly human reproductive behaviour, I think there must be few men – or women – who have not, under this capacious definition, broken this one of the Ten Commandments. But what is important about this statement is the spirit it conveys and its setting. Delivered in circumstances that recall the giving of the Law to Moses, Jesus seems to be saying that our duty is neither to disregard the Law nor to observe it slavishly, but to go beyond it. Jesus seems to be asking his followers to remain strictly within the Law, but also to think about its fuller implications and nuances.

We see the same message when we look at Jesus' comments concerning the laws of cleanliness prescribed to Jews. The Pharisees were appalled when they saw Jesus and his followers eating bread together without first observing the *mitzvah* commanding a Jew to wash his hands. Jesus' attitude is described in the paragraph which follows: 'There is nothing from without a man which entering into him can defile him; but the things which can come out of him, those are they that defile the man' (Mark 7: 14, 15). Far from rejecting the laws of Judaism, Jesus was emphasizing the parallel between them and the individual's spiritual life: 'For from within, out of the heart of men, proceed evil thoughts, adulteries, fornications, murders, thefts, covetousness, wickedness, deceit . . .' (Mark 7: 21).

To his death, Jesus remained an observant Jew, but his central message was that observance *alone* was not enough. Ritual actions were meaningless, 'making the word of God of none effect through your tradition', unless conducted in an awareness of what the ritual meant for one's morality. Similarly, in striking down the tables of the money-changers, Jesus was not seeking to abolish Temple sacrifice but, like the prophets six hundred years earlier, arguing that it was pointless when performed as an end in itself.

It is possible that this message of Jesus was misunderstood by many at the time, just as it has been since, and is now. As far as many within the Jewish and the Roman authorities alike were concerned, here was a *tsaddik* (righteous person) from the troublesome province of Galilee who was gathering an increasing following of alarming proportions. He had Zealots and knife-wielding *sicarii* among his followers, and was prepared to risk popular riot by overturning the tables in the Temple during a major festival. It was necessary to act swiftly to neutralize him. After his death, it was not possible to keep his cult alive using only this rather abstruse, Jewish message for Jews; so the writers of the gospels and Paul modified the message, presenting it as a substitute for the Jewish Law that offered salvation for all.

Of course, it is impossible to distinguish what Jesus the man said from the glosses and asides put in later. We can go some way towards this by analysing the language of the gospels: sentences which could be conveyed only in Greek but not in Aramaic are obvious later additions. We can also discount, on grounds of his very Jewishness, 'quotes' that make Jesus out to be an enemy of Judaism. But the real Jesus will always be something like the image on the Turin Shroud – a ghostly impression with mysterious origins.

One Christ or many?

It has often been remarked that, where painting and sculpture were concerned, the ancient Egyptians were far in advance of the rest of the world. But where religious beliefs were concerned, they seemed to be on another planet. They had no concept of linear time, and therefore none of history. They viewed heaven and earth as being the same. They seemed to see few distinctions between the living and the dead, or between men, plants and animals. Their leaders, the pharaohs, were gods. Other gods took the form of falcons, cows, crocodiles and cats, which were adorned with bangles and jewels, fed on luxury rations, and mummified when they died. Ancient Egypt had no concept of abstract law, no written codes of conduct – all legislation came from the god-king. Judaism, with its one God in heaven, its sense of history, its detailed laws, seems more familiar to our modern, western viewpoint. But there are also some interesting similarities between the various beliefs that thrived in the Near East at the time of Christ and long before.

For some people, the obscurity surrounding the historical Jesus suggests that his story is nothing other than a myth, a myth common to several Near Eastern civilizations – albeit a deeply meaningful myth. This is the premise of *The Pagan Christ* by the Canadian theologian Tom Harpur, a professor of Greek and New Testament at the University of Toronto, and a man who spent many years as an Anglican priest. Harpur, noting the poverty of historical evidence for the life of Jesus, points to previous versions of the legend, most notably from ancient Egypt, where the idea of a divine spirit being incarnate in a human body was current as early as 4000 BCE. 'The truth is', says Harpur, 'that the Gospels are indeed the old manuscripts of the dramatized rituals of the incarnation and resurrection of the sun god Osiris/Horus.'

Drawing on previous work by the biblical scholar Alvin Boyd Kuhn, as well as nineteenth-century Egyptologists, Harpur presents a startling list of similarities between the

Jesus of the gospels and Osiris, who became Horus in Greek mythology. They include:

- Horus was baptized by a god-figure called Anup the Baptizer (who was later beheaded, as was John who baptized Jesus).
- Horus walked on water, healed illness and exorcised demons.
- Horus was crucified between two thieves, buried in a tomb and resurrected on the third day after his death.
- Horus, like Jesus, was supposed to reign for a thousand years.

Harpur suggests that the identity and story of the Christian Jesus had been prefigured by Egyptian mythology in the incarnation, the resurrection, and even the specific details of miracles like the raising of Lazarus.

Given his vigorous religious faith and his former career as an Anglican priest, Harpur's conclusions are remarkable. They may say rather more about him that they do about Christ; after all, statisticians pick holes in these kinds of coincidence as part of their academic bread and butter. Be that as it may, Harpur proposes that Christianity took a wrong turn when the early church presented archetypes, metaphors and a reworking of previous mythology into concrete historical fact. Those ancient stories contained within them some great spiritual truths and moral guidance; but interpreting them literally, as though all these events happened to a single person in a single lifetime, was absolutely wrong. For Harpur, myth should be neither dismissed as archaic fancy, nor cherished as history, but valued correctly as stories rich with spiritual meaning. 'Through prolonged meditation and inner searching, the ancients discovered archetypal symbols that corresponded to what they saw and experienced in the natural world. They came to know, for example, the reality of the Christ (or atman, or soul) within by inward exploration. They discovered what God had already in compassion planted there: the divine, God's own image.'

Myth, for Harpur, has even more value than history – which, he argues, makes ancient wisdom commonplace. The precipitation of hard fact out of ancient mythology does a great disservice to the scriptures, which represent a body of oral tradition, ritual, allegory and myth forged over thousands of years among the ancient civilizations of the Middle East. The stories of Jesus are much more telling when considered as symbols of some greater truths rather than when taught as fact. Literalism, claims Harpur, has actually been incredibly damaging. Most atrocities committed in the name of religion have been inspired by it – both violence against non-Christians, and violence between different groups of Christians. It is perhaps fair to say that a completely literal reading of any text – religious or otherwise – is not conducive to an especially tolerant outlook on life.

In the light of these theories, then, what should an objective observer make of the Jesus narrative? Should we write it off as a corruption of ancient mythology, suitable only for those who need literalism to pin down an otherwise untamed religious symbolism? And where does that leave the living, breathing Jesus of history?

Some Christians, or ex-Christians who want to retain some element of faith, are happy to accept Jesus as mortal. They see him as a prophet, a healer and a teacher, but not as the incarnation of God. But for most Christians this version of Jesus of Nazareth is completely inadequate. C. S. Lewis, the English novelist and writer, converted to Christianity in middle age (he was riding to a zoo in his brother's motorcycle sidecar at the time). In a series of BBC radio broadcasts in the 1940s, later published as a book entitled *Mere Christianity*, Lewis expressed his belief that a kind of Christianity dependent on a mortal Jesus is not a religion worthy of the name:

> A man who was merely a man and said the sort of things Jesus said would not be a great moral teacher. He would either be a lunatic – on the level with a man who says he is a poached egg – or he would be the devil of hell. You must take your choice. Either this was, and is, the Son of God, or else a madman or something worse. You can shut Him up for a fool or you can fall at His feet and call Him Lord and God. But let us not come with any patronizing nonsense about His being a great human teacher. He has not left that open to us.[10]

Believing in Jesus as a mortal makes him mildly interesting at best, and at worst a lunatic, though surely not 'the devil of hell'. But in either case, Jesus the sage or moral teacher would have been only a footnote in the history of God instead of towering like a Colossus over the past two millennia. The full-throttle Christian rendition of Jesus Christ as God incarnate underpins Christian theology and most Christians' faith. It is that vision, of Jesus the Son of God, whose death invites all men into an everlasting life, that makes him relevant for over two billion people on this planet. Whether that rendition is based entirely on fact or partly on myth perhaps, in the end, does not really matter.

The vagueness of the Jesus portrayed in the gospels, and to some extent the puzzle of the Trinity, might also be a significant contributory reason for Christianity's success. The more explicitly one depicts a man, the more rigid his portrait, the more likely it is that some people will find faults in him or feel unsympathetic. What people are drawn into through Jesus' story is a universal theme, rather than the events occurring in a particular life; for the tale of an innocent person, tried and executed by the powerful, strikes a powerful chord. The theme of suffering, death and rebirth is visible and poignant to anyone who has considered the nature of existence for even a few moments – you need not be a Buddhist or a Baal worshipper to notice that everything decays, only to be born anew. Lines like 'turn the other cheek' and 'blessed are the meek' are arresting to the human brain because they run so contrary to the laws of nature – when attacked, we fight; when strong, we triumph. Jesus the mild has proven to be a far more vivid and powerful figure than any gun-toting revolutionary.

Indeed, so arresting has Jesus' story been that he is considered of significance in other religions as well. For the Muslims, Jesus is recognized as 'the breath of God' and considered a prophet. Muslims also accept that he was born of a virgin and that he performed miracles, although they deny that he was crucified, arguing that someone else was executed in his place. Even certain Jewish movements maintain that Jesus was more than a misguided prophet. In the mid-nineteenth century, certain European Jews who had converted to Christianity felt that they had lost touch with their culture as a result. They also noted that, before Paul, the followers of Jesus had been observant Jews. This gave rise to the birth, in 1866, of the Hebrew Christian Alliance and Prayer Union of Great Britain. A number of related groups have come into operation since, all of them united by an acceptance of Jewish tradition on the one hand, and a belief in Jesus as the Messiah on the other. There are currently around 100,000 Messianic Jews (or Jewish Christians, as some prefer to be called) in the United States, as well as congregations in Canada, Australia, Britain and Israel. It must be noted that the wider Jewish community views these movements, such as Jews for Jesus, with grave suspicion, considering them to be propaganda devices funded by evangelical Christians. This is mostly, I think, because they seek to convert Jews – always a threat to the non-proselytizing Jewish tradition. But this controversy merely attests to the power of the Jesus story, surely one of the most influential manifestations of the Divine Idea ever to rise in human consciousness.

6

Muhammad and Islam

Anyone who has ever visited the Middle East will have experienced the vibrancy setting it apart from most of Europe. Streets teem with activity long into the heat of the evening; transactions are conducted with a great deal of vigorous banter. The call of the muezzin, the rich scent of spices, piles of colourful fruits and live animals on their way to market create a heady sensory experience. Life often seems frenetic, balanced on an edge. The geographical location of the Middle East, a sort of saddle between Europe and the Orient, between Africa and Asia, gives it the flavour of a crossroads, a region where ideas were once traded as much as goods. The Middle East is where, it seems, humans first found an environment that favoured settlement; this in turn led to written language and hence the beginnings of what we understand to be civilization. It is surely no surprise that three of the world's most widespread religions sprang from this rich soil.

Sadly, one of these world religions, despite having the clearest links with the religions of neighbouring peoples, despite embodying a passion for the highest ideals of

ABOVE: **Grotte de Gargas**
Many of these hand-prints, stencilled over 27,000 years ago, mysteriously have fingers missing. Do they represent early humans' attempts to reach out to God? (page 55)

BELOW: **Jean Clotte in Grotte de Gargas**
Jean Clotte discovered slivers of bone pushed into these cracks – perhaps messages like those pushed into the Western Wall of the Temple in Jerusalem.

ABOVE LEFT: **Luxor**
Rameses II ordered the building of Thebes. My ancestors may have been among the slaves who worked these pillars into position.

ABOVE RIGHT: **Mithraic altar**
In the crypt of San Clemente Church in Rome, the Mithraic temple icon represents the pagan worship that was once celebrated on that spot (pages 211–212).

BELOW: **Teotihuacan, the Pyramid of the Moon**
These pyramids are notorious for the human sacrifices which took place at their summits. (Reproduced with kind permission of Eddie Bowman, FRPS.)

OPPOSITE PAGE:
TOP: **The golden calf**
Local legend has it that this rock formation at the base of Mount Sinai is the golden calf made by the Israelites when Moses went up the mountain to receive the Ten Commandments (see page 159).

BOTTOM: **The Western Wall, Jerusalem**
Worshippers at the 'Wailing Wall', the only part of Herod's Temple (pages 205–6) left standing, frequently push paper messages into the cracks asking for divine intervention.

RIGHT: **Hindu temple**
Some Hindu traditions suggest that there are 333 million different gods. Just a few of them are celebrated on this temple façade in Colombo, Sri Lanka.

BELOW: **Reclining Buddha, Polonnaruwa, Sri Lanka**
This famous eleventh-century reclining Buddha is 14 metres long. The Buddha is at the moment of throwing off all desire, all illusions, and achieving nirvana; variations in the rock colour convey the idea of ripples encroaching on human consciousness.

ABOVE LEFT: **Fire Temple, Isfahan, Iran**
The fire has burned continuously in this Zoroastrian temple for 1,700 years (pages 135–136). When the priest goes behind the glass barrier to add fuel, he avoids contaminating the purity of the flame by wearing a mask.

ABOVE RIGHT: **Ruwanweliseya stupa at Anuradhapura, Sri Lanka**
Brick-built and covered with coral, this 2,000-year-old stupa or commemorative monument has a chamber containing sacred relics of the Buddha. King Dutugemunu built it, being inspired by a bubble floating on water. One interpretation suggests the dome represents heaven; the conical spire, enlightenment.

BELOW: **The Emperor Constantine**
Constantine was the first Christian Roman Emperor (pages 215–216). This massive bust is in the Capitoline Museum in Rome.

LEFT: **Raphael's God**

The Raphael print from my collection of biblical prints shows God's hands and fingers outstretched and new creatures emerging from the ground (page 327).

BELOW: **Ceiling in the Church of San Clemente, Rome**

Carlo Fontana designed this ornate carved ceiling. How much artistic endeavour has been devoted to attempts to connect with God?

LEFT: **St Peter's, Rome**

March 2005: Easter Mass being celebrated in Pope John Paul II's hearing for the last time.

LEFT: **Church of St Bernadette, Lourdes**
The cave over which the church is built is bottom right.
St Bernadette, who had nineteen visions of the Virgin,
died of tuberculosis in 1879. Her body has been
exhumed on three occasions and, according to the
Church, has shown no sign of decomposition.

BELOW: **Ladies (Shaykh Lutf Allah) Mosque, Isfahan**
This mosque, in the great maidan in Isfahan, shows the
glorious flourishing of Muslim abstract art in its tilework.
Shah Abbas built it in the seventeenth century.

RIGHT: **Highgate
cemetery**
The rituals of human
burial are regarded
as evidence of
spirituality in early
humans. Respect
for dead bodies has
persisted throughout
time.

Large Hadron Collider, CERN
Standing in the largest magnet in the world. When the Hadron Collider is eventually switched on in 2007 (see pages 390–391), we may finally know whether the Higgs boson actually exists.

justice and equality, despite sharing the same moral values as other great faiths, such as a deep respect for the sanctity of life, has been vilified and slandered in recent years. Islam, founded by an Arabian called Muhammad in the sixth century CE, is often perceived as violent, repressive and backward-looking, the spiritual basis for terrorism, public amputations and the suppression of women. In contemporary Britain, right-wing parties no longer single out the Jews as the target of their aggressive contempt, but centre on our Muslim communities, depicting them as divorced from mainstream life and breeding grounds for extremism. As we have seen before, ideas about God can act as a source of conflict. They can also be profoundly misunderstood by people from other traditions, and this is particularly true for how Islam is too often seen in the West.

This misunderstanding frequently emanates from the very people who should know far better. Some time after the attack on the World Trade Center in New York on 11 September 2001, the Reverend Franklin Graham, the well-known American evangelist preacher and chief executive officer of the Billy Graham Organization, said: 'The God of Islam is not the same God of the Christian or the Judeo-Christian faith. It is a different God, and I believe a very evil and a very wicked religion.' And the Reverend Moody Adams, another well-known American evangelist, tried, in a media interview, to show how open-minded he is as a Christian preacher by saying: 'I like Muslim people. Those that I have known. I think they are very nice people.' (Do I hear an uncomfortable echo here? As a Jew myself, I just remember the phrase 'Of course, some of my best friends are Jews.') He continues: 'I think they are being victimized by a very, very dangerous book – the Qur'an. When a Christian kills, he's disobeying Scripture, and he's refusing to follow the example of his leader, Jesus Christ. When a Muslim kills, he's obeying his Scripture. He's following the example of his leader, Muhammad.'

Such dark prejudice, particularly from people who claim to be God-fearing, is very disturbing indeed. But one problem is that so many of us who are not Muslims do not understand the religion and are, to some extent, prevented from getting that knowledge. As a non-Muslim, I am not allowed to visit the holy city of Mecca (Makkah in Arabic), though I would very much like to. Above all, it would be an important experience to see what Muslims regard as the centre of the world, the most holy building that they venerate.

An empty cube in the desert

The Ka'ba (or Cube) is not of a particularly adventurous design. It is a plain, flat-sided structure which has been restored repeatedly over the centuries. Muslim tradition says that the original cubic building on this site was as old as time itself, having been erected by Adam. It is towards this refurbished structure that modern Muslims still pray each day. Ibraham renovated it, building a shrine with the help of his son Ismail, whom he had offered as a sacrifice on God's request. The shrine is where Ismail, together with his mother, Hajara, wandered in the desert looking for water after being thrown out of Ibrahim's household because of Sara's jealousy. This shrine is called Baitullah, the House of Allah. The parallels with Jewish scripture are very striking. 'Ibrahim' is, of course, 'Abraham' of the Old Testament, 'Ismail', 'Ishmael' and 'Hajara', 'Hagar'. Baitullah is very close to the Hebrew Beit El, also meaning the House of God (and from which, of course, we get Bethel, a very common name for evangelical churches in the United States). In the Jewish Bible, though, it was Isaac that was almost offered up as a sacrifice, not

Ishmael. Moreover, in the biblical account, Hagar was thrown out of the household for mocking Sarah's infertility. Even so, the similarities are strong; and I mention them here because, although Judaism and Islam are so frequently seen as being at loggerheads, there are many beliefs, practices, moral attitudes and philosophical elements within the two faiths which show the strongest relationship. These two great religions, like Christianity, stem from the same root.

After Ibrahim's death and before the arrival of the Prophet Muhammad, the Ka'ba was a small sanctuary with low walls in the middle of the arid southern Arabian desert. It seems that, in time, the local Arabs forgot the teachings of Ibrahim and used the Ka'ba as a pagan shrine, filling it with hundreds of different idols – a kind of pantheon. None the less, before the arrival of the Prophet, it had become an important centre of religious life, towards which many of the desert peoples came to pray. No trace of any of these pagan religions now remains. When Muhammad returned here from his exile in Medina, he cleansed the Ka'ba and destroyed the idols. The modern Ka'ba is a more substantial building, measuring 10.5 by 12 metres in area, and standing an impressive 15 metres high on a marble base. It has been rebuilt several times since the Prophet's day, but essentially to the same design. The corners are aligned so that each points approximately towards one of the four points of the compass. On the east of the door, the holy Black Stone lies in three large pieces with some smaller fragments, tied together with a silver band. According to at least one tradition, the Black Stone is a meteor that fell to earth. The historian Ibn Sa'd says that it shone like the moon for the people of Mecca until the pollution of impure people turned it black. For most of the year, the Ka'ba is covered by a black curtain of Egyptian fabric, yielding the striking image seen in photographs well known to non-Muslims.

Perhaps one of Islam's difficulties in the modern world is that it began its life as a reaction against urban living. Muhammad ibn Abdallah was an affluent merchant in the city of Mecca, a descendant of the Quraysh tribe. His profound break with tradition took place during the month of Ramadan in the year 610 CE, when he withdrew from the city to Mount Hira for a spiritual retreat. Muhammad was deeply troubled by what was going on in Mecca, which in a matter of decades had become a flourishing mercantile centre at the hub of a nexus of trade routes. Mecca owed much of its prominence to the fact that the Ka'ba was regularly visited as a place of pagan prayer, and, as a place of pilgrimage, generated substantial revenues, to the benefit of the ruling clans in the city. Desert tribes like the Quraysh had been partially responsible for this success and they were now extremely wealthy. But wealth brings its own problems in any age, and men like Muhammad felt that the traditions of their lean, ascetic forefathers had been rejected in favour of a gluttonous pride in personal wealth.

Desert life was tough, certainly, with its twin imperatives of fending off starvation and fighting one's corner in the constant inter-tribal warfare. To adapt to these conditions, nomadic tribes like the Quraysh had developed a strong ethic of co-operation. The survival of the tribe was the most important thing, and therefore every member took care of everyone else. There were few possessions anyway, but what they had, they shared. Hospitality was shown to strangers, even if they were enemies, because no-one knew if they might one day find themselves in similar circumstances, wandering lost in a barren land. Generosity was considered vital; the stockpiling of individual wealth, unthinkable. Trade had now assured tribes such as the Quraysh of economic success, but there was still no stability in the region. Surrounded by the vast empires of Persia and Byzantium, the Arabs adhered to codes of conduct which forged unity within, but not between, tribes. Christian ideas may also have had an unsettling effect. For the Arabian tribes, the only immortality one could achieve was in the survival of the tribe; yet Christianity argued that salvation was an individual undertaking.

The Islamic world refers to this period immediately preceding Muhammad's revelation as *jahiliyah*, the time of ignorance. But it was not godless. The Arabian tribes had their pagan gods, with worship centred on the Ka'ba shrine in Mecca. And though we now think of the Hajj,[1] the pilgrimage to Mecca, as a uniquely Islamic tradition, it is actually far more ancient. The end of the *jahiliyah* period was a time of longing for many Arabs. The Christians and the Jews had been sent prophets by their God, to guide them towards righteousness and/or salvation. No such favour had been shown to the Arabs. Neither did they have a sacred book like the Torah or the gospels. Their neighbours viewed them as spiritual inferiors. The Arabs seem to have felt this inferiority themselves, and it contrasted uncomfortably with their pride in their codes of conduct, their stern submission of individual needs to the needs of the tribe, and their ability to survive in the desert. One story describes a member of Muhammad's tribe, Zayd ibn Amr, leaning against the Black Stone of the Ka'ba during the traditional ceremonies and declaring, almost forlornly, 'Oh God, if I knew how You wished to be worshipped, I would so worship you; but I do not know.'

Muhammad's experiences met this need in a most vivid manner on the seventeenth night of Ramadan in 610. He was woken violently from his sleep by an angel who commanded him: '*Iqra!*' – 'Recite!' Like Moses before him, Muhammad said he could not. And, just as YHWH had done with so many of the biblical prophets, the angel persuaded Muhammad with violence, squeezing the air out of his body in a sort of transcendent bear-hug that left him weakened and terrified. Eventually, words bubbled out of him and he uttered the first of God's revelations to the Arabs.

It is interesting that so many figures, throughout the history of the Divine Idea, have experienced the call to prophecy in this way. You might think that being chosen as God's messenger would be a matter of pride and joy. Instead, men and women have felt it as a shattering, humbling assault that leaves them reeling. For some people, like the theologian Rudolf Otto,[2] this universal experience of fear and fascination provided the only evidence we have for what God is – a mighty presence, something entirely 'other'.

Certainly Muhammad was far from overjoyed. In his particular background, prophets were of a status well below that of an affluent merchant. Women consulted them to find out about women's matters; they were involved in curing toothaches and finding lost camels. He fled to his wife and told her what had happened to him. This first wife, Khadija, an older woman who had been married once before and was a rich and successful merchant in her own right, listened to him carefully. With a shrewdness that should silence those who say Islam has no regard for women, Khadija suggested they consult her cousin, who had become a Christian. The cousin confirmed that Muhammad had received a revelation from the God of Abraham and Moses. God was at last speaking to the Arabs – and Muhammad was his messenger. Over the next twenty-three years God revealed the Qur'an to Muhammad, line by line.

The Mother of Books – or 'toilsome reading'?

The Qur'an was never fully written down, or collated, in the Prophet's lifetime. Rather, it was recited and memorized by a group of scholars, the Qurra, who were diligent in preserving Muhammad's words and were personally supervised by him. Just as with the Hebrew Bible, it is not clear when the text was first codified. This may have been done by Abu Bakr, the caliph who succeeded Muhammad, or Uthman ibn Affan, the third caliph.[3] In any event, the definitive collation was complete by around 650 CE. Just as Jews regard the Torah as the actual explicit word of God, Muslims view the Qur'an as the word of God explicitly communicated to Muhammad, who was its conduit – much as Moses was the conduit for the Torah. The Qur'an is God's speech, His communication with humanity.

Many years ago, having become fairly fluent in classical Hebrew and gained a smattering of Aramaic, I tried very hard to learn Arabic. I found it much more difficult than I had expected, even though it has much in common with Hebrew; I wish I had persisted. When I was on *Desert Island Discs*, I chose the Qur'an as the book to take to my island. I supposed that, given long isolation, with no distractions, I might have familiarized myself with the language. This Mother of Books, as Muslims call it – *Umm al-Kitab* – is widely agreed to be expressed in the most perfect poetic Arabic. It is accepted as the fount of Arabic grammar. It is, as Reza Aslan has written, 'to Arabic what Homer is to Greek, what Chaucer is to English: a snapshot of an evolving language, frozen forever in time'.[4]

The words of the Qur'an have to be written in Arabic. Indeed, every Muslim recites them in the original language, because to translate these words risks corrupting their meaning. No translation can do justice to the text of the Qur'an, it is frequently said. This is, I think, why Islam can be so difficult for the non-Muslim to understand fully. This position is not substantially different from that of Jews who view with suspicion, disdain or sadness loose translations of the Torah. Precise interpretation of a passage in the Torah can be very difficult for those who do not have a sound basis in its original language. So often the motives and views of Jews have been misleadingly or downright falsely described, usually to the detriment of the Jewish people, because of some false translation of the Jewish law.

So it is with Islam. In any translation, the Qur'an is a difficult book – for Judaeo-Christian readers, at least; and as a consequence, to this day, people tend to invent theories about the religion rather than see for themselves what it is that Muslims believe. The Bible, like the Qur'an, needs careful interpretation; but the Bible is above all a story. By contrast, the Qur'an is a series of *suras*, or chapters, subdivided into *ayahs*, or verses. The *suras* cover a range of subjects and are disposed in the order in which they were revealed to the Prophet. Some have odd-sounding names, like al-Baqarah (the Cow), which is about religion, al-Alaq (the Clot: about man's beginnings and limitations) and al-Ankabut (the Spider; dealing with unbelievers). In many cases, even parts of the same *sura* do not seem to have a consistent theme. So the Qur'an certainly cannot be read like a novel.

I must say it is as toilsome reading as I ever undertook. A wearisome confused jumble, crude, incondite: endless iterations, long windedness, entanglement; most crude, incondite – insupportable stupidity, in short! Nothing but a sense of duty would carry any European through the Qur'an.[5]

So fumed the Scottish historian Thomas Carlyle in the

nineteenth century. Many others have admitted, in more considered fashion, to finding it perplexing. One of the major reasons for this is the difficulty of translating Arabic into other languages. Even the everyday Arabic utterances of politicians, or secular literature, can sound weird and stilted when translated. In addition, the Qur'an was intended to be recited, not read privately. The rhythms of the language may affect the audience as much as, if not more than, the actual content. The Arabic of the Qur'an is considered a sacred language, and when Muslims approach it they feel they are entering into a spiritual experience, in the same way as Jews do when they hear the Torah. By contrast, Christian attempts to render the Bible into everyday language in order to spread the Word far and wide have sometimes diminished that aura of spirituality surrounding the text. Accordingly, some Christians have felt surprise and bewilderment, even a sense of anti-climax, when reading the sacred books of others in translation. Also, in doing so they do not have the benefit of over a thousand years of stories, commentaries and analysis of the Qur'an, which shape the context in which it is approached by modern Muslims.

The tradition of memorizing and reciting the Qur'an, established by the Qurra, evolved into *tajwid* – the art and science of its recitation. There are formal instructions as to when and where one may breathe during a recital, which syllables may be stressed, when and where the reader may stop, when a congregrant may prostrate himself. Along with the use of pictorial art, Islam shuns the use of formal music during any form of worship. But the reader of the Qur'an, in congregation, does use spontaneous inflection and melody, and these chants can be very beautiful indeed. I suspect that many Qur'an readers are a little like the *hazanim*, the cantors in formal Jewish worship. *Hazanim* are notorious for being a bit like operatic *prime donne* – sometimes self-absorbed and occasionally a touch vain about their performance. When I was recently in the Al-Aqsa Mosque in Jerusalem, I listened to the reader, an elderly man of very serious mien, recite the ninety-nine names of Allah from the Qur'an for my benefit. As we walked out of the mosque together some moments later, he asked rather plaintively for reassurance: 'Was it not beautiful? Did it not sound very good?' I replied that I thought it the most beautiful sound – as, indeed, it was.

As the direct speech of God, presented without any chronological framework, the Qur'an is not always easy even for Muslims to fully understand. So, as Jews developed the Mishnah and the Talmud, Muslims also evolved an oral tradition of numerous commentaries on the matters within their divine text, formulated to embellish, underline and expound Allah's intentions. These commentaries are called the Hadith.

Most of Muhammad's contemporaries and successors were deeply impressed on hearing the Qur'an. Some were converted on the spot, feeling certain that only God could have been responsible for such a work of beauty. One of Muhammad's most vociferous opponents, Umar ibn al-Khattab, was converted to Islam when his sister was overheard reciting it in secret. His first response was to strike her to the ground. But then he picked up the book and, being literate, began to read it out loud. He immediately surrendered to Islam – and indeed, that is what the word *islam* means: 'surrender' or 'submission'.

A modern-day imam (the leader of worship in the mosque) writes of the standards of conduct required while reading from the Qur'an:

Reading the Qur'an is not like reading an ordinary book. It has some etiquette and manners. Imam Al-Ghazali mentions that one should perform ablution, be soft spoken and quiet, face the Qiblah [the direction of the Ka'ba], keep the head lowered and should not sit in

The Qur'an must be treated with the utmost reverence even when it is not in use. The Qur'an would never be placed on the floor, for example. It may be wrapped in a cloth to avoid its coming into contact with dust; it is often placed on a top shelf so that no other book can be placed above it; and when a volume of the Qur'an is present in a room people are expected to behave with care and not act cruelly or indecently. When it is being read aloud, people do not talk, eat or smoke, or make distracting noises. All these matters of respect are very similar indeed to the way Jews treat the Torah. Jews, for example, avoid touching the parchment of a scroll of the Law – and, for that matter, do not place any object on a printed Pentateuch. My grandfather, a rabbi in west London, once had a sleepless night. He came downstairs and, wandering around, found that he had placed a copy of the Pentateuch on the dining-room table upside down – with the end of the book facing upwards. He turned it over, climbed the stairs to his bed, and fell into a deep sleep immediately. I suspect that a Muslim would have no trouble understanding this.

Against this background, it should be easy to understand how shocking desecration of the Qur'an is for Muslims. In May 2005 the Arabic television station Al-Jazeera reported that one detainee in the US prison camp at Guantanamo Bay, Abdul Rahim, said that abuse of the holy book was routine during the interrogation of Muslim prisoners there by American forces: 'Abuse of the Qur'an was done routinely, particularly in the early days of detention . . . They would throw the holy book on the ground, trample upon it and tell the prisoner under interrogation no-one could stop them from doing that. The news of sacrilege sent shock waves among the prisoners and all of us went on a hunger strike.' If such allegations are really true, as the International Red Cross has stated, then non-Muslims should be equally outraged.

What, then, is the message of the Qur'an? Any paragraph-length summary must be simplistic to the point of glibness, but its core message is not unlike that of Jesus. Muhammad's audience had placed personal wealth and worldly success at the centre of their world. They had forgotten that Allah, God, was the creator of everything. The Qur'an does not see itself as inventing a new religion, but as reminding the Arabs of a truth to which they had become blind. It points to various aspects of existence, as if to nudge its audience into realizing, with their own powers of observation, that of course Allah is the bedrock of it all. *Suras* frequently begin with challenges, such as 'Have you not seen . . .' or 'Have you considered . . .'. The core principle being asserted is the centrality of God to everything. Like Jesus, Muhammad in no way sought to overthrow the traditions of the past, but said that his people had forgotten the meaning of these traditions. The signs of God's goodness and benevolence were everywhere to be seen. It was up to men to reproduce this goodness in their own society, otherwise they would be out of step with the true nature of creation. This is why the faith of Muhammad is called 'surrender': it refers to the voluntary act of surrendering one's will to Allah, of recognizing his supremacy and importance.

The Qur'an does not waste much time speculating about the nature of Allah – in fact, it dismisses this as *zannah*, navel-gazing of the worst kind. Accordingly, Muslims have always found it hard to understand the Christian's obsession with the nature of the Trinity – a perplexity expressed, comically, by the fourth-century Christian bishop Gregory of Nyssa:

> Everything is full of those who are speaking unintelligible things –
> streets, markets, squares, crossroads. I ask how many oboli
> [currency] I have to pay; in answer they are philosophizing on the
> born or the unborn; I wish to know the price of bread; one answers,
> 'The Father is greater than the Son'; I ask whether my bath is ready;
> one says, 'The Son has been made out of nothing.'[6]

For Christians, Paul's attempts to make Jesus into a God resulted in considerable confusion. How could God be present as an omnipotent Creator, appear in space and time as Jesus, and also be present everywhere, any time, in the Holy Spirit? Did that mean there were, in effect, three gods? And if so, which was greatest? What may seem like pointless speculation to us generated bitter division and antagonism in the early Christian church – the arguments at the Council of Nicaea, as we saw in chapter 5, being a good example.

Islam suffered from no such confusion. In the Qur'an, Allah is given ninety-nine names, all of which emphasize His superiority to the created world: they include, for example, al-Ghani, rich and infinite; al-Muhyi, giver of life; al-Alim, knower of all things. Many of His names contrast with one another – he is both giver and taker-away, exalter and he-who-lays-low. All of this contradiction serves to remind Muslims that the God they worship is simply beyond human thought and language.

Allah: a God for everyone

Initially Muhammad's message, like the teaching of Jesus, was tolerated. Groups that did not participate in the commercial success of Mecca – slaves, women, the under-privileged – grasped it enthusiastically. The haughty aristocracy had little time for a faith that urged them to distribute their wealth and bow down on the ground, but they allowed it to carry on. However, their 'do it by all means but don't scare the camels' attitude shifted when Muhammad's message became more strictly monotheistic. When he began to tell people that the old gods were a pointless illusion, and that worshipping them was a sin – 'Have you then considered al-Lat, al-Uzza, as well as Manat [older deities] . . . ? These beings are nothing but empty names which you have invented – you and your forefathers – and for which God has bestowed no warrant from on high' – the break with tradition was seen as deeply threatening. Time and time again in the history of the Divine Idea we see the same drama played out. There is something richly appealing about a god with a shrine, a god with whom one can engage in simple transactions to gain what one desires. The shift to an unknowable One, pleased only by a turning around of one's whole life, rather than simple actions, is painful.

Muhammad found himself at odds with the ruling classes of the Quraysh clan. In 622 he and his followers sought a haven in the northern settlement of Yathrib, later known as Medina. Yathrib was sympathetic to his teachings – it had been torn apart by inter-clan warfare, so was welcoming to an idea that offered unity. The presence of a number of prominent Jewish tribes in Yathrib had also opened people's minds to the concept of the One God. Muhammad, in line with his conviction that Allah was the God of Abraham, had taken steps to bring Islam closer to the Jewish faith, allowing inter-marriage and urging his followers to pray three times a day, like the Jews, facing Jerusalem.

But eventually there was a break between the Jews and the Arabs. The Jews expected a Messiah, but they had seen Jesus and others fail, and they were sceptical towards prophets of all kinds. In 624 Muhammad directed his followers to pray towards Mecca instead of Jerusalem. This break with the Jews lent Islam a distinct advantage among Arabs, for many who had previously seen it as a dangerous offshoot would have been impressed by its open continuity with the tradition of the Ka'ba, the Arabs' most holy shrine. A decade of warfare and struggle ensued, but by the time Muhammad died in 632 most Arabs had accepted Islam – in many cases, including that of the city of Mecca itself in 630, without violence. In the final year of his life, Muhammad conducted the venerated Hajj pilgrimage and incorporated it as one of the pillars of the Islamic faith.

The special nature of God as one transcendent being is one of the key teachings of Islam: 'Allah (the Supreme), the One. Allah is Eternal and Absolute. None is born of Him. He is Unborn. There is none like unto Him' (*sura* 112). God is referred to in the masculine as 'He' and is thought of more as the 'Creator' of humankind than as a 'father'. The concept of 'father' seems to imply a totally different role for God and to Muslims, it is very important to avoid risking the division of the nature of God – the sin of *shirk*. The Muslim God also knows everything, and therefore everything is predestined and in His hands; there is no such thing as a random or chance event in human life.

A second key teaching is that Allah is compassionate and merciful. Muslims are expected to emulate this aspect of divine behaviour in their conduct with their fellow men. This notion of *imitatio dei* is very similar to a concept found in both Judaism and Christianity. Thus, for example, when a Jew acts as a physician he is 'playing God', and this is a laudable concept – providing he imitates Him and does not attempt to supplant Him.

A third teaching is that humanity is the highest physical creation of Allah. Each of us has a soul, and each person has an allotted time on earth that is predestined. The notion of an immaterial soul is held in common by all of these world faiths, though Jews consider that the human, though unique, is certainly not necessarily the pinnacle of creation. Muslims also believe that humans have free will and exercise a choice between good and evil. The idea of free will is clearly difficult to reconcile with the notion of predestination, but it seems that in Islam life itself is a form of test, in which all humans are given choices.

A further feature feels rather more strange to some Christians and much more strange to Jews. The vast universe of God's creation contains mystical unseen beings – angels, the devil and *jinns*. Each person on this earth has two guardian angels, who record that person's actions and are the agents who prick their conscience. The devil, or Shaitan, is chief of the *jinns* and is the enemy of humans, determined to lead hearts and minds away from God. *Jinns* are non-physical beings that can assume human shape and may require exorcism – a belief which has echoes in Christianity. *Jinns*, however, are not always bad spirits, though they may be malevolent if they enter a human body. Mainstream Judaism shuns the idea of spirits or demons (though the Talmud does refer to spirits quite often) and even dabbling in the occult is seen as reprehensible. Nor is a belief in Satan truly acceptable. Although the book of Job mentions Satan, this Satan is almost certainly just another manifestation of God's many facets. The occult has always been viewed with suspicion by Jews, though the Kabbalists of the Middle Ages certainly seemed to engage with the supernatural.

One important Muslim teaching is that humans live only once on this earth; after death, each individual faces judgement and eventually is committed to heaven or to hell. It seems that the Qur'an describes heaven and hell in poetic, symbolic terms, rather than suggesting they are actual places where physical pleasure or physical torment will be experienced. This is in contrast with a mainstream Christian tradition in which a vividly rendered physical hell often features strongly – indeed, one of the most significant pieces of Christian poetry ever written is the *Divina Commedia* of Dante Alighieri, where heaven, hell and purgatory are described in great detail. Muslims also believe in resurrection: at some point in time, God will resurrect all people, irrespective of how they died. Jews mention the resurrection of all in their prayers, but it is very unclear whether the new lease of life means much more than a single day spent out of the grave – a rather unattractive proposition, this, I feel.

Islam requires adherents to observe a number of signs of commitment, referred to as the five pillars of faith. The first pillar is the need to bear witness to the faith. Once a person has sincerely taken the decision to become a Muslim, his first act is to declare his belief in Allah as the sole God, and Muhammad as His Prophet; this declaration is made in front of two witnesses. This bearing witness is called *shahadah*, and the same form of words is used by the muezzin for the call to prayer five times each day. Worship, *salah*, is the second pillar of Islam. The prescribed ritual is a mixture of words and movements, such as kneeling to face Mecca. The five daily prayers are Fajr, the morning prayer, which is recited at dawn; Zuhr, at the height of the sun; Asr, when the afternoon shadows appear; Maghrib, after sunset; and Isha, at night. Jews also pray at prescribed times – normally three times a day with additional prayers on the Sabbath and festivals. By contrast, very few Christians pray at such regular intervals, except in institutions such as monasteries or nunneries, where there is still the same notion of a discipline of regular prayer at set times during each day. Orthodox Muslims are expected to pray in a congregation at least once a week, at midday on Fridays, when all work ceases.

Like the tithe given by both Jews and Christians, almsgiving is a key pillar of Islam. *Zakah* is the obligation of all and ideally should be anonymously given. *Zakah* helps the poor, but also helps the rich perform their duty to be responsible people. The fourth pillar of Islam is *sawm*, fasting, undertaken as an exercise in spiritual and physical discipline to cultivate a peaceful frame of mind. Fasting is done at varying times during the year, but a key fast observed by all is Ramadan, a month during which food may be eaten only between sunset and sunrise. Most Muslims read through the whole of the Qur'an during Ramadan, in spells of around two hours each night.

The Hajj

The Hajj, the pilgrimage to Mecca, is the fifth pillar of Islam. This is a very serious undertaking and, for most Muslims, involves a complete upheaval in an individual's life for several days. It is compulsory for every adult who can afford to go to Mecca. Sometimes people save up most of their lives to achieve one pilgrimage because it is regarded as so important. Money earned dishonestly may not be used to fund a person's Hajj. Children are welcome, but their attendance on a Hajj pilgrimage does not in any way exempt them from making their own pilgrimage once they become an adult.

The rituals of the Hajj can seem complex and bewildering to outsiders. The pilgrim's first act is to don a white 'death shroud', the *ihram*, to symbolize his departure from the mundane world. He then heads for the Ka'ba, which for the faithful sums up in its emptiness the unknowable nature of Allah, his absolute otherness, without shape, without colour. After kissing the stone, or reaching out in its direction – for the vast crowds make it impossible for many to come close to the Ka'ba itself – the pilgrim walks around it, anti-clockwise, seven times. He then conducts prayers in the eastern corner of the complex, the 'station of Abraham'.

After this, the pilgrim moves on to a well known as Zamzam, to drink three draughts of its water, before making seven trips between the hills of Safa and Marwa. This re-enacts an episode in the Arabian–Jewish legend of Abraham, who left his concubine Hagar with her son Ishmael in order to return westwards. In desperation and searching for water, Hagar ran seven times between the two hills. After this is completed, the pilgrim has his head shaven (women usually cut an inch or so off their hair), and puts his workaday clothes back on. But the full pilgrimage is far from over.

The Hajj takes place over six days, between the 7th and 13th days of the month of Dhul-Hijja. Each day has its own particular rites. Those of the 7th are described above. On the 9th day, for example, the pilgrim stands on the plain of Arafat from dawn until sunset, begging for forgiveness. On the 10th day, he casts pebbles at three stone pillars, a symbolic stoning of Shaitan, the devil.

So holy are the rituals of the Hajj that non-Muslims are not even permitted to enter the city of Mecca. It is, in any case, notoriously hard to enter the kingdom of Saudi Arabia in the first place. In making the television series which accompanies this book, we found it impossible to get visas to film anywhere in Saudi Arabia on two separate occasions – which is sad, because my essential motive was to try to contribute to a better understanding in the west of the great worth of the Islamic world. Non-Muslims can only rely on the testimony of the pilgrims to discern the meaning of this sacred journey. The anthropologist Victor Turner, whose work we discussed in chapter 2, noted that in various cultures pilgrims separate themselves from daily life and enter a different state of being. In the Hajj, the white shrouds donned by the pilgrims, the shaving of the head, the prescribed activities – as well as the very fact of the arduous journey undertaken – all contribute to a sense that they have visibly separated themselves from their former existence.

Turner felt that pilgrimages are one way in which societies reinforce their unity. When, in entering his new, altered state, the Hajj pilgrim dons a uniform, it makes him resemble all other pilgrims, stripping them of individual differences. Through practices such as these, Turner argued, the dominant mood becomes what he called communitas – an essential brotherhood and equality.

Malcolm X, the legendary American black activist, certainly experienced this shift in perception on his Hajj in 1964. Until that time, he had been a devotee of the teaching of Elijah Muhammad, a black Muslim leader who advocated segregation between blacks and whites. For Malcolm X, the experience of becoming a pilgrim, of joining thousands of people of different backgrounds and colours in a unified experience, led him to shed his segregationist views. The Iranian writer Ali Shariati explains the Hajj in similar terms: 'Everyone melts himself and assumes a new form as a "mankind". The egos and individual traits are buried. The group becomes a "people" or an "Umma". All the Is have died . . . what has evolved is We.'[7]

To this day, the Hajj remains the most vivid and most readily recognized symbol of what it means to be a Muslim. Ask the average non-Muslim about the five pillars of the faith, and they will probably not know about the one-ness of God, the duties of prayer, alms-giving and fasting. But the image of a sea of white-clad pilgrims circling around a huge black monolith has an arresting power, for Muslims and non-Muslims alike.

We humans are intrinsically social animals; we usually become depressed in isolation. Religions make use of this impulse towards community in many rituals. Anyone who has ever been part of a large crowd, like a protest march or a football match, can relate to this experience and it's not hard – for an Arsenal fan like myself, at least – to see why it feels like something transcendent or holy. Evangelists hold total sway over massive congregations; the Catholic faithful fill the space outside St Peter's in Rome at every Easter celebration. The experience of being in a large group with a common set of values, or at least a common aim, is richly liberating – the mind, as it were, is allowed to soar above the limits of the individual body. In experiencing unity with our fellow humans, many of us take a step towards perceiving a greater unity that lies behind everything.

Judaism also makes use of congregations in quite a similar way. Although only ten men are needed to form a religious quorum (a *minyan*), there is an undoubted requirement to pray with others if at all possible; this increases spiritual power. Indeed, there are many Jewish prayers which can be recited only when a *minyan* is present. And communities have always got together for the main festivals and many events, such as weddings. However, in past times – ever since the destruction of the Temple – Judaism was somewhat wary of large gatherings. To an extent, this caution reflects the character of a people who lived in exile, often in environments that were hostile to them. Mass gatherings were dangerous; they both attracted unwelcome attention and provided an opportunity for aggressors to attack with maximum potential for damage. The synagogue is a hub of Jewish life, certainly, but the home is a place of equal religious value.

Without doubt, in according respect to the powerful and ancient ritual of the Hajj, Muhammad knew what he was doing. His religious quest began as a reaction to disunity among his people, and ever since, Islam has contained the capacity to unite disparate groups in a common cause. It has also, since its earliest days, been a highly tolerant faith. The Qur'an did not see its revelations as cancelling out the messages of previous prophets, but as adding to the tradition of humans, in different ages and cultures, experiencing God. Tradition says that there have been 124,000 prophets – a number intended to suggest an infinity, rather than a certain limit.

Islam and modernity: *jihad* and *shari'a*

Jihad, although not a pillar of the Muslim faith, is a key idea of great importance. It is also a concept very frequently misunderstood by non-Muslims as well as unfortunately misrepresented by some within the faith. *Jihad* is certainly not 'holy war' to convert non-believers. This is a stereotype which was promulgated by Christian crusaders, and does not truly reflect the nature of the Islamic idea. Indeed, Islam has a proud history of being tolerant towards other monotheistic faiths, and defends individual liberties. Faith is a matter of choice, and while Muslims encourage conversion, it is never regarded as acceptable to force a person to accept Muhammad's message. In the Qur'an, believers are required to be patient with non-believers: ' "Bear with patience what they say, and when they leave give a courteous farewell," Allah requested' (*sura* 73: 10); 'Deal gently with un-believers; give them enough time [to change their minds]' (*sura* 86: 17).

Jihad is often taken to mean military action. But in most contexts *jihad* means a struggle – and that struggle in turn is most usually that of the soul attempting to over-come the obstacles which prevent an individual from getting closer to God. Islam is no more a warlike faith than Christianity, even though it has frequently become depicted as being in favour of wars (although the word *islam* means 'surrender', it actually has the same root as *shalom* in Hebrew, meaning 'peace'). According to the strict rules laid down by the Prophet, *jihad* may be declared only in defence of Allah, and not for conversion or conquest. 'Fight in the way of God those who fight you, but do not begin hostilities; God does not like the aggressor' (*sura* 2: 190). This type of *jihad* may be used to restore peace and freedom of worship, but must be led by a spiritual leader and judge. Hostilities must cease when an opponent surrenders, and women, children, the old and sick must not be damaged; nor may trees or crops be destroyed. Consequently, it is a profound misrepresentation to believe that *jihad* covers wars of aggression, border disputes, the intent to colonize, or – particularly relevant today – acts of terrorism or indiscriminate bombings. The Qur'an repeatedly makes this abundantly clear, but perhaps it is worth quoting one verse (*sura* 8: 61): 'If they seek peace, then you seek peace. And trust in God – for He is the One that hears and knows all things.'

It is very unfortunate that in recent years some political, religious and academic leaders in Muslim countries have rewritten the rules of *jihad* to suit their own political ends. In doing so, they have greatly damaged the standing of Islam and caused huge turbulence in the world. Thus, in Iran, the Ayatollah Khomeini emphasized the need for a militant approach to *jihad*. Having first used this tactic to underpin the Islamic Revolution of 1979, in which he assumed supreme power, he then committed Iran to the terrible eight-year-long war with Iraq – all in the name of *jihad*. These actions provided the fuel for the establishment of the terrorist group Hizbollah, whose subsequent promulgation of suicide bombing has resulted in appalling crimes against humanity which go completely against all Islamic morality. And we have recently seen in both Iraq and London how such terrible crimes involve the murder of the totally innocent – small children, Muslims as well as non-Muslims. The foundation of Hamas, another such terrorist organization which has been highly active in Palestine and Israel, was a response to the teaching of at least one academic in Saudi Arabia. There, Professor Abdullah Yusuf Azzam carries great responsibility for persuading disaffected young men to adopt a most aggressive and violent approach to the idea of *jihad*: '*Jihad* and the rifle alone – no negotiations, no conferences, no dialogues.' And from this stable, too, came Osama bin Laden, so destructively commencing his immoral, notorious – and misnamed – *jihad* against the west.

Shari'a, the body of Islamic canon law, divides human behaviour, both religious and secular, into various categories, attaching to some of them prescribed reward or punishment. First are actions which are expected of a believer – their undertaking is rewarded, their omission is punished. Second are actions which are seen as good deeds – these are laudable and may be rewarded, but if not done there is no specific punishment. Then come neutral actions – neither punished nor rewarded. Then there is conduct which is considered bad or reprehensible, but not in itself sufficiently grave to warrant a punishment. And lastly are those actions that are proscribed by law and punishable.

In Islam, all people are equal in the eyes of the law. No matter how humble the citizen, he or she is worthy of protection. A person has a right to a fair trial and a right to legal defence. A judge is expected to be truly independent and without bias. *Shari'a* requires that no judge may try a case while he is in the wrong frame of mind – angry, tired or hungry, or distracted by other matters. No person can be imprisoned unless legally convicted by a properly convened court of law. It is a serious offence to convict an innocent person. One aspect of *shari'a* which is very frequently misunderstood is the requirement that a legal sentence be carried out publicly.

This is not done to please a bloodthirsty or sadistic people, but to ensure that justice is seen to be done and that punishment does not exceed what has been prescribed. As the Qur'an says: 'O believers, be staunch in justice, witnesses for God, even though it be against yourselves, your parents or your kindred, whether the case of a rich or a poor person – for God is nearer to both . . . if you lapse or fall away, then lo! God is ever informed of what you do' (*sura* 4: 135).

One problem with *shari'a* as a code of law is that its traditional observance accords ill with our modern notions of human rights. Most Muslim countries have attempted to reconcile *shari'a* with a more modern legal system, but the marriage is not entirely happy and the results are certainly not completely successful. In Egypt and Pakistan the general approach has been to accept *shari'a* as a source of civil law only, for cases involving the family or inheritance. In Saudi Arabia, and currently in Afghanistan, little has been done to modernize *shari'a* law, and the governments there seem to have largely ignored internationally agreed standards of respect for human rights. In Iran there has been a more progressive attitude, and considerable headway has been made in trying to reconcile democratic values and liberal attitudes with the traditional approach of *shari'a*. I was impressed deeply during my visits to Iran, and feel that, although massive changes are still taking place there, the west might do rather better to encourage Iran in its progress forwards than to attempt to isolate it, as is currently largely the case.

While I do not pretend to be an apologist for *shari'a* law, it is no more than fair to observe that western reporting of *shari'a* in the popular press does not always represent the true conduct of Islamic law. Also, a number of quite barbaric Muslim governments have occasionally invoked the law or used it in ways that seem at variance with the path that Muhammad intended; sometimes, indeed, the laws of Islam are actively flouted when it is claimed they are being represented. Having said that, some punishments which are handed out seem barbaric to western eyes. In the modern world it seems impossible to accept the damage caused by amputation of a hand, even when a culprit has continued the most persistent thieving. But it is worth recalling that in the New Testament (Mark 9: 43), Jesus said: 'If your hand causes you to sin, cut it off. It is better to enter life maimed, than with two hands go to hell – into the fire that never shall be quenched.' Flogging for drunkenness seems extraordinary to our society; but the rules about how and when flogging is carried out are certainly intended to temper this unacceptable punishment with mercy. Much has been written about the law of execution for adultery. But under strict *shari'a* law the death penalty cannot be enacted unless the act of adultery has taken place in public and been seen by four reliable witnesses – in actual practice, conditions which should make it largely impossible for such punishment to be delivered.

The status of women

Before Muhammad, women in Arabia had an unenviable status. They could not own property and were not even considered competent to inherit it from their husbands. Indeed, women were themselves considered as property, and their menfolk could always take control of any property which came into a widow's or a female orphan's hands. Muhammad was intent on changing the status of women and gave them a level of independence and a degree of equality largely unprecedented in that region of the world. Women were given the right to inherit and to hold property, and to keep their dowries if they chose to withhold them from their husbands once they were married. Indeed, a man was forbidden to take possession of his wife's dowry and was expected to provide adequate support in every way for his wife. Muhammad went further. Much to the dismay of many men, he stipulated a limit on the number of wives a man could marry and granted the right of divorce to women.

It is not surprising, then, that, like early Christianity, Islam had a distinct appeal to women – a fact which may surprise those westerners who see the religion as oppressive and male-dominated. During the recent military campaigns against Afghanistan and Iraq, the image of women in the *burqa*, a long, tent-like garment covering the entire body, has been used by the media to suggest that Islam uniformly shutters up and confines its women. But nowhere in the Qur'an is the veil generally enjoined on women. Veiling and seclusion of women was common before the establishment of Islam, and was never seen as an essential religious requirement. The universal veiling of women and their separation from the world of men were, in fact, customs prevalent in Persia and Byzantium which later entered into Islam. The one mention of the *hijab* in the Qur'an applies to Muhammad's wives alone:

> O Ye who believe! Enter not the dwellings of the Prophet for a meal without waiting for the proper time, unless permission be granted you. But if ye are invited, enter; and, when your meal is ended, then disperse. Linger not for conversation. Lo! that would cause annoyance to the Prophet, and he would be shy of [asking] you [to go]; but Allah is not shy of the truth. And when ye ask of them [the wives of the Prophet] anything, ask it of them from behind a curtain. That is purer for your hearts and for their hearts. (*sura* 33: 53)

Muhammad's house was the centre of the community, and this was a necessary way of assuring the Prophet's household of some privacy. Since then, the term for this curtain, the *hijab*, has come to mean any form of the veil, and its use in those various forms has persisted throughout the Muslim world. Sometimes governments or religious leaders have imposed these forms of female dress with extreme strictness – as was the case at the start of the Islamic Revolution in Iran. More recently, any visitor to Iran can see that the veil is worn by very many women with increasing liberality. In Afghanistan, the strict regime of the Taleban has caused a strong hostile reaction. Women there were clearly oppressed when the Taleban were in power. But it has to be said that many Muslim women elsewhere see the custom in a more positive light. They note that wearing 'Islamic' dress gives them a sense of identity with their tradition and frees them from being treated as objects of desire by men.

While Muhammad clearly wanted to give women a better status within his society, Islam's history has not always been liberal. One of Muhammad's companions – Umar, of whose conversion the Prophet was particularly proud, not least because he was a member of the Quraysh tribe – was an aggressive alpha male. A noted warrior, he was zealous in ensuring the moral rectitude of the true faith. It is told that he had a fiery temper and often expended his violence on women. He was one of the early leaders who tried to ensure that women were confined to the home and even argued that they should not be allowed inside the mosque. He made sure that women could not pray with men and, after Muhammad's death, forbade his widows to go on pilgrimage. He also established a number of penal sentences which were designed to keep women in severe submission. It was he who instituted the stoning of female adulterers, which is certainly not part of the Qur'anic tradition.

Nevertheless, over time, women were increasingly allowed an education, and a number of famous women have contributed greatly to the oral law. For example, one of Muhammad's wives, Aishah, was the author of a substantial part of the more highly regarded Hadith. Fatima bint Ali and Karima bint Ahmad, both of whom lived in the eleventh century, are two of the most significant teachers and scholars in Islamic history, and Zaynab bint al-Shari, who lived over a century later, was noted for her great expertise in exegesis.

It must be observed that Islam, like any other faith, was greatly influenced by the cultural norms of the societies in which it arose. The desert societies never considered women as members of the tribe equal with men, either before Muhammad or after Islam was established. The Qur'an clearly reflects these attitudes – and it is only recently that Muslim women have started the struggle for real equality with men, in part because Islam, which has always been cut off to some extent from the west, was not subject to the extraordinary influence of intellectual movements such as the Reformation.

The struggle for the caliphate and the great schism in Islam

There can be no denying that Islam had very warlike beginnings. Natural death among its early rulers was not a normal event. The unity that Muhammad's faith gave to the warring Arabian peoples was, unfortunately, only transitory. After his death, as happens in many human stories, a power struggle emerged between rival successors: Islam was fashioned only after a long and bloody struggle. The details are complex and what follows is only a bare summary – but a summary which is needed if we are to understand how this great religion developed.

A new leader was clearly essential if the Prophet's

mission was to live on after his death. Among the key roles to be played by this leader, or caliph (the word means 'successor'), would be to exemplify the way of life Muhammad had espoused and to pass on authentic Hadith. Abu Bakr was a strong candidate because he was Muhammad's first convert; he was also the father of the Prophet's youngest wife, Aishah. Given that Muhammad had on his deathbed requested Abu Bakr to lead the congregation in prayer, his appointment as Muhammad's successor was almost inevitable. Ali, the Prophet's son-in-law, would have been an obvious alternative, but he was passed over and Abu Bakr became caliph in 632 CE. He led Muhammad's followers for just two years, building up a substantial military and ideological force, his main preoccupation being to ensure that Islam was protected against the local warring Arabs. When Abu Bakr was dying, he nominated Umar as his successor, without reference to the community.

Umar was caliph for ten years – a period within which Islam continued to spread with extraordinary rapidity. He oversaw the capture of Jerusalem in 638 CE, forbidding any harm to its Christian and Jewish population. Umar was a moderate who, though warlike in expanding the territory of Islam, never forced conversion on the peoples who came under his dominion. It is said that in 644 CE a soothsayer came to Umar and quoted verses in the Jewish Torah which forecast his imminent death. At this time, a Persian Christian captive called Firoz was working in Medina as a carpenter, blacksmith and painter – the sort of odd-job man (perhaps fortunately) difficult to find in London today. Firoz sought audience with Umar to complain that his master was taking too much money from his wages. He asked the caliph to reduce this levy. Umar enquired about the work he did. Firoz explained, saying that he could also make windmills. Umar told him that because the jobs he did were lucrative the levy did not seem excessive, but he did promise to contact his employer. Firoz, dissatisfied, left sulking. Later, when Firoz returned, Umar suggested that, seeing as he made windmills, he might make one for him too. Firoz is reported to have replied: 'Verily I will make such a mill for you, that the world would talk about it.' On 1 November 644, at morning prayers, Firoz went to the mosque. When the congregation stood, Firoz rushed from a hiding place and stabbed Umar repeatedly. He died three days later.

Uthman was the next caliph. He was a Qurayshi and, in a move perhaps characteristic of the ambitious nature of this family, appointed a number of his own clan to govern the territories which Islam was in the process of conquering. These continued to multiply hugely. Under Uthman, the Muslims beat the Byzantines in North Africa and drove back the Romans from the Mediterranean. Arabia, Yemen, Persia, Iraq, Armenia, Azerbaijan, Syria, Palestine, Jordan, Egypt, Libya, Algeria, Tunisia and Morocco were now effectively under one ruler, and the Islamic state was larger than the mighty Byzantine or Persian Empires had been. Although an attempt to invade Spain was unsuccessful, parts of East and Central Africa started to practise Islam. But this rapid expansion led to growing instability and civil war closer to home. There was increasing resistance to Uthman's caliphate, and eventually he too was assassinated at prayer, at the age of eighty. When the insurgents who had been besieging his house in Medina broke in and brutally attacked the caliph, his wounded wife, Nailah, tried to shield her husband; in the process her fingers were chopped off. She sent them, together with a plea for help, to her husband's cousin, Mu'awiyah, in Syria.

Ali, Muhammad's son-in-law, having waited in the wings ever since the Prophet's death, finally became caliph in 656, succeeding the murdered Uthman. He found it difficult to bring the assassins of the third caliph to justice and there was a feeling that in some way he was responsible for the murder. Once in power, he immediately faced a challenge from Aishah – the widow of the Prophet – who led an army against him at the Battle of the Camel. The two armies met just outside Basra, now in Iraq, on 4 December 656 CE – the first time that Muslims had fought Muslims. Ali gave orders that his men were only to defend themselves, not initiate attacks. No wounded were to be slain, no fugitives pursued, and no plunder taken. Aishah's forces inundated the air with arrows, but Ali refused to return shot. Eventually, as the camel bearing Aishah advanced, a heavy pitched battle developed round the animal. Ali recognized that so long as the camel remained standing, the battle would continue, with unnecessary loss of life; so a soldier was deployed under the camel to amputate its legs. The soldier positioned himself behind the camel, swung his sword and the unfortunate animal slumped to the ground. What had started as a skirmish resulted in carnage: Muslims say that 16,796 men of Aishah's forces and 1,070 of Ali's died in the battle. Once Aishah was captured, fighting ended fairly rapidly.

Even after defeating Aishah, Ali still faced resistance, now headed by Mu'awiyah from Syria. Uthman's shirt, besmeared with his blood, and the chopped-off fingers of his wife, Nailah, were exhibited from the pulpit in Damascus. Soon, Mu'awiyah had fomented the whole of Syria into hostility against Ali – who, for his part, had little alternative but to meet this growing threat and advanced into Mesopotamia with an army of 80,000. He planned to invade Syria from the left bank of the Euphrates. After various skirmishes with the Syrians he reached the plain of Siffin, where Mu'awiyah's forces were waiting. The Syrian positions controlled the water supply of the whole valley, and there was no access to the river for Ali's men, who were now suffering extreme thirst. Ali had no time to lose. He attacked Mu'awiyah's army at full gallop, turned their flank, and gained the river bank. Now it was Mu'awiyah's turn to experience thirst. But Ali allowed the Syrians free access to the river, because he believed it was unethical to fight men who were dying of thirst. With the two armies opposite each other, negotiations continued for nearly four months. There were skirmishes every day, but eventually full-scale battle broke out on 26 July 657. The fighting continued overnight. The eighteenth-century historian Edward Gibbon wrote:

> The Caliph Ali displayed a superior character of valour and humanity. His troops were strictly enjoined to wait the first onset of the enemy, to spare their flying brethren, and to respect the bodies of the dead, and the chastity of the female captives. The ranks of the Syrians were broken by the charge of the hero, who was mounted on a piebald horse, and wielded with irresistible force, his ponderous and two edged sword.[8]

Ali, appalled at the bloodshed incurred in the clash of armies, sent a message to Mu'awiyah, challenging him to single combat: whoever won should be caliph. Gibbon says: 'Ali generously proposed to save the blood of the Muslims by a single combat; but his trembling rival declined the challenge as a sentence of inevitable death.' Having lost his nerve, Mu'awiyah devised a trick. His soldiers tied copies of pages of the Qur'an to their lances and raised them aloft. Ali's men, all believers, refused to fight; and Ali, with victory so close, was forced to retreat – with terrible consequences. Many of his warriors were shocked at this 'surrender' and around 12,000 men mutinied, deserting his army on the march back through Iraq. The seceders, the so-called Kharijites, then fought skirmishes against Ali's army and gradually collected widespread support. Rebellion broke out across Asia

Minor and Egypt. Ali, increasingly beleaguered over the last years of his life, was in turn assassinated in a mosque in 661. Philip Hitti, the famous historian of the Arabs, writes: 'Valiant in battle, wise in counsel, eloquent in speech, true to his friends, magnanimous to his foes, Ali became both the paragon of Muslim nobility and chivalry and the Solomon of Arabic tradition, around whose name poems, proverbs, sermonettes and anecdotes innumerable have clustered.'[9]

Ali's death left his two sons, Hasan and Hussain, as potential claimants to the caliphate. Grandsons of the Prophet, both were regarded as *imam* – leader of the congregation. When Ali was assassinated, Imam Hasan was acclaimed caliph by the people of Kufa, the city south of Baghdad where Ali had had his headquarters. But by now a strengthened Mu'awiyah in Damascus had ambitions for himself. He prepared for war again, this time gathering 60,000 troops from Syria, Palestine and northern Arabia. Mu'awiyah fully expected to intimidate Hasan, given the ebb in support for his family. If Hasan gave way as his father had, Mu'awiyah's cause would be legitimized. If not, he would attack Kufa head on. Soon after, Hasan left Kufa with his main force to fortify his position at nearby Madain – where, even today, Muslims are still fighting each other, Shi'ite against Sunni.

Here Hasan faced a serious situation. Some of his troops defected to the enemy, and when the Kharijite contingent joined them, Hasan was fatally weakened. Realizing that his position was hopeless with the rump of four thousand troops left at his disposal, Hasan made peace with Mu'awiyah, who entered Kufa saying: 'I am the first king of Islam.' Thus the Umayyad dynasty was founded, with Mu'awiyah at its head and its centre in Damascus. Hasan left for Medina to live out his life in prayer and piety. Sadly for him, his continuing existence still represented a threat to Mu'awiyah's ambitions. There are confusing tales about Hasan's death; possibly his wife poisoned him; possibly Mu'awiyah had offered her money and a new marriage to his son, Yazid. Whatever the truth, her new match did not materialize.

Mu'awiyah died after ruling for nineteen years, determined that Yazid should succeed him. Ali's second son, Imam Hussain, in a very weak position when his father died, had fatally agreed to waive his immediate right to succession to the caliphate, and had previously given his allegiance to Mu'awiyah. Like some other misguided aspirants to kingship before and since, Hussain agreed to delay his succession, with the promise of becoming caliph on Mu'awiyah's death. But Yazid, with his father dead, and power in his hands in Damascus, issued orders to the governor of Medina to exact homage from Hussain.

The stage was now set for one of the great religious schisms, a rift whose implications still reverberate in world politics 1,400 years later. Imam Hussain, with members of his family, made his way to Kufa, where he expected to raise support. But Yazid's four thousand soldiers intercepted him on the plain of Karbala, just 25 miles short of Kufa. They were under the command of Ibn Sa'd and they controlled the banks of the Euphrates. Imam Hussain had a mere seventy-two supporters with him and was virtually encircled; nevertheless, amazingly, he held off an initial attack. Thereafter, negotiations lasted eight days, during which time Ibn Sa'd tried to persuade Hussain to bow to the inevitable and accept Yazid's caliphate. After five days, a message came from Damascus to deny Hussain access to water, rendering his predicament increasingly serious. A daring sortie by his younger brother Abbass succeeded in filling a few water-skins, but eventually Ibn Sa'd advanced towards Hussain's camp in full force. Hussain assembled his followers, trying to persuade them to flee. Tabari, the tenth-century Arabic historian, quotes him as saying:

> I give praise to God ... I know of no worthier companions than
> mine; may God reward you with all the best of His reward. I think
> tomorrow our end will come. I ask you all to leave me alone and to
> go away to safety. I free you from your responsibilities for me, and I
> do not hold you back. Night will provide you a cover; use it as a
> steed. You may take my children with you to save their lives.

All the relatives and followers of Hussain refused to
decamp. They spent the night in prayer, recitation of the
Qur'an, worship and meditation. In the morning, Hussain
drew up his small band, some of whom were only fourteen
years old. Hopelessly outnumbered, Hussain's followers
fell, one by one. When Hussain held out his infant son to
invoke the pity of the attackers, the baby was pinned by
an arrow through the neck to Hussain's arm.

Hussain received numerous wounds before he finally
died. His decapitated body was trampled into the ground.
A few women from Hussain's family were given safe
passage under guard, together with Hussain's desperately
wounded elder son, Ali Zayn. Edward Gibbon may have
the last word: 'In a distant age and climate, the tragic
scene of the death of Husayn will awaken the sympathy of
the coldest readers.'

The news of the massacre of the Prophet's family at
Karbala was like an earthquake which rumbled through
the Muslim world. After the battle, Yazid had Hussain's
son Ali Zayn tied to the back of a camel and paraded as a
trophy through the streets; and, to reinforce the new
caliph's authority, his father's severed head was displayed
to the crowds in Kufa. Far from quelling dissent, this
brutal display provoked rebellion and civil war. The
Kharijites set up an independent government in Persia and
the Arabian peninsula, the people of Kufa revolted, and in
Mecca one of Aishah's supporters at the Battle of the

Camel declared independence from the Umayyad rule in Damascus. Responding to the potential break-up of the extended empire, Yazid besieged Mecca and Medina and, after extensive devastation, these cities surrendered.

Four years after Karbala, a group of the faithful calling themselves 'penitents' (*tawwabun*) gathered in sackcloth and ashes at the field of that battle to mourn the death of Hussain and his family – partly in homage to the imam, and partly in penitence for not having left Kufa to come to his aid when he most needed help during the battle. Their collective lamentations and their notion of atonement through sacrifice were to become the central elements of the new sect of Shi'ism. The ideas still reverberate in the notion of a pious believer who, in the struggle for justice against tyranny and oppression, is prepared to accept martyrdom.

Karbala has become a holy place, with a shrine where the imam is buried. Ashura, the modern Shi'ite festival which commemorates Hussain's death, is a period of mourning which lasts for ten days. Throughout its duration, Shi'ite Muslims weep for the triumph of the evil Yazid over the good Hussain. They perform passion plays that represent the details of the fighting at Karbala. The daily gatherings end spectacularly on the last day with the funeral procession of fasting men, who beat themselves with chains – and sometimes slash themselves with knives – in memory of the wounds suffered by their martyred Prophet, Muhammad's grandson, Imam Hussain.

Nearly 20 per cent of the world's Muslims are Shi'ites. Most of them live in Iraq and in Iran. They consider that they are the only authentic faithful, believing that Sunni Muslims (those descended from the Umayyad line) are the creation of a lax and corrupt time over 1,300 years ago; Sunnis, for their part, believe that they are the representatives of the true faith – tolerant, compassionate and upholding the consensus of the community – and see

Shi'ites as rigid and extremist. Shi'ites believe that their own imams are not just religious leaders but highly spiritual men with exclusive authority to interpret the Qur'an. They believe that these special people are direct descendants from Ali, divinely inspired, infallible, sinless and preserving the Prophet's message.

Islam, with all its internal tensions, has become an increasingly important influence in the modern world. Although there are no precise figures, around one billion humans are Muslims. They come from all social, economic, linguistic and ethnic backgrounds, and from all human communities. Europe and North America, for example, have an increasing, vibrant and influential Muslim community – it is said that in the next century over half of all births in Europe will be to Muslim parents. Islam was born out of conflict and, as we have seen, its early history was frequently turbulent and violent. Today the Islamic community is still riven. Just like some very religious Christians and Jews, there are some very religious Muslims who see the modern world, dominated as it is by American and west European values, as a deeply corrupting influence. These Muslims abhor many of our liberal moral values and the nature of our pluralistic society.

But they forget that Islam was founded with peaceful values and is centred on a deep respect for human life. And Islam has in its time also been a most tolerant religion, supporting the notion of religious pluralism in a way that few other monotheistic religions have. The Qur'an (2: 256) itself asserts firmly that 'There can be no compulsion in religion,' recognizing the nature of human diversity, and perhaps the diversity of Islam itself.

Of course, the world continues to face conflict, much of it based on religion and some of it sadly centred on Islam. The horrific tragedy of 11 September 2001 is not an isolated incident, and the world will undoubtedly see more similar violence. But there is surely cause for real optimism. A groundswell of more sophisticated thinking and of reform is slowly sweeping the Muslim world and we must all help to nurture this. Struggle will continue, but eventually the real values of Islam – tolerance, mercy, respect for life – must surely triumph. These are values which ultimately cannot be hijacked by a few people who have confused bigotry and fanaticism with religion, and blind denial of moral behaviour with God.

7

Heresies and Schisms

In the centre of Kiev there is a giant statue of a man in
armour astride a huge, prancing horse. This effigy, rearing
up on a large rock in the main square, is of Bogdan
Chmielnicki – to Ukrainians, a great nationalist hero; but
to Jews, the incarnation of evil, the wicked butcher who
massacred Jews and Poles without mercy. It was he who,
in the spring of 1648, led a Cossack revolt against the
Polish occupiers of the Ukraine and conducted atrocities
throughout the territory. His uprising, given bloody
support by Cossacks and Tartars from the Crimea, was
aimed chiefly at the landowners and to some extent
the Catholic Church. But it was for the most part the Jews
who found themselves on the receiving end of the rebels'
violence. Their roles as middlemen for the Polish land-
lords, and as moneylenders, naturally made them targets,
as did the wealth of a few of their communities.

Chmielnicki ran amok through every major town, and
many small villages. Jews were invariably singled out.
Rabbi Nathan ben Moses Hannover, a witness to some of
the Chmielnicki massacres, wrote:

Some of them [the Jews] had their skins flayed off them and their flesh was flung to the dogs. The hands and feet of others were cut off and they [their bodies] were flung onto the roadway where carts ran over them and they were trodden underfoot by horses . . . And many were buried alive. Children were slaughtered at their mothers' bosoms and many children were torn apart like fish. They ripped up the bellies of pregnant women, took out the unborn children, and flung them in their faces. They tore open the bellies of some of them and placed a living cat within the belly and they left them alive thus, first cutting off their hands so that they should not be able to take the living cat out of the belly . . . and there was never an unnatural death in the world that they did not inflict upon them.[1]

In many cases, the terrified Poles used the Jews as bargaining material – in the town of Tulchin, for instance, where thousands of Jews had sought refuge, the authorities delivered them up to the Cossacks in return for their own lives – only to find that many of the Polish leaders were killed anyway as soon as the Jews had been dispatched. At Bar, the Poles refused to admit the Jews into the fortified town. At Nemirov, the Cossacks invaded the precincts disguised as Polish soldiers and killed six thousand Jews. One eye-witness account of events is that of a rabbi who survived, Shabbetai ben Meir HaCohen:

They arrived as if they had come with the Poles . . . In order that he open the gate of the fortress . . . and they succeeded . . . and they massacred 6,000 in the town . . . and drowned several hundred in the water and by all kinds of cruel torments. In the synagogue, before the Holy Ark, they slaughtered them with butchers' knives . . . after which they destroyed the synagogue and took out the Torah books . . . they tore them up . . . and they laid them out for men and animals to trample on . . . and they made sandals of them and several other garments.[2]

Similar atrocities were perpetrated throughout the

region, with synagogues burnt, sacred books defiled, and Jews murdered in a variety of gruesome ways. It is not clear to this day how many lost their lives. Contemporary Jewish accounts put the number of dead at 100,000, but many serious historians believe the real figure was closer to 200,000. Whatever the number, the consequences for Europe's Jews were serious, as refugees poured out of the Ukraine, with tales of their ordeal at the hands of a godless horde.

And what of Chmielnicki? In spite of his declared hostility to the church, and even though the Pope commended those who bore arms against him, it is said that the Archbishop of Corinth girded him with a sword which had lain in the Holy Sepulchre, and the Metropolitan of Kiev absolved him from all his sins, without the usual preliminary of confession, before he rode forth to his battles.

Kabbalah, mysticism and messianic expectation

It is no coincidence that this turbulent period in eastern Europe and Russia was marked by a revival of interest in mysticism among Jews. When the world becomes unstable, people seek solace in the other-worldly. As we have seen, prophecies about the Messiah first emerged during the time of Babylonian exile. The Roman occupation, culminating in the destruction of the Temple, was another time of chaos for Jews. And in the fifteenth century, the expulsion of the Jews from Spain and Portugal uprooted a deeply established community: suddenly, Iberian Jews who had lived and prospered alongside Christians and Moors alike for several hundred years found themselves largely destitute and forced to travel far and wide simply to survive.

This latest upheaval sparked a popular interest in the

Kabbalah, a collection of medieval writings which, like the Gnosticism of the first centuries CE, taught that the created world was inherently flawed, having come into being as the result of a disaster in heaven. Once a highly esoteric text, originating probably in Provence and Spain in the twelfth and thirteenth centuries and circulating only among small circles of initiates, it was now taken far and wide by the expulsions. The invention of the printing press was a major force in its dissemination – technology advancing religious ideas, rather than destroying them. Soon competing versions of the Kabbalah, edited and adapted for popular readership, were being printed in high volumes throughout Europe. During this process, the content of the Kabbalah itself changed. New sections were added, concentrating on the notion of Zion as the Promised Land, and on the imminent coming of the Messiah. Both features reflected the insecurity of the Jewish communities who read the book, fearing that every New Year might bring another expulsion or massacre.[3]

Many Jews emigrated to Italy, Greece, Egypt and North Africa at this time. Others decided to 'return' to Palestine in this period, where some poor Jewish communities had always lived. One person who was drawn in this direction was Rabbi Joseph ben Ephraim Caro, from whom I am a direct descendant. My family was expelled from Toledo (the name Caro is still inscribed on the wall of a disused synagogue there) in 1492 and moved to Portugal. Thrown out of there in turn in 1497 (when Joseph was still a child) they took ship across the Mediterranean to Turkey. From Constantinople, the family went to Adrianople, Nikopol and then Thessalonika in northern Greece. In 1536 Joseph finally settled in Palestine, eventually founding a famous synagogue in Safed in what is now northern Israel. That community is still there and flourishing.

Caro is interesting because he was one of the leading medieval codifiers of Jewish law: he wrote the standard legal text which is still used worldwide by all Orthodox Jews today. But – extraordinarily for a lawyer, someone whose whole professional life was bound up in rationality – Caro was also a Kabbalist, who believed himself to be visited every night by a heavenly mentor. Caro was known for having the faculty of 'automatic speech', by which the mentor's voice came out of his mouth and was heard by witnesses. Caro was not in a trance when this happened because he could remember many of these messages afterwards and wrote them down in his mystical diary. Parts of this diary have survived in manuscript and some of these were subsequently printed, in Lublin in 1646 and (a fuller version) in Amsterdam in 1708.

It was, however, not Caro but another Jewish mystic who formed the actual school of Kabbalah at Safed around this time: Rabbi Isaac ben Solomon Luria. Luria's biography bears some resemblance to that of your average mystic. Born in 1534 and raised in Egypt by an affluent businessman uncle, Luria specialized in the trade of pepper and corn. He was just fifteen when he married his cousin; thereafter he moved to a remote island in the middle of the Nile owned by his wealthy father-in-law. Here he lived alone for seven years, apparently leaving the island only to spend the Sabbath with his family, to whom he spoke Hebrew not Arabic – and then only through his wife, who acted as his mouthpiece. He is remembered for invariably wearing pure white during the Sabbath. In his early twenties he claimed he was visited by the prophet Elijah, who told him to leave Egypt for Palestine.

Luria left little written work (some people claim that this is because he was preoccupied with his business affairs, though this seems doubtful – more likely he was not particularly interested, or skilled, in writing); nevertheless, his influence on others in Palestine was profound. A circle of disciples grew around him, some believing him to be the Messiah. The names of about thirty of them are still on record nearly five hundred years later. It was said that he could talk to the birds. He often conversed with the prophets. He would stride around Safed with his loyal band of pupils, pointing out the graves of various holy men. But he remained a trader to the last, compiling his final set of accounts just three days before suddenly dying in an outbreak of plague. When he passed away in 1572, legends immediately sprang up relating to his resurrection.

Luria's power lay in his ability to teach people how to achieve states of ecstasy swiftly. He provided exercises in breathing control and concentration, the latter being based on rearranging the letters of the name of God. He also related the mystical contents of the Kabbalah to the very particular circumstances of the Jews in exile. He taught that, while the created world was an error, it nevertheless contained *tikkim*, or sparks of the essence of God. The Jews themselves, imprisoned in a hostile world, were a symbol of this separation of God from Himself. But they were also active agents: through their piety, their prayers and their close observance of the Law, they had the power to release the sparks from their material prison. When this process had been completed, the Messiah would arrive.

These ideas were attractive to a community that found itself powerless. Throughout the Dark Ages and medieval period, Christian rulers had used the Jews for their own political ends. First admitting them to stimulate trade, then persecuting them to ensure their own popularity with the church, occasionally the rulers of Europe shunted whole communities around as if they were goods. But Luria's ideas taught the Jews that they were not merely stateless chattels – they had an active part to play in their own destiny, and in the destiny of Creation itself. Unsurprisingly, this idea was warmly received, and by the mid-seventeenth century, the Lurianic Kabbalah was being read and studied across a wide area from Turkey and eastern Europe to France.

In his mammoth study of Jewish mysticism, the historian Gershom Scholem has pointed out that the seventeenth century was a time of enormous instability and change, which made the soil especially fertile for mystical ideas. The massacres in the Ukraine were followed by war. Even Jewish communities in unaffected areas experienced upheaval as refugees poured in, bringing new ideas, different traditions and an expectation of disaster. By the 1660s there was a widespread notion that the process described in Luria's mystic text had begun. The signs were everywhere – the recent Chmielnicki massacres being just one example: the Messiah was undoubtedly on his way.

Enter a saviour

In due course, one appeared: Shabbetai Zevi, proclaimed God's messenger in Gaza in May 1665. The tale of this seventeenth-century Messiah is perhaps the most startling example of a Jewish heresy, one which caused uproar, panic, bitter divisions and disillusionment. It begins with another Palestinian mystic, called Nathan of Gaza. Nathan had been born in Jerusalem, and as the son-in-law of a wealthy merchant had plenty of time to devote himself to scholarship. He had quickly mastered the techniques of Luria's Kabbalah, and by 1665 was able to experience prolonged mystical states and visions. Nathan knew of Shabbetai Zevi, but initially, along with most other people, regarded him as someone you should cross the street to avoid. However, as Nathan's mystical experiences deepened, he became convinced that the true Messiah was right around the corner.

Zevi would probably, in modern times, be diagnosed as suffering from bipolar disorder. His periods of exaltation and enthusiasm were interrupted by long spells of existential despair. In his manic phases, he had a rather troubling tendency to flout the Jewish Law and fly in the face of custom. He pronounced the forbidden name of God. He 'married' the Torah scrolls under a wedding canopy. He also married a refugee girl with a dubious reputation called Sarah, echoing the actions of the prophet Hosea. The Ukrainian massacres of 1648 prompted him to start proclaiming that he was the Messiah. Jewish communities viewed him askance, not least in Constantinople where he celebrated the three main festivals of the year – Pentecost, Tabernacles and Passover – in one week, declared the abolition of the Ten Commandments, and pronounced blasphemous blessings to 'Him who allows the forbidden'. He was expelled from there to his home town of Smyrna, from where he went to Rhodes and Cairo and then Jerusalem, where he led a rather ascetic life for a while.

Like so many people suffering from mental illness, Zevi had periods in which he was capable of a great deal of insight into his condition. In bipolar depression in particular, there is a midway point free of mania or depression, when the mind may be calm and balanced. In these phases, Zevi expressed distress at the things he had done and was highly genial. Moreover, as a deeply learned scholar – the son of a successful merchant, he had been singled out from among his brothers for a rabbinical education – he was also widely respected when he was his 'normal' self. A modern doctor might prescribe lithium or carbamazepine to help stabilize his patient's moods – but in 1665 the recognition of such disease was a long way off. In April that year, Zevi heard that there was a remarkable man of God in Gaza who could get to the secret root of the soul. Believing himself to be possessed by demons, he took the two-day journey there to meet this skilled Kabbalist who he thought would be able to cure him.

As we see so often in the story of God, movements are sparked into life when a few random coincidences hit a specific set of circumstances. In this case, a great swathe of the Jewish population was receptive to mysticism and believed in the Messiah. Meanwhile, Nathan of Gaza was already experiencing visions of an imminent Messiah. Then Zevi turned up seeking help. The stage was set.

Rather than exorcising Zevi's demons, Nathan turned out to be something of a Svengali figure, convincing this troubled individual that he wasn't sick, that he really was the Messiah; and on 31 May 1665 Shabbetai Zevi formally announced he was indeed the Messiah. Thereafter, Nathan stage-managed the whole affair. He led Zevi around Gaza on horseback and despatched ambassadors to summon all the tribes of Israel. He took him to Hebron, to Abraham's tomb, on a pilgrimage and then organized a tour through Jerusalem, where he symbolically circled the city seven times. These look like the actions of an impresario: but we shouldn't take Nathan's organizational expertise as agent, tour operator and publicist as a sign that he wasn't completely convinced. This man, like a great number of his contemporaries, fervently believed in the coming of the Messiah; and he thought he had found him.

Initial reactions were mixed. A great many rabbis in Gaza joined the cause. Elsewhere, particularly in Jerusalem, some rabbis were sceptical, particularly as this Messiah hadn't performed any miracles. But they none the less considered it wise to hedge their bets, and issued no formal condemnations. Meanwhile, believing, or perhaps having been made to believe, that he was going to inherit the throne of Turkey, Zevi began a carnival-like journey northwards, first to Syria and thence towards Constantinople.

As this manic Messiah moved through the cities and towns on his route, hysteria followed him like a royal train. Zevi continued to shock and surprise – eating pork, for example, shouting out the forbidden name of God and urging others to follow him. Any rabbi unwise enough to protest found his house burned down by the zealous mob of followers. In Smyrna, Zevi attacked the doors of the synagogue with an axe when the authorities refused to admit him. Once inside, he dismissed the rabbis as unclean animals, danced with a scroll of the law in his arms, sang a Spanish love song, announced the date for the Redemption of the World as 18 June 1666, and proclaimed the downfall of the Turkish sultan. When challenged by a rabbi, Zevi excommunicated him and, terrible to hear, led the crowd in chanting the forbidden name of God out loud.

As Zevi outraged and fascinated crowds throughout the Balkans, Nathan of Gaza was acting as his Alastair Campbell and Max Clifford rolled into one: propagandist and fixer, printing and distributing texts which announced Zevi's mission. These were enthusiastically received in communities as far part as Palestine and Amsterdam. Their overall effect was a sort of mass penance on the part of Europe's Jews. In Prague they fasted; in Frankfurt they took continual ritual baths to ready themselves for the Redemption. Jews lay down naked in the snow. Many sold all their possessions and returned to Palestine. Some shady businessmen battened on to the hysteria, selling non-existent package deals to the Holy Land.

On his way northwards, Zevi stopped in Aleppo. There he created a wave of excitement – a revivalist atmosphere. There were reports that the prophet Elijah had reappeared; funds were set up to help those who would be adversely affected economically by the cessation of commercial activity that he caused. Back in Smyrna that December, Zevi caused a huge commotion in the Jewish community as it was swept off its feet. By now he had dumped Nathan and was acting on his own, wearing royal clothing and conducting ecstatic prayers.

Unsurprisingly, the Turkish authorities were fairly troubled by Zevi's swelling wave of followers, particularly as they were now advancing towards Constantinople, again declaring that the end of the Sultan's rule was imminent. The news of Zevi's impending arrival in Constantinople was causing intense excitement, not only among the Jews, but also within the non-Jewish population. At this point, just as state power had intervened in the story of Jesus, so it did with Zevi. Fortunately for him, the Grand Vizier Ahmed Koprulu was a wise and able statesman and acted with considerable restraint. The death penalty, which was the normal way of dealing with treason, was not invoked. Instead, Zevi's boat was stopped in the Sea of Marmara, where he was arrested and put in chains. Imprisoned at Gallipoli, Zevi sat in cheerful majesty, charming the Grand Vizier with his genial manners – but refusing to take back a word he had said.

Across Europe, concerned communities waited for news. The Jews of Venice discreetly sent messages to their brothers in Constantinople asking for a progress report. They received encouraging news, cleverly disguised from prying Turkish eyes as a trade report. 'We . . . examined the merchandise of Rabbi Israel . . . We have come to the conclusion that they are very valuable . . . but we must wait until the great day of the Fair arrives.'[4] The Turks themselves seem to have been hedging their bets, waiting to see what would transpire on the foretold day of 18 June. When that day came and went, the Sultan moved swiftly against Zevi, summoning him to the palace. As the ruler listened from a secret chamber, Zevi promptly denied everything he had claimed. He was offered the choice between converting to Islam and death. The Sultan's doctor, a Jewish convert to Islam, urged him to choose life. Zevi, always an impressionable soul, agreed. He adopted the name Aziz Mehmed Effendi and took a job as Keeper of the Palace Gates. It was said that the Sultan had a soft spot for him.

Elsewhere, people were less favourably disposed towards this religious cross-bencher. The news of Zevi's sudden conversion was met with dismay and uproar. Communities were split between those who had been swept up by the hysteria and the smug 'told you so' types. Rabbis feared that the Turks, hitherto tolerant of other faiths, would begin to persecute Jews in their empire and force them to convert, as Zevi had done. Official records were collected up and destroyed in an effort to expunge the whole shabby Shabbetai affair from recent history.

But the failure of a prophecy does not necessarily result in the end of the prophetic belief. The Jehovah's Witnesses, for example, have twice predicted the end of the world, but maintained their status as a major movement by carefully reformulating their beliefs after the foretold event has failed to happen. A similar regrouping was effected in seventeenth-century Judaism, largely attributable to the skill and scholarship of Nathan of Gaza. He claimed that, rather than being a betrayal, Zevi's conversion was a supreme act of sacrifice. The Messiah had 'descended into evil' in order to enact the trickiest tasks in the overhaul of the cosmos. Nathan reminded his audience that Zevi had always done strange things whose meaning was often hidden to those around him. Donning a turban and calling himself Effendi was merely the latest in a long line of signs and wonders.

Zevi did not live out his days reading the Qur'an and twisting his worry beads. He continued, during his manic phases, to declare himself the Messiah, and his sexual exploits scandalized his neighbours. Eventually the Sultan was persuaded to exile him to the barren, mountainous territory of Albania, where Zevi died in 1676. Still undaunted, Nathan of Gaza declared that the Messiah had been absorbed into heaven. He formulated a clever explanation, which took into account not only Zevi's career but any future events. There was not one set of divine sparks, as the Kabbalah had stated, but two – a good set, and an indifferent set inclined to be bad. Creation occurs as an interplay between these two sets of sparks, and the Messiah plays an important role in this process, often appearing to be evil in order to achieve good.

Jerusalem Syndrome and strange alliances

As I have suggested, Shabbetai Zevi was probably psychotic. Indeed, modern science may be able to offer a particular explanation of his condition. Quite recently, a group of psychiatrists working at Kfar Shaul Mental Health Centre in Jerusalem published a riveting paper.[5] They noted that in the previous thirteen years over 1,200 tourists in Israel had been referred to their psychiatric hospital, and 470 of them – all visitors to Jerusalem, a huge percentage – required admission. Jerusalem, the authors point out, conjures up a unique sense of 'the holy, historical and heavenly', holding attraction for people from a wide variety of faiths. When people dream of Jerusalem, they observe, 'they do not see the modern, politically controversial city, but rather the holy biblical and religious city'.

These psychiatrists described several forms that what they call 'Jerusalem Syndrome' may take. One example was the case of a forty-year-old American tourist who had had some psychological problems. He began working on his body image, using weights in the gym and body-building, in the framework of a rehabilitation programme. Eventually he felt the urge to get on a plane and fly to Israel, having begun to believe that he was the biblical character Samson and feeling compelled to go to

Jerusalem. His intention was to move one of the giant stone blocks (each of which weighs many tons) that form the Western Wall of Herod's Temple. In his opinion, these stones were not in the correct place.

Once in the Temple area, he tried to remove one of these huge stones. Readers may recall that when the Israeli Prime Minister, Ariel Sharon, trespassed in the same Temple area he created an international incident which sparked off the *intifada*. So it is hardly surprising that, in this most sensitive part of the city, the American visitor's action caused a terrible commotion: the police arrived in droves and the tourist was carted off to the Kfar Shaul Hospital. Very unfortunately, the trouble really began there because the duty psychiatrist – against all accepted clinical practice – challenged the American, telling him he could not possibly be the genuine Samson because, according to the book of Judges, Samson never visited Jerusalem.[6] The American was enraged, got very aggressive, broke a hospital window and escaped through it. Eventually a student nurse found him in the middle of Jerusalem, standing patiently at a bus-stop. (Sadly, the authors do not tell us whether it was the stop for buses to Gaza, the Philistine capital.) The nurse showed great presence of mind, and with commendable wisdom told him that he clearly had demonstrated qualities which were those of Samson and that he could now return to the hospital if he wished, which he duly did. Once his psychosis had settled down with drug treatment, he was released and flew back to the United States.

It does seem strange that, at the right time and in the right place, deranged people might be sufficiently persuasive to recruit other, presumably sane, people as their disciples. In the seventeenth and eighteenth centuries, the cult of Shabbetai continued to draw converts; it even gave rise to a new Messiah in the form of a Balkan trader called Jacob Frank. Frank declared that he had inherited Zevi's soul and led an underground movement in Poland, whose practices probably included sex acts forbidden by the Jewish Law. Excommunicated in Poland, he fled to Turkey and became a Muslim. His supporters, meanwhile, appealed to the Catholics for help, proclaiming themselves 'contra-Talmudists' and exaggerating the aspects of their belief that chimed with Christianity. The Catholics, thinking them ripe for conversion and perhaps also spying a handy opportunity to divide the Polish Jewish community, lent their support and ordered copies of the Talmud to be publicly burned.[7] However, one of the key Catholic supporters of the movement, Bishop Dembowski, had the bad luck to die on the day of the book-burning. The Orthodox rabbis saw this as a sign of divine intervention and persecuted the Frankists with even greater zeal. As an act of defiance, Frank became a baptized Catholic in 1759, along with his followers. But, like Zevi before him, his sexual antics (he surrounded himself with a bevy of beauties known as the twelve 'sisters') raised eyebrows, and he eventually found himself in prison. Undeterred, he then converted to the Russian Orthodox Church.

Curiously, there are still a few Shabbeteans living today. According to recent reports, around eight thousand adherents of his brand of 'Judaism' live in Izmir (formerly Smyrna) and Istanbul, speaking Turkish and Ladino – which is a little like Yiddish in that it is a mixture of Hebrew with Castilian Spanish and, in this case, Arabic and Turkish.

It is perhaps significant that nearly all these messianic movements were forced to turn to other religions in order to have any chance of survival. This reflects the circumstances of the Jews in Europe as a stateless people squeezed between the major political powers of the day, which were themselves fundamentally associated with Christianity and Islam. It is also worth noting that, despite committing the most shocking infractions of Jewish Law and causing widespread upheaval, neither Zevi nor Frank was threatened with death. Judaism is no more and no less prone to offshoots and factions than any other religion. But in its exiled communities in Europe, the punishment for dissent was *herem* – excommunication – not execution. Excommunication was of course a severe penalty in its own right, particularly in conditions of exile, for the community offered the means of livelihood and protection against the often hostile Gentile world. An excommunicated Jew might have little option but to convert to another religion if he wished to survive. None the less, Judaism's response to religious rebellion was markedly different to that of Christianity, where the authority of the state was bound up with that of the church. One of the greatest paradoxes of the Divine Idea is its capacity sometimes to permit or even motivate acts of great cruelty. Nowhere is this more apparent than in Christianity, where belief in a loving, forgiving God has been enforced with torture, branding, amputations and death by fire.

A corrupt world: the faith of a thousand heresies

It should not be surprising that Christianity, which began as an offshoot of Judaism, has itself divided into a kaleidoscope of differing, sometimes warring sects. What does seem surprising is the frequency with which men and women have been persecuted or put to death over what seem like tiny points of theological detail. The Roman and Eastern churches, for example, split several centuries ago in a feud over the nature of the Trinity – more precisely, over whether the Holy Spirit is produced by the Father alone, or comes from the Father and the Son. In eighteenth-century Holland, there were bitter feuds between the Supralapsarians, who claimed that God had decided who would be saved before he created the world, and the Infralapsarians, who said that God had made his decision after Adam was ejected from Eden. Many of these differences seem obscure and pointless to modern eyes. As David Christie-Murray notes in his *History of Heresy*, 'if the pagans of the first century were amazed at the love which Christians bore for one another, those of later centuries could have been equally astonished at the loathing and intolerance the upholders of the loving God ... displayed towards their associates whose formulae for defining the indefinable differed from their own'.[8]

Throughout human history, people have been inclined to rally around a common banner even though they have a range of interests and complaints, giving an appearance of unity to an underlying diversity. In recent years, demonstrations against the poll tax, the abolition of hunting and the war in Iraq have been attended by people with a host of very different political views and grievances against the government of the day. We shouldn't suppose matters were any different in the past. It is safe to assume, when we read of people rioting because of a line or two in a heretical book, or the words of an outspoken preacher, that they were partly motivated by other concerns. For a long time in European history, important religious positions were bought and sold, and a bishop or a cardinal was little more than a nobleman with a job. Attacking churches, or defying the orders of the church, became a means by which the poor and the downtrodden could express their anger at the ruling classes. A Marxist would, of course, say that this is the whole meaning of 'heresy' – one group seeking to overthrow the power of another.

However, heresies also seem to have a lot in common with scientific inquiry, in that both often represent courageous attempts by people with questioning minds to defy accepted tradition and contemporary assumptions, and to ask about issues that go right to the heart of existence. Throughout the history of Christianity, many arguments seem to revolve around the same set of themes: How could a good God create a world of evil and suffering? Do we really have free will, or has God already ordained all of our actions? How can we reconcile the humility and generosity of Jesus with the power and wealth of the church? These questions continue to preoccupy people who believe in God. The fact that, in former ages, people were prepared to undergo torture and execution for the right to ask them means that those enquiring individuals deserve our attention.

One of history's most disturbing heresies came from Egypt some 150 years after Christ. Despite its obscurity and extremism, it managed to convert some respected Christian thinkers to its cause. The driving force of the movement was a prophet called Montanus, who believed that the early church had already become corrupt and immoral, and was unrepresentative of the examples of Jesus and the Apostles. This theme, of return to some older, leaner, truer set of values, runs throughout the story of heresy: the men and women who promote upheaval are frequently, in fact, deeply conservative.

Montanus believed that the Trinity consisted of only a single person (with three attributes or aspects), in contrast to orthodox Christians, who maintained that the Trinity is one God of three persons. He seems to have been a priest of Cybele – an oriental fertility cult – before his conversion to Christianity. His followers encouraged ecstatic prophesying and speaking in tongues, in marked departure from the more sober and disciplined approach of mainstream Christianity. The sect also believed that Christians who fell from grace could not be redeemed; so rigid adherence to the rules was important. These were ascetic in the extreme. Montanists observed a strict discipline of celibacy, amounting to complete abstinence from sexual activity; they fasted for two whole weeks before Easter, and permitted themselves only dried food the rest of the time. Montanus was accompanied by two prophetesses, Priscilla and Maximilla – the latter specializing in delivering prophecies that never came true. Others of the fold claimed Jesus had appeared to them in the form of a woman, and that the soul was a physical, man-shaped object that could be grasped with the hand.

Despite the strangeness and severity of Montanist beliefs, the group received a major fillip when the respected Christian theologian Tertullian converted to their cause in 207. Disillusionment with the pride and wealth of the mainstream church may have motivated him to take this step, although it seems odd that this learned lawyer, who had played such a significant role in mainstream Christian theology, could not have found a better means of opposing immorality than joining an extremist suicide cult.

The Montanists were widely scorned by the Christian establishment, but nevertheless clung on for three centuries or more, achieving particular popularity in North Africa. Their success was due in part to their organization: Montanist missionaries conducted door-to-door conversion work in much the same manner as the Jehovah's Witnesses still do today. However, a movement like this was destined ultimately to have a limited shelf-life. The Montanists' belief in Jesus' imminent return, heralding the dawn of a new age, resulted in a marked lack of planning for the future. Their rigid discipline, un-appealing diet, and stress on martyrdom and chastity also made it quite hard to win converts among the ordinary folk, as well as keeping their numbers from growing internally. The last remaining Montanists set light to themselves in their churches towards the end of the sixth century CE, after sustained persecution from the Byzantine Emperor.

Damnable doctrines: the two Gods and the virtuous Cathars

Like the doctrines of the Jewish Kabbalah that influenced Nathan of Gaza and his manic-depressive protégé Zevi, many Christian heresies drew their inspiration from Gnosticism, a secretive, esoteric movement whose adherents claimed to possess the hidden knowledge that ensured salvation for all humankind. Gnostics believed that creation had been an error, and that the material world was essentially evil. Man, whose inner essence was the same as God's, was stranded in this corrupt world of flesh, from which he could attain freedom only through the magical practices of gnosis (which means 'knowledge' in Greek).

Various early heretics used the ideas of gnosis to explain the problem of suffering. The sailor-turned-bishop Marcion, for example, argued in the second century that there were two Gods. The cosmos was torn between a wicked God, who was identical to the YHWH of the Old Testament and who had created this world, and a good God, who had sent his son Jesus down to earth to combat his opponent and guide men to the truth. Our knowledge of Marcion's beliefs comes chiefly from the attacks of his opponents – ironically, it was Tertullian, later to become a heretic himself, who was Marcion's most vigorous critic.

Offshoots within early Christianity tended to be tolerated as long as they did not contradict official church teaching. But Marcion's views were quite clearly beyond the pale. For a start, to say that there were two Gods was a denial of monotheism, and also meant that the Christian God was not all-powerful. Montanus also denied that

Jesus had had a material body, and therefore that the crucifixion and resurrection had really taken place. Those baptized into Marcion's strand of the faith were also forbidden to marry – meaning once again that it was extremely hard for the sect to grow, except by winning converts; and, in an age when to profess radically unorthodox beliefs could mean death, conversion was never going to be that popular.

Although Marcion's sect had died out by the fifth century CE, various of his ideas survived in other forms. The notion of two Gods of equal power, one evil and one good, was central to the beliefs of the Cathars, a sect that flourished in southern France in the twelfth and thirteenth centuries. Few heresies have been surrounded by such romanticism and mystery as these world-rejecting ascetics, who retreated to mountain strongholds in the face of severe reprisals from the church and Catholic nobles. Their castles – Peyreperteuse, Puivert and Carcassonne, among others – situated in remote and lonely spots around the Languedoc and Roussillon regions, are a striking testimony both to the power of their beliefs and the power of the forces that were set against them.

Much of the modern interest in the Cathars stems from a book written in 1982. *The Holy Blood and the Holy Grail* advanced the theory that Jesus had had a wife and children, whose descendants became the Merovingian dynasty in France. These descendants, and the whereabouts of the Holy Grail, the cup from which Jesus drank at the Last Supper on the eve of his crucifixion, were protected by sects like the Cathars and the Knights Templar. The idea is currently receiving a resurgence of attention in Dan Brown's *The Da Vinci Code*, a controversial bestseller that has been condemned by the Vatican with a force not dissimilar to the tone in which medieval popes reacted to various heresies – though admittedly without the physical violence.

In reality, the Cathars were just a particularly persistent, largely pacifist sect who turned away from the pomp and ceremony of the Catholic Church, and who maintained that the created world was the product of an evil God. They aspired to the simple life, were notably honest and tolerant, repudiated luxury, and sought direct contact with the divine. Moreover, they were seen as representing an alternative model to the clerical arrogance and corruption of Rome, whose depredations on the church's congregations were increasingly difficult to bear. So their neighbours tended to see them as highly virtuous. One deposition now in the Library of the Vatican gives an eye-witness account:

> They are the only ones who walk in the ways of justice and truth which the Apostles followed. They do not lie. They do not take what belongs to others. Even if they found gold and silver in their path, they would not lift it unless someone made a present of it. Salvation is better made in the faith of these men called heretics than in any other faith.[9]

But the Cathars altered history, because the reaction to them set the tone for the way the Catholic Church was to respond to all future deviations from official doctrine. Up to that point, the outright killing of people whose views differed from the norm had been frowned on by many orthodox churchmen. In 1145 Bernard of Clairvaux, himself an enthusiastic persecutor of heretics, nevertheless maintained that they should be won over to the truth through reason and not by force, a view that many shared. In 1184 the Holy Roman Emperor Frederick Barbarossa – who ruled over territories roughly corresponding to modern-day Germany – declared that heretics should be punished by banishment. However, the Cathars became too numerous and powerful to reason with, forgive or expel. At a certain point, it seemed likely that the whole of southern France was about to split away from Catholicism, and offshoots of the Cathars were also flourishing in Italy, Spain and as far away as Constantinople. At this juncture the official church embarked on a response whose violence marked its history for centuries thereafter.

Why did the official church take such exception to heresy? The papacy was, in many respects, a worldly power, and one of the ways it enforced its dominance was through a rigid body of doctrines and practices. To use a modern metaphor, we might say that the Catholic Church held the patent for the way men were to enter the Kingdom of Heaven, and imitations or adaptations were slapped down just as I would be if I began to market a 'McWinstons' burger. To challenge a single semi-colon of church doctrine, or question a single practice, was to challenge a superpower to war. Christian rulers, who paid hefty taxes to the Pope, were nervous of incurring his wrath, in case he exhorted their land-hungry neighbours to mount a crusade against them.

But it was the church itself, in its widespread worldliness and degeneracy, that actually caused much heretical activity. Much of the Cathars' popularity stemmed from the laxity that was spreading throughout the established church. Many of the clergy were illiterate and ignorant, some had wives in defiance of the rules, and there was a roaring trade in the sale of relics and important clerical positions. In the south of France particularly, the corruption of the church was profound – there were churches where no Mass had been said for decades, where priests completely ignored their pastoral role and conducted all kinds of businesses, where clerics feasted, flagrantly kept mistresses, employed huge retinues of servants and travelled in gilded coaches. The Cathars (or Cathari, a name from the Greek meaning 'Pure Ones') by contrast were ascetic and driven individuals, whose moral superiority to the clergy was self-evident.

The Cathars believed that man was originally a spiritual being, who had been tricked into becoming imprisoned in the flesh by a wicked God. The soul had to be liberated from its fleshly prison through a life of self-denial. Every concession to the will of the flesh imprisoned the soul further, whereas every act of self-denial weakened the bars of the prison. This philosophy lent them considerable strength in resisting the pleasures of the world and facing down opposition from their persecutors. In modern times, Islamic terrorists present a serious threat because of their readiness to die in the cause of their faith. The Cathars' disregard for their own physical bodies gave them a similar advantage.

They had no churches and, because they denied that the crucifixion had really taken place, used no crosses in their worship. They adopted a 'pick and mix' attitude to the scriptures, rejecting those passages they disagreed with. They also believed that the Catholic Church was itself a creation of the evil God. To put it relatively mildly, the established church viewed them askance. The Dominican order of monks was founded in 1206 specifically to undertake to convert Cathars back to the true faith. They might have had some success in peaceful persuasion – but in 1208, a representative of the Pope was killed at the court of Count Raymond of Toulouse, a wealthy Cathar sympathizer. From then on, the response to the heresy would be a violent one.

The Pope, Innocent III, called on Raymond's neighbours to join an undertaking against him known as the Albigensian Crusade (after a local name for the Cathars, the Albigenses). In 1209 some 200,000 crusaders rallied to the Pope's call, gathering on the Rhône. They were led militarily by a great English bully, the ruthless Simon de Montfort, who held a minor French barony, and spiritually by the fanatical papal legate, the Cistercian Arnaud of

Cîteaux. They were lent support by many northern French nobles who were perhaps less concerned with stamping out heresy than with making the most of an opportunity to lay hands on some of Raymond's huge fortune. Christians against Christians, the followers of the Pope were given an implicit licence to plunder and pillage, and to steal whatever property they chose.

Twenty years of bloody persecution followed, during which entire towns, like Béziers, were put to the sword. The massacre at Béziers was particularly infamous. Once the inhabitants realized they could not defeat their persistent attackers, they had to decide whether to submit to a siege or surrender. They decided to hold out. The first act of siege they resisted was a demand by the crusaders to the good Catholics inside the city that the city authorities give up the Cathar heretics among them. A list of 222 names was sent under a white flag to the Bishop of Béziers, but neither the bishop nor the citizens would denounce or imperil their heretic neighbours.

Béziers was too well-built a fortress town to be easily taken very quickly, so for a while it looked as if the siege might fail and the besieging force would just move on to more easy pickings. The defenders gained in confidence – and then some young hotheads made a serious error of judgement. Familiarity, possibly mixed with wine, breeding contempt, a small number of the inhabitants of Béziers made a foray from the city walls on the day of the feast of St Mary Magdalene. Sauntering almost at leisure, they met and killed a lone French crusader, throwing his body into the river. Then, from a respectful distance, they foolishly taunted his comrades as they strolled around. But they strayed just a bit too far from the city gates. A sizeable group of papal soldiers made to cut them off. Suddenly they were seized with terror and fled up the hill leading to the city gate, but in their panic managed to let some of the besieging troops through as well. Immediately scaling ladders and siege engines were drawn up and, with help from those troops who had managed to get inside, the city's defences were broken. Within minutes, as the army poured into the town, the place was awash with blood: a massacre had begun, and within around two hours the city was completely overrun. The surviving citizens fled to take sanctuary in the cathedral. The crusaders broke down the doors and started indiscriminately to put everybody to the sword. The wanton bloodthirstiness they showed is best summed up by the reputed words of Bishop Arnaud of Cîteaux. When some soldiers asked how they might distinguish the Catholic defenders from the heretic Cathars, he is said to have replied, 'Tuez-les tous. Dieu reconnaîtra les siens' ('Kill them all. God will recognize his own'). Twenty thousand people died in this one episode alone.

In 1229 Raymond accepted defeat and renounced his loyalty to the Cathars. He then demonstrated his recovered allegiance to the Pope by burning eighty heretics alive. The remaining Cathars fled to the Balkans, where in time their beliefs became absorbed into the more tolerant Islam of the area. In France no traces of Cathar belief remain, except perhaps a linguistic one: in the Languedoc area, people still refer to 'Good God' rather than simply using the word 'God'. The Catholic Church had survived another assault on its authority; but the lesson it drew from these years of bloodshed was that a specific strategy was needed to target heresy. It was partly in response to the challenge of the Cathars that the infamous Inquisition was formed in 1230.

God's constabulary

It had become apparent to the leadership of the Catholic Church that the threat of damnation alone was not enough to stop people from asking questions. They therefore decided to create a crack squad of officials whose sole task was to root out and punish dissent, by whatever means proved necessary. This was not without difficulties, however. Even in the ruthless past, holy men were not supposed to wield red-hot pokers against their fellow humans. Pope Innocent IV therefore added torture to the list of their powers, giving Inquisition officials the power to absolve one another for their crimes.

The severity of the Inquisition's rule doubtless contributed to the eventual breakdown of the Catholic Church's monopoly on Christian belief and practice. The inquisitors operated on the basis that any man accused of heresy was guilty until proven innocent – and, for many, proving one's innocence was an impossible task. They conducted their investigations in a way that encouraged a culture of fear and blame. People could give evidence against a heretic in secret. Lying to the Inquisition about potential heresy was permitted if done in zealous concern for one's faith. Servants and employees were forbidden to be loyal to their masters if to do so would hinder the Inquisition's investigations. People testifying against heretics were rewarded with 'indulgences', in other words, a cancelling-out of their sins. Torture was officially permitted only once, but enthusiastic sadists got round this by declaring that they used only one torture session, with 'intervals'.

From the thirteenth century to the sixteenth, resentment towards the Inquisition, whose methods stood in such frank opposition to the doctrine of a loving, merciful God, grew; however, most were understandably reluctant to challenge this religious police force openly. Then, towards the end of this period, the circumstances of Christian Europe changed. During the fifteenth and sixteenth centuries, there was an upsurge in nationalism, as discoveries of new territories in Africa, the Americas and the

Far East gave nations access to unprecedented wealth and they became more concerned with building up their own power. Rulers became less interested in obeying the orders of the Pope, who was after all merely an Italian prince, and resented the idea that a foreign force had the power to police their own lands. The stage was prepared for the last major heresy to emerge within the Christian church – one which would divide Europe and shape the world we live in today.

Angry priests and wealthy poverty: the Protestant Reformation

At the beginning of the sixteenth century a young university student was walking in the countryside near the town of Erfurt in Germany. A storm began and lightning flashes tore the sky apart. A bolt plunged to the ground, narrowly missing the young man. Martin Luther decided that God had spared him and, in thanks, he became a monk.

Luther was prone to bouts of depression. One of these hit him in 1510, while he was on a trip to Rome. It is a feature of certain types of depression that the world we perceive begins to seem ugly and revolting – music pains us, the television seems like a lot of nonsense, people's conversation bores and drains us. It was no different for Martin Luther, who gazed with a jaded eye upon the immense wealth and lavish opulence of this supposedly holy city. Cardinals dressed in lavish robes and rode round like princes. Mighty buildings demonstrated the power of the church. Everything contrasted so violently with his own upbringing in a simple peasant family as one of seven children. He, after all, had been leading an ascetic life as a monk following the example of Christ, a wandering holy man who had said that only humility would allow a rich man to enter heaven. The experience motivated Luther to formulate the most successful assault on the Catholic Church, in the end forming an entirely new branch of Christianity.

It would be wrong to claim that Protestantism came about solely because of a depressed monk's experiences in Rome. As we see throughout history, great changes occur when random events coincide with unrelated but propitious circumstances. And for centuries before Luther – indeed, ever since the Christian church became wedded to the mighty Roman Empire – people had been objecting to its power and its wealth.

In the eleventh century, an Antwerp man called Tanchelmus began to preach throughout the Low Countries. He had started professional life as a diplomat, but abandoned this to become a kind of wandering free-lance minister for the peasants, wearing monk's clothing. He rapidly gained a huge following, which included twelve 'apostles', one of whom was a woman he called Mary. He persuaded some wealthy followers to seek salvation by giving him large sums of money, having first advised them to stop their regular tithes to Rome.

Tanchelmus was the ultimate paradox. His sudden wealth enabled him to adopt a luxurious lifestyle. He was surrounded by many admiring followers, and lived opulently wearing fine apparel. Eventually, Tanchelmus asserted that he had become betrothed to the Virgin Mary and organized an elaborate wedding at which he took a statue of the Virgin Mary as his wife. Wedding presents and jewellery were showered on the happy couple by the local population (it is unclear whether the marriage was actually consummated). Yet his message was that the official church was corrupt, and its officials therefore had no power to absolve people from sin or administer the body and blood of Christ.

Tanchelmus was arrested by the Bishop of Cologne and

imprisoned; he died during an escape attempt, but his message was echoed throughout Europe. It seemed self-evident to many Europeans that the character and behaviour of priests, archbishops and the like had an effect on the duties they performed. How could you intercede with God the Almighty if you carried on like a drunken lord, if you took money for indulgences, if you didn't even know your Bible very well? Various groups emerged, flourished and faded. The Petrobrusians, founded by Peter de Bruys in about 1110, expressed their contempt for the official church in extreme fashion, smashing churches and burning crosses. A couple of decades later another sect, the Henricians, declared that the sacred rituals of the church were utterly pointless if they were not performed by celibate priests who lived in poverty, as Christ had done.

England had its own pre-Protestant rebel in the form of John Wyclif, a fourteenth-century Oxford don who, having taken holy orders as a priest himself, began to criticize the church for its wealth and formal hierarchy. Images, icons, crucifixes and the like were pointless, he asserted, as was the practice of paying monks to sing Masses for the dead. Wyclif formed a movement of 'poor preachers', who wandered England in their distinctive uniform of plain tunics and wooden staffs. He also translated the Bible into English, maintaining that every person had the duty to know the truths of scripture for themselves. All of this may seem mild stuff to modern eyes, but it was a huge affront to the medieval church. Wyclif escaped persecution in person because of his connections to royalty; but after his death the Council of Constance banned his profuse written work, and Pope Martin V ordered that his books be burned and his bones be pulled out of his grave.

But it was Martin Luther who made the greatest mark. He directed his initial anger at the sale of indulgences, a practice which had become a nice little earner for the church. At the time he was mulling over his experiences in Rome, the market in indulgences was booming. In 1516 the Pope granted indulgences to everyone who paid to see a collection of holy relics. In 1517 a Dominican friar called Tetzel had a travelling road-show throughout Germany, selling indulgences in order to raise money to build a cathedral and pay off his archbishop's debts. Holy relics – purporting to be fragments of the cross or remnants of Christ's clothing, with a provenance no more reliable than a Cartier watch from Korea – were on sale throughout the land. It is certain that some monasteries even churned them out on an industrial scale, living off the proceeds. Simony, the practice of buying and selling important church positions, was another established practice – Tetzel's boss, the Archbishop of Mainz, had incurred his debts buying his job. All of this struck Luther as more than just absurd, but absurd it certainly was: 'When a coin was placed in the chest for purgatory, as soon as the penny fell ringing upon the bottom, the soul immediately started for heaven.'[10]

The enraged Luther drew up a list of ninety-five 'theses' or propositions about the regrettable practices of the church, and on 31 October 1517 famously nailed them to the door of Wittenberg Cathedral. This may seem to our modern eyes as a dramatic, even histrionic flourish. But in the Middle Ages this was a widely established method for people wanting to start a debate. That it certainly did; indeed, it initiated one of the greatest religious upheavals in European history. Almost five hundred years later, those following Luther's protest – Protestants – remain divided in certain parts of the world from Catholics, Northern Ireland being a poignant example. The Church of England, whose head is appointed by the Queen, owes its existence to the protest of this disillusioned monk (by way, it is true, of Henry VIII's need for an heir and a great deal of money). The patchwork character of Christian belief in modern times – Pentecostalists, Unitarians, Methodists, Baptists – was made possible by this sixteenth-century schism. Protestant beliefs drove the industrial revolution and motivated settlers to found new communities in America.

Luther's challenge to the established teachings was centred on three main points. First, he believed, on the basis of a text in St Paul's Letter to the Romans, that faith alone was the quality that assured people of salvation. Priests, monks, bishops and the like were no more holy than the simple peasant who had faith. It followed from this that elaborate rituals, images, bowing down, purchasing indulgences or relics did nothing to assure people of salvation. *Sola fide*, 'faith alone', was what mattered. Given this, it was possible for men to serve God through the joyful fulfilment of their duties on earth, not just through ritual and prayer. Hard work, whether that be through bringing home the harvest or trading commodities, was a valid way of worshipping God.

Luther also maintained that the scriptures were the only source of Christian knowledge. Individuals had the right and duty to read them and think about them for themselves. Priests and the whole official machinery of the church, right up to and including the Pope, were not ultimately necessary, because salvation was a matter for the individual, his conscience and God.

Third, Luther argued that there was nothing magical about rituals like baptism or the consumption of the bread and wine. These were simply symbols of faith. Catholic doctrine teaches that the priest has the power to transform the bread and wine into the actual body and blood of Christ during the Mass, in a process known as transubstantiation. Luther dismissed this, instead arguing that the Holy Spirit is present in the bread and wine as fire is present in a red-hot poker. Thus he set the stage for later

Protestants, like Calvin, to claim that the Mass was nothing more than a symbol.

None of Luther's ideas was necessarily revolutionary, or even exclusive to Christianity. As we have seen in earlier chapters, the prophets of ancient Israel railed against the pomp and ceremony of the Temple sacrifice, and urged people to serve God by being good to one another. Mystical sects like the Kabbalists believed that the world could be repaired through simple adherence to Jewish Law, which covered the whole of daily life. Muhammad railed against the use of images in worship, and twelve hundred years before him, Siddhartha Gautama, the Buddha, had taught that shrines and offerings were point-less, and that salvation was an individual path. It seems that wherever the Divine Idea tends towards a hierarchy of priests and officials and elaborate ritual, others will claim that individual piety is of greater importance. Wherever the Divine Idea attempts to separate worship from the world, others will argue that everything is sacred, including the most mundane actions of our daily lives.

But there were certain historical reasons why Luther's own protest became a battle-cry for so many. The spread of literacy and the availability of cheap printed books, working hand in hand, helped his ideas to move from one country to another. The authorities could not hope to quash them simply by burning books, as the new tech-nology meant that more could swiftly be printed. Renewed interest in the writings of the ancient Greeks and Romans bred a humanism which celebrated man's powers of reason above all else, and encouraged sceptical enquiry. Humanist translations of the Bible from Hebrew and from Greek allowed people to question official church inter-pretations of key texts and passages. Scholars like Erasmus (1466–1536) used satire as a weapon against ignorance and superstition. Books like his *Praise of Folly*, or Sebastian Brandt's *Ship of Fools*, had the readers of Europe roaring with laughter at their attacks on the church and other previously unassailable institutions.

Protestantism also became successful because it was ideally suited to changes in the pattern of the economy. New technologies, the discovery of fresh sources of riches in the New World, and improved methods of communication and transport made it increasingly possible for people to trade goods far and wide, and to amass fortunes doing so. The family unit, producing its own food and clothing, became less self-sufficient, with people trading their labour in return for the ability to buy the things they needed. This new capitalist way of life had its drawbacks, as Karl Marx pointed out in a later century. Those without money were powerless, forced for all time to sell their labour and buy their necessities from the same people who exploited them for maximum profit. But Protestantism, in a sense, made this more meaningful. The idea that God could be served through daily life meant that work was itself a spiritual activity. In its dislike of luxury and opulence, and its insistence on simple, hard-working piety, Protestantism encouraged capitalists to acquire more and more profit, but also to reinvest it in their businesses rather than squandering it on showy displays of wealth and status. Protestantism did not fuel capitalism, nor vice versa, but there was a perfect fit between the two ideas, and both benefited from it.

The emergence of nationalism also had an important role to play in the early spread of Protestantism. For European rulers who had long resented the dominion of a few leading Italian families over the papacy – and resented even more having to fund costly crusades and pay hefty taxes – Luther's rejection of the Pope's authority was an attractive idea. Luther himself, excommunicated but enjoying the protection of his own sympathetic prince, worked out the structures for a Protestant way of life. He was a prodigious writer, and as well as publishing many books wrote hymns and composed services of Eucharist and baptism. He also came up with the principle of *cujus regio, ejus religio*, by which religious differences were to be tolerated, but each country should follow the religion of its leader.

As national leaders took sides in the conflict within the church, wars broke out that would rage for decades – conflicts that had far more to do with territory and wealth than whether Christ was really present in the bread and wine. The same is true for modern-day Protestant–Catholic struggles. In Northern Ireland, communities are polarized by issues like employment and rights to housing and schooling, rather than debates about theology. As we have seen throughout this book, people's ideas about God, and how He is to be worshipped, tend to be 'borrowed' by other causes and exploited to confer respectability, to allow disparate followers to display unity and to lend legitimacy to competing claims.

By allowing individual rulers to decide the faith of the territory they ruled, the Protestant heresy ultimately resulted in the end of heresy itself within the church. A certain strand of religious belief and practice can only be heretical if the power exists to suppress it. Religious battles were to continue, on and off the battlefield, for many more centuries, but the Catholic Church received a severe blow to its authority in 1555, when the Peace of Augsburg granted liberty and tolerance to the Protestant subjects of Protestant rulers, enshrining in law the principle of *cujus regio, ejus religio*.

In its stress on the individual's responsibility for his or her own salvation, Protestantism eased the passage into light of ideas that form the basis of modern society. When, in August 1789, French revolutionaries proclaimed that 'Men are born and remain free and equal in rights', this declaration of the universal rights of man drew on a spirit that was partly brought to life by Protestantism. The individual, free and equal to all other individuals, had become of paramount importance – an importance surpassing that of the church. The Age of Enlightenment taught that each individual's capacity for reason was God's true gift to humanity. In his famous 1784 essay, 'What Is Enlightenment?', the philosopher Immanuel Kant wrote:

> Enlightenment is man's leaving his self-caused immaturity. Immaturity is the incapacity to use one's own understanding without the guidance of another. Such immaturity is self-caused if its cause is not lack of intelligence, but lack of determination and courage to use one's intelligence without being guided by another. The motto of enlightenment is therefore: *Sapere aude!* Have courage to use your own intelligence!

Individual choice, for the good or the bad, is one of the hallmarks of modern life. Timothy Leary's exhortation to the acid-loving hippies of the 1960s, 'Turn on, tune in, drop out' was possible partly because, four centuries earlier, men had risked death by denying that priests and popes held ultimate authority. Margaret Thatcher could not have claimed that there was 'no such thing as society', had Protestantism not prepared the ground by teaching that individuals were responsible for their own salvation. Modern commentators often refer to religion in the modern world as being like a supermarket, in which the individual is free to pick and choose from an array of different faiths. The term 'Christianity' itself now covers a myriad beliefs and forms of worship; but today these are seen as paths to the same goal, rather than heresies to be stamped out.

Modern splinter groups: malicious animal magnetism and a babel of tongues

One Christian splinter group that continues to enjoy support today is that of the First Church of Christ, Scientist. It has no official clergy, but its website claims two thousand branches in eighty countries. Like many of the faiths dubbed as heresies in previous centuries, Christian Science draws quite heavily on Gnostic ideas, considering the material world to be a corrupt illusion.

The sect began in the 1870s after a Bostonian lady, Mrs Mary Baker Eddy, was cured of her headaches and depression by a healer called Phineas P. Quimby. Quimby believed that ill-health was the result of 'wrong thinking' – an idea not that different from those of Freudian psychiatrists, who take the view that many bodily disorders reflect conflicts in the mind. Mrs Eddy took Quimby's ideas and, taking them as her basis, developed her own doctrine, publishing a tome called *Science and Health* in 1875. The core of her belief was that the material world does not really exist, and that Jesus' mission, through his life and death, was to bring us to realize this. Evil and illness were the products of erroneous thinking, or, as she called it, 'Malicious Animal Magnetism', on the part of humans. *Science and Health* sold widely and Mrs Eddy, a shrewd businesswoman, invested the profits well, netting a small fortune for herself in the process.

Modern Christian Scientists remain true to Mrs Eddy's original teachings, considering her book to be a vital text. Mrs Eddy herself kept the movement unified for a long time, expelling anyone who questioned her teachings and insisting on the same order of service in every Christian Scientist community, including a fixed order of readings from *Science and Health*. There are no sacraments and the Bible is seen purely as a book of allegory. Christian

Scientists abstain from tea, coffee, tobacco and alcohol, and they also refuse medical treatment. Since the flesh is not real, they argue, we doctors are an unnecessary breed.

Movements like Christian Science – and Montanism long before it – are inevitably going to have a restricted appeal. For most people, the requirement to live a life of total abstention from sex and most food, or to choose a painful and maybe lingering death over medical treatment, is asking too much. But some splinter groups enjoy success with certain groups of people precisely because of what they demand.

The movement we call Pentecostalism emerged in 1900, when a Kansas Bible student began to speak in tongues. This phenomenon, known as glossolalia, is a common feature of people experiencing states of religious ecstasy. Believers see it as a sign of God's blessing, or of baptism by the Holy Spirit. People who have studied glossolalia point out, perhaps unnecessarily, that the outpourings of the blessed, or afflicted, are characteristically rhythmic and bear no relation to any human language.

Pentecostalism is a powerful force in modern Christianity, being a popular form of worship even in strictly Catholic countries like Portugal. Its sheer numbers – some estimate it as having ten to twelve million adherents – have led people to call it a 'third force' alongside Protestantism and Catholicism. Its worship is spontaneous and joyful – people sing, speak in tongues or exclaim enthusiastically in response to a preacher. Churches are simple affairs, with no icons or images. Pentecostalists believe that we are all sinners, in need of a direct conversion experience through personal spiritual contact with Jesus Christ. Such experiences announce themselves through manifestations such as fainting, or speaking in tongues, and it is our duty to anticipate them by attending Pentecostalist services, called 'waiting meetings'.

Pentecostalism has always been popular with under-privileged minority groups – for instance, poor black communities in the Southern states of America, or Afro-Caribbean immigrants in British cities. It is not hard to see why, because the experience of religious ecstasy is both open to all and at the same time gives worshippers a kind of status. Mainstream Protestantism, lacking the comforting rituals of Catholicism, and with its stress on inner piety, can be a tough discipline. Faiths like Pentecostalism offer a degree of controlled rebellion against a lowly status in the social hierarchy. Not only are worshippers permitted to shout out loud, fall down, weep, laugh and sing – as they would not be in daily life – but these actions are also outward signs of being blessed by God. Through simple religious enthusiasm, one claims membership of a kind of spiritual aristocracy within one's own community.

This phenomenon has been noted in other 'ecstatic' kinds of movement as well. Possession by spirits is particularly common in the Islamic parts of Somalia. The somewhat controversial anthropologist Ioan Lewis noted that the evil spirits, known as *sar*, seemed to show a marked preference for visiting women, who have a low status in society.[11] When they were under the influence, so to speak, the possessed women spoke and acted in ways that would have been normally forbidden to them. They could demand luxuries and foods, and use kinds of language, normally reserved for men – and it was the duty of their husbands to provide what the demon wanted in order to send it away.

Showgirls and saints

Forms of worship that begin as simple and humble gatherings to serve the spiritual needs of the oppressed do not always remain that way. The skyline of 1920s Los Angeles was dominated by a vast white dome-like structure called the Angelus Temple, which could accommodate up to five thousand believers at any one time. The Temple, which cost $1.2 million to build, was the creation of one Aimée Semple McPherson, a former missionary to China, who had moved west from New York in an automobile painted with the slogans 'Jesus is Coming, So Get Ready' and 'Where Will You Spend Eternity?'

McPherson's rise to fame began in 1921, when she helped a crippled woman rise from her wheelchair at a prayer meeting. The audience declared her a healer, but McPherson modestly refused the title. 'Jesus is the healer,' she said. 'I am only the little office girl who opens the door and says, Come in.' Such modesty was not reflected in the organization she built. The Angelus Temple, opened in June 1923, was topped by a rotating cross visible for 50 miles, and boasted a vast gospel choir composed of buxom young women, a brass band and a pipe organ. A broadcasting station sent the message far and wide and a Miracle Room displayed row upon row of crutches, wheelchairs and leg braces, the discarded debris of McPherson's miracle cures. In her love of spectacle, and her deliberate courting of publicity, McPherson quite consciously aped the tactics of the neighbouring Hollywood film industry to win converts.

Her career hit a peculiar point in 1926, when she went missing while swimming in the sea. For a long time, she was presumed to have drowned. But thirty-two days after her disappearance McPherson knocked on the door of a farmhouse in Mexico, claiming she had been kidnapped. Critics thought, as they still do today, that she had staged a publicity stunt. Certainly there were a few bum notes in her alibi: although she claimed to have walked across the burning sands to the farmhouse, her shoes were unscuffed; and it was noted that during her absence the handsome, muscular chief engineer of McPherson's radio station was also nowhere to be seen.

Whatever the truth, McPherson's vanishing act did her no harm. In June 1926 a crowd of around 50,000, including politicians, sports stars and matinée idols, turned out to welcome her home. But, perhaps because of her immense popularity, speculation about her disappearance continued to rumble. Chambermaids and room clerks were produced who claimed that McPherson and the engineer regularly met for illicit liaisons. She also endured a very public battle with her mother and sister for control over her wealthy Foursquare Gospel Church – a battle she won, but which divided her from her family until her death. She continued to be active throughout the Depression years of the 1930s, preaching hope for the future, and dispensing food and healthcare to the poorest. Married three times, widowed once and divorced twice, Aimée Semple McPherson died in September 1944 in a hotel room in Oakland, California; the coroner recorded a verdict of accidental death by sleeping pills. A monument to her remarkable career, the Foursquare Gospel Church continues to be active today, with around 100,000 members in the United States and branches in a range of countries, including Britain. Today the building in downtown Los Angeles stands near Sunset Boulevard. The organization's budget in 1993 was said to be well over $300 billion.

McPherson was an adept impresario, and the popularity of her movement shows that she met a definite need. Throughout the story of God, various offshoots and schisms occur precisely because people want to worship through joy and dance and song, rather than with formal prayer and prescribed ritual.

If you should visit the borough of Stamford Hill in north London on a Friday evening, you could be forgiven for thinking you had entered a time warp. Groups of men, dressed in frock coats and wide-brimmed hats, walk together to the synagogue. Many sport long beards, even fur-brimmed hats that owe more to the fashion of eighteenth-century eastern Europe than that of modern-day London. Each variation of this self-consciously outdated garb is a badge, marking out a local community from one of many different regions of Poland over four centuries ago. These are Hasidic Jews, or Hasidim (*hasid* means 'pious' in Hebrew). Although they have large communities in many major cities across the world, such as Antwerp and New York, these Hasidim keep themselves separate from mainstream life, and even from other Jewish communities. They tend to cluster around certain professions – the diamond trade in London, for example, and the camera trade in New York – but devote much of their time to study and prayer. Central to their belief is the idea that the Messiah is on his way, and different sects of Hasidim sometimes regard their own charismatic leaders, or rebbes, as the ones specially chosen to announce his arrival. Anachronistic and exclusive they may seem, but in fact their movement has much in common with the ideas that fed Protestantism and the later Enlightenment. Their rather severe appearance also disguises the fact that they worship in a uniquely enthusiastic and joyous fashion.

I have shown elsewhere how enthusiastic and spontaneous forms of worshipping God can sometimes be a response to oppressive conditions. This was certainly the case in eighteenth-century Poland. Not only were the Jews oppressed from without, by the Catholic majority, but their own society was hierarchical and strictly conformist. The show was run by an intermarried caste of rich merchants and rabbis. This elite had formidable powers within the community. If a poor man had a grievance against a rabbi, he had little chance of winning the dispute. The Jewish family was also a unit of strict control. A father was entitled to use force to teach the

Law to his son once he was twelve. From the age of thirteen, the youth was treated as a man, but the ancient Law of the Rebellious Son, derived from Deuteronomy 21: 18–25, theoretically dictated that delinquents could be stoned to death:

> If a man has a wayward and defiant son who does not heed his father or mother and does not obey them even after they discipline him, his father and mother shall take hold of him and bring him out to the elders of his town at the public place of his community. They shall say to the elders of this town, 'This son of ours is disloyal and defiant; he does not heed our voice. He is a glutton and a drunkard.' Thereupon the men of his town shall stone him to death. Thus you shall sweep out evil from your midst: all Israel will hear and be afraid.

The Talmud points out that this never happened, either in biblical or in post-biblical times, and it certainly has never happened since; but the mere existence of such a law might have influenced in some measure the interaction between fathers and sons in some families.

Nevertheless, Jewish tradition has always contained within it a certain amount of room for people who refuse to fit in with the mainstream. From the sixteenth century onwards, as we have seen, there was a great deal of popular interest in magic and mysticism, even though all kinds of magic are condemned in Orthodox thought. Holy men, who wandered Europe dispensing magic amulets and casting spells, were an accepted feature of the Jewish landscape. There was a tradition, harking back to God's creation of Adam out of clay, that certain gifted men possessed the power to create a being known as a *golem* from dirt. Once activated, by recitation of the divine name, the *golem* could perform a range of tasks for its master, for instance, fetching water or even doing battle against Gentile enemies. Popular mythology teemed with tales of heroic *golemim* – and also of some, like Dr Frankenstein's monster, which ran amok and caused mayhem.

In the 1740s, a German Jew called Jacob Hayyim Falk fled Westphalia. Falk was a Kabbalist who claimed to perform miracles, demonstrations of which he gave before Christian notables in Germany and elsewhere – leading to accusations of wizardry, arrest, and sentence to death by burning. Having escaped the German authorities, he arrived in England (at that date somewhat more welcoming to asylum-seekers) in 1742, and spent the rest of his life in London, with occasional trips to Paris. In London, he impressed a number of Jews and Christians alike, acquiring a considerable sum of money in the process. He was vehemently attacked by most of the Orthodox rabbis, some of whom denigrated him as a disciple of Shabbetai Zevi, but was friendly with the Chief Rabbi, Tebele Schiff. He established his own synagogue in Wellclose Square and, his interest in magic undiminished by his narrow escape from the stake, set up an alchemist's laboratory at London Bridge. There is a story, almost certainly apocryphal, which suggests that he saved the Great Synagogue from burning by writing magical inscriptions either side of the door.

As Falk conducted his experiments, another mystic was attracting popular support in Poland. Born around 1699 as Israel ben Eliezer, and referred to as the Besht (from the initial letters of his adopted name, Baal Shem Tov, meaning 'Master of the Good Name'), he was by no means a member of the Jewish elite. He seems to have worked in the slaughterhouse, as a watchman and as an innkeeper. Portraits depict him smoking a pipe. He was certainly a man of the people. He left no writings. Instead, he wandered the land, dispensing amulets, casting out demons and performing miracles. None of this was unusual at the time; as noted earlier, itinerant mystics were no rarity among the Jews of eighteenth-century Poland.

But the Besht possessed certain special qualities that led to the formation of a movement around him. He was clearly a charismatic leader, drawing disciples to his fold who eagerly copied down his utterances. He was also a creative thinker, putting a fresh spin on two long-standing Jewish religious traditions.

Since ancient times, the Jews had cherished the notion of the *tsaddik*, the righteous man who was slightly above mere mortals because of his exceptional closeness to God. In a way, a *tsaddik* is the Jewish equivalent of a saint; but saintliness, in this context, carries no special certification by the rabbinate or any other Jewish authority – a *tsaddik* is a *tsaddik* by popular acclaim. The Besht, coming to prominence after the ignominious failure of men like Shabbetai Zevi and Jacob Frank, gave a new meaning and significance to this idea. The public conversion of these men to other faiths, and the obvious falsity of their prophecies, had left the Jews needy and bewildered. The *tsaddik*, a figure exemplified by the Besht himself, filled a gap; not a Messiah, not an ordinary rabbi, this was a new type of religious personality, who, according to the Besht, had descended into the world in order to bring people to the truth. This, in itself, was slightly similar to the way many Christian sects viewed Jesus Christ. But since the *tsaddik* was not claiming to be God's exclusive messenger, there could be many of them – an idea that in turn led to the spread of the movement. The various thriving Hasidic sects today are still centred on individual *tsaddikim*, both living and dead, and these men are deeply revered.

The Besht also revolutionized the idea of prayer, stressing it as both a physical and a mystical discipline, to be undertaken with full concentration and energy. The doctrines of the Kabbalah taught that men could repair the fractured cosmos through their piety and prayer; the Besht espoused this mystical idea but made it more accessible, providing a technique by which ordinary people could achieve states of ecstasy through prayer, and thus feel that they were active agents in the salvation of the world.

The worship of the Besht's followers, the Hasidim, was a turbulent, enthusiastic affair – quite different from the staid Torah recitations of the established synagogues. In their own prayer houses, men smoked and drank to achieve altered states of mind. They prayed at the top of their voices, often swaying, dancing and clapping their hands. The Besht, in one of his sayings, likened prayer to the experience of orgasm, so we can imagine that it was a noisy and joyous activity. But prayer was only one feature of the Hasidic approach. Followers also believed that they could release sparks of the divine light, trapped in creation, through the joyful fulfilment of daily life – eating, sleeping, working and sex all had the potential to be holy activities if undertaken in the correct, devoted way.

In this respect, Hasidism shared much in common with the mind-set that gave birth to Protestantism. Once again, stress was placed upon the individual's own ability to achieve salvation – he was not powerless, but an active agent in the repair of the cosmos. Hasidism was popularist, a religion of poor men at the bottom of the hierarchy of Jewish society. Protestantism similarly appealed to the poor because of its attack on the pomposity and wealth of the church. Both religions also taught that it was possible to worship God in and through daily life and, in doing so, they reflected a quite radical shift in thought.

We have seen before that the Divine Idea often involves a separation of existence into sacred and profane compartments. Certain foods, objects, buildings, activities are sacred or holy – and, in the case of Catholicism, it was the church alone that provided access to this sacred sphere. Movements like Protestantism effectively stormed the garrison, taking the sacred out of the hands of the elite and placing it in the hands of the individual, irrespective of status.

But this idea, ironically, also created a serious threat to the whole basis of religion. In making the sacred the possession of all people, priests and paupers alike, it can be argued that it forfeited some of its power and mystery. The sacred, by its very definition, can be sacred only by being set apart from daily life, being something 'other'. Only a minority of spiritual athletes are likely to be particularly good at conducting their daily affairs as if they were worshipping God. For most of us, rituals and symbols are necessary to induce that state of mind. We can feel peace, and a connection to God, in the tranquil setting of church or synagogue; it's far harder to find these spiritual states in the hubbub of the market place or stuck in a traffic jam.

Movements which emerged to keep religion alive might ultimately have led to its retreat. The belief in the individual as responsible for his own salvation led to the Enlightenment idea that the individual, with his or her own powers of reason, was of paramount importance in the scheme of things. It became possible not just to choose whether you wanted to be a Catholic or a Lutheran, but also to use your reason to decide whether you wanted a religion at all. Reason, and its application to the material world in the form of science, became a means of abolishing God altogether.

8

God in Retreat

Among my more prized possessions are some seventeenth-century engravings illustrating Old Testament stories. They are dedicated to Raphael, the great biblical painter, being copies of what he had painted some one hundred years earlier. In one glorious drawing, God is striding about his newly created earth, his cloak billowing behind him. The intense look on his face shows he is involved in some deep spiritual effort. His arms are outstretched either side of him, forefingers pointing across the horizon. And he is surrounded by newly created animals: a lion, a lobster, an elephant, a deer, an ostrich. There is even a unicorn prancing around in the background. Other animals – a leopard and a horse – are still coming out of the ground where they have been formed together with various creeping things, including lice. This is the Creation, described as 'the finger of God' (Exodus 8: 19).

These furrows on the Creator's brow and the outstretched fingers are repeated in quite similar ways throughout Renaissance art – in frescoes, in paintings, in sculpture and on glass windows. Very often God's fingers are symbolized by the rays of light which emanate from the heavens in these images. Perhaps the Renaissance glass-makers thought of this metaphor when they saw the moving rays of light refracted or recoloured through their stained glass onto the brick walls and tiled floors of the Florentine and Roman cathedrals. Whatever the case, this understanding of the natural world and its various depictions are the culmination of millennia of human belief. They are not simply confined to Christianity, though it is the western Christian world that has come to dominate so much of our scientific thinking and technology.

St Ambrose, who died in 397 in Milan, vividly describes how Christians have tended to think about the Creation: 'We must remember that the light of day is one thing and the light of the sun, moon, and stars another – the sun by his rays appearing to add lustre to the daylight. For before sunrise the day dawns, but is not in full refulgence, for the sun adds still further to its splendour.' Light was divine and pervaded the universe; this was the prevailing view, which lasted until the time of the Renaissance. Its influence can be seen in early Renaissance painting, even as more scientific ideas gained impetus. It is interesting to see how the idea of the division of light from darkness persisted in all sorts of ways. One set of stage directions for a mystery play says of the stage performance re-enacting God creating the world: 'Now a painted cloth is to be exhibited, one half black and the other half white.' An example of the same idea given more permanent form appears in Michelangelo's 1511 fresco of God in the Sistine Chapel. The Almighty, frowning deeply, extends his arms and forefingers, separating two huge discs in the sky. One dark disc represents night, the other day. Around him, amazed cherubim wonder at what he is doing.

Creation: what, where and how?

The *matter* of which the universe is made has always engrossed Christian theologians and scientists. What existed before the world was created? Did God do something unique – not merely make matter, but create 'something out of nothing'? Humans are capable of making things, but, according to Jewish tradition, only God can create. In Hebrew, there are various verbs meaning 'to make', but in Genesis, one verb describes specifically what God did: *bara* – 'he created'. God uniquely creates, but humans use the God-given tools – the earth's natural materials, with which they have been provided at the time of creation – and their God-given intelligence to research and make new things. In the Jewish view, this is science; and it is 'good'. *Imitatio dei* is something essentially expected of humanity. Hence it is incumbent on us to 'play' God – to heal the sick, to change our environment, to explore the natural world.

Quintus Septimus Florens Tertullianus – Tertullian – is widely regarded as the first great Latin Christian writer. A Roman convert to Christianity around 200 CE, he was a somewhat controversial figure. As we have seen already, he joined the Montanists before eventually forming his own breakaway movement. He started an argument which continued to occupy Christians for the next two thousand years. If there had been any pre-existing matter out of which the world was formed, Tertullian declared, the scriptures would have mentioned it; so, by not mentioning it, God has given us a clear indication that there was no such thing. To us, now, it might seem odd that such a question was of such basic importance – but then, this issue still dominates much of modern particle physics. Tertullian insisted on his view; any other was, in his opinion, heretical. Some two hundred years later, one of the most influential Christians of all, St Augustine, put forward a refinement of the argument. After all, he said, the matter out of which the universe was formed must have pre-existed in some way: 'although the world has been made of some material, that very same material must have been made out of nothing'.

For the next thousand years, Tertullian's opinion dominated Christian thinking, becoming part of that unquestionable certainty which characterized so much of medieval theology. Eventually, the church got organized. The Fourth Lateran Council in 1215 declared firmly that God created everything out of nothing. (This was the same gathering, by the way, that required Jews and Muslims to wear a special dress to enable them to be distinguished from Christians. It also enforced its authority by insisting that Christian princes must take measures to prevent blasphemies against Jesus Christ.) These views of creation persisted not only among Catholics, but later among Protestants, too. Prominent Reformers such as Philipp Melanchthon, Martin Luther's great friend, insisted that the universe was created out of nothing and in a mysterious way, in an instant *and* in six days, citing the text: 'He spake, and they were made.'

Theologians also spent a huge amount of effort working out the length of time that God took to complete his greatest works. The biblical account must have puzzled very many early thinkers. How could it be possible, for example, that a tree, which takes so long to grow, could be created in a single day? But the book of Genesis is unambiguous. The whole enterprise took a working week, with the seventh day being the day God rested. And the biblical account is quite precise in its description of what was created on each of the six days. Some philosophers, such as Philo – a Jew born in 20 BCE – were not content with this bald description. Philo decided that each act of creation must have been instantaneous, because a latter passage in Genesis says: 'He spake, and they were made; he commanded, and they were created.'

Philo was a Hellenized, secular Jew, and his view seems never to have greatly concerned most mainstream Jewish thinkers. Indeed, there is reason to believe that he was not fluent in Hebrew and did not read Genesis in the original language. For the most part, Jewish scholars did not linger over interpretations of this aspect of Genesis. But Christians did: they agonized about this aspect of God's work. Christians began to affirm that it was safer to believe *both* statements; that mysteriously, God created the universe in six days, yet produced it in an instant (as Melanchthon asserted). So the science of sacred mathematics was ingeniously developed to reconcile these two accounts. The creation of the heavenly bodies took place on the fourth day because of 'the harmony of the number four'; of the animals on the fifth day because of the five senses; of man on the sixth day because of the same virtues in the number six which caused it to be set as a limit to the creative work of God. Says St Augustine:

> There are three classes of numbers – the more than perfect, the perfect, and the less than perfect, according as the sum of them is greater than, equal to, or less than the original number. Six is the first perfect number: wherefore we must not say that six is a perfect number because God finished all his works in six days, but that God finished all his works in six days because six is a perfect number.

At the pinnacle of this line of argument was God's repose on the seventh day – explained by the vast mass of mysterious virtues inherent in the number seven. At the other end of the spectrum, the number two was often regarded as evil – for that was the day that, according to Genesis (said St Jerome), God did not declare of his creation 'it was good'. Similar ideas were reiterated centuries later by Bede.

It would become excessively tedious to spend overlong

on the calculations surrounding the time of Creation; to us the thinking behind them now seems rather convoluted. But St Thomas Aquinas, who died in 1274, and who was undoubtedly one of the keenest thinkers of the Middle Ages, also spent considerable time expounding on the mysteries of creation. He taught that God created the substance of things in a moment, yet gave six days to the work of separating, shaping and adorning this creation. Even the early Reformers did not deviate from this view. Luther especially showed himself equal to the challenge, boldly declaring that Moses 'spoke properly and plainly, and neither allegorically nor figuratively', and that therefore 'the world with all creatures was created in six days' – and then went on to show how, by a great miracle, the whole act of creation was also instantaneous.

To the medieval church these matters were of key importance, because the creed of the church took the description in Genesis as its starting point. Both in Rome and, later, in Augsburg and Westminster, divines were drawing up their Confessions of Faith – the declarations of the core beliefs of their churches. They specifically laid it down that it was necessary to believe that all things visible and invisible were created not only out of nothing, but in exactly six days.

The time the Creation took is only part of the story. The time *since* the Creation also exercised many minds. As I write, the Jewish year is 5765; by Jewish tradition the world is 5,765 years old. Relatively few Jews take this seriously now, though Jews, too, have the odd creationist. Even in the Middle Ages, as we shall see, Jews preferred a figurative rather than a literal translation of the beginning of Genesis; and even among Christian scholars of the time, only a few went quite to the detailed lengths of the earnest Dr Lightfoot. It was in the Jewish year 5400 (1640 CE) that Dr John Lightfoot, Vice-Chancellor of the University of Cambridge and Master of

St Catherine's College, Canon of Ely Cathedral, a scholar renowned for his extensive knowledge of Hebrew and his Aramaic studies, made his remarkable calculations. By studying in great detail both the biblical text and the Talmud and engaging in some divine mathematics, he arrived at the conclusion that 'heaven and earth, centre and circumference, were created all together, in the same instant, and clouds full of water', and that 'this work took place and man was created by the Trinity on October 23, 4004 B.C., at nine o'clock in the morning'. This, it turns out, was a Sunday. Work it out – apart from the small matter of a 250-year discrepancy this is quite close to the Hebrew year at the time, at least.

I recently visited Trinity College, Dublin. If you get the chance to do likewise, above all visit the great Ussher Library. Among its prized possessions is the fine Book of Kells – a ninth-century manuscript of the gospels, renowned worldwide for its rich illustrations. The old Ussher Library is a long hall, with a high, arched ceiling and seventeenth-century galleries, crammed with ancient books – all sorted on the shelves Irish fashion: by size. James Ussher, Archbishop of Armagh and Primate of All Ireland, who died in 1656, was another scholar who worked assiduously to uncover the mysterious dates in the Bible. In addition to establishing the date arrived at by John Lightfoot, he went one better. He calculated that Adam and Eve were driven from Paradise on Monday, 10 November 4004 BC – having had a taste of paradise for just seventeen days. And the Ark touched down on Mount Ararat on 5 May 1491 BC – 'on a Wednesday'. His *Annals of the World*, published posthumously in 1658, are worth a brief skim:

For as much as our Christian epoch falls many ages after the beginning of the world, and the number of years before that backward is not only more troublesome, but (unless greater care be

taken) more lyable to errour ... Now if the series of the three
minor cicles be from this present year extended backward unto
precedent times, the 4713 years before the beginning of our
Christian account will be found to be that year into which the first
year of the indiction, the first of the Lunar Cicle, and the first of the
Solar will fall. Having placed there fore the heads of this period in
the kalends of January in that proleptick year, the first of our
Christian vulgar account must be reckoned the 4714 of the Julian
Period, which, being divided by 15. 19. 28. will present us with the 4
Roman indiction, the 2 Lunar Cycle, and the 10 Solar, which are the
principal characters of that year ...

... and so on.

Sadly for Dr Ussher, within two hundred years of his
death, the wonders of ancient Egypt would be uncovered.
Theologians then had to grapple with incontrovertible
evidence that, at the supposed time when God had been
creating the earth, a powerful, sophisticated state had
built huge temples along the River Nile; had enjoyed the
grain, the fruit, the wine of a highly developed civilization;
had built the pyramids at Giza in Egypt; and had
invented a sophisticated form of writing. And quite
similar advances had been made by other humans during
the same period in Mesopotamia, India and China.

Jews, Muslims and science

Until quite recently, the vast majority of the Christian
faithful – whether Catholic or Protestant – were still
taught that the universe was created out of nothing, in an
instant, and yet also in six days. Jews, on the other hand,
for the most part tackled the biblical text in a quite
different way. There were many shades of opinion
about the precise nature of creation, but none warranted
the accusation of heresy. Moses Maimonides, one of the

greatest of all Jewish geniuses – doctor, rabbinical scholar, commentator, philosopher, scientist – wrote mainly in Arabic, having been born as he was in Muslim Spain. He died in 1204, having written extensively about the nature of existence and belief. He never fell into the trap of the pre-Copernican belief that man was the crown of earthly creatures and the centre of God's universe. In his *Guide for the Perplexed* he wrote:

> The major source of confusion in the search for the purpose of the universe as a whole, or even its parts, is rooted in man's error about himself and his supposing that all of existence is for his sake alone. Every fool imagines that existence is for his sake . . . but if man examines the universe and understands it, he knows how small a part of it he is.

And later in the same book he writes: 'All mankind and certainly all other species of living things are naught in comparison with all of continuing existence.'

Maimonides wrote a whole book on creation, *Maaseh Bereshit* ('The Work of Creation'). Creation in the Bible, he wrote, 'has been treated in metaphors, in order that the uneducated may comprehend it according to the measure of their faculties and the feebleness of their comprehension; while the educated take it in a different sense'. Clearly, then, interpretation of the text is required; and, as we saw in chapter 4, although Orthodox Judaism requires scrupulous observance of exactly how the Torah is written, Jews are expected, indeed required, to interpret it. This is, in effect, a form of support for what we would later see as scientific thinking. Eight centuries ago, before so many of the Christian church's accusations of heresy about how Genesis was interpreted, Maimonides pointed out that the prevailing scientific opinion about the universe was that of Aristotle – the pre-eminent Greek philosopher, who exercised a pervasive influence well beyond his own time. Aristotle had written that the universe had existed for ever. It was fixed and unchanging and all matter within it was conserved. This notion was a complete contradiction of the biblical account. But Maimonides argued that Aristotle had not proved conclusively that there was eternity of matter. Since, philosophically, the concept of eternity of matter and the concept of divine creation were equally acceptable alternatives, Maimonides said, he preferred the account in Genesis. But nevertheless, he said, if new evidence came along which proved Aristotle correct, then he would have to re-examine the biblical text and reinterpret the verses to accommodate the scientific theory. That flexible attitude has been a consistent feature at the core of how Judaism deals with scientific exploration.

Nachmanides, another great biblical commentator, who died in 1270, went somewhat further in stating that, in his view, the story of the Creation in the Bible also spoke only in general terms – its primary purpose being not so much to tell a story as to state that nothing was created except at God's command. So Jews, who have always tended to take the text of the five books of Moses very literally, never had quite such a problem with the Creation story as Christians.

It is surely no accident that so many Jewish philosophers of this period lived and thrived (at least for a while) in Muslim Spain. Indeed, although they were nearly all fluent in many languages, most of the early medieval Jewish scientists, sages, poets and rabbis wrote their books in Arabic, rather than Hebrew. Muslim cities like Córdoba and Toledo, as well as Baghdad, Damascus and Cairo, were important centres of civilization. These cities flourished during a period of stability sometimes lacking in much of the rest of Europe and Asia Minor. Muslims saw Europe as backward, disorganized, strategically insignificant and essentially irrelevant – while aloof Christian Europe, with its increasingly entrenched clerical hierarchy, returned the compliment, regarding Muslims as culturally inferior infidels. So initially, the new scientific discoveries made in the Islamic lands before the twelfth century were largely ignored by Christian Europe.

Córdoba was the capital of Muslim Spain – a centre for enlightenment and learning attracting scholars and students from various parts of the world. Many manuscripts were deposited there. Nowadays, we judge academic success by impact, reflected in the number of publications a scholar or institution produces. In these terms, ninth-century Europe offers an interesting comparison. It is said that the library of the monastery of St Gall (in what is now St Gallen in Switzerland) was the largest in Christian Europe, boasting thirty-six volumes. According to some Muslim sources, the library of Córdoba contained over half a million.

Jews studied in small *kehilot* (communities), Christians in monasteries. Muslims formed the idea of colleges in the late 600s and early 700s. The earliest European colleges are those established at the universities of Paris and Oxford, founded in the thirteenth century. They were funded as trusts, rather like those in Islamic countries, and legal historians have traced the concept back to the Islamic system. The organization of these European colleges was also similar to those in Islam – the distinction between the graduate (*sahib*) and undergraduate (*mutafaqqih*), for example, seems to have been derived from Islamic practice.

In 762 the Caliph al-Mansur built Baghdad and established it as his new capital. In those days, the religious climate there was tolerant, secularized and relatively intellectual. Given improving trade with the east, as well, it was natural that Arab scholars should look far over the horizon. They studied classical Greek from Byzantium and mathematics from India and China, and improved their medical knowledge through contacts with Persia. Jews were tolerated, and a number became physicians and scientists at court. They adopted the Chinese abacus and used the Hindu system of numbers and decimals. Trigonometry was used to build buildings, which were decorated with the geometric designs that can be seen in so many stunningly beautiful mosques today. Ptolemy's system of astronomy was important because it helped to identify the correct times for prayer and a calendar for festivals. In the absence of a magnetic compass, too, astronomy could be used to determine the direction to Mecca from any town in the Muslim world.

In mathematics, the Arabic notion of the number zero (derived from the Hindus) and the decimal system were gradually introduced to Europe and eventually became the basis for our scientific revolution. Europeans adopted Arabic numerals because they made mathematical tasks much easier. Problems that had taken days to solve could now be solved in minutes. If you doubt this, try adding up your supermarket bill using Roman numerals. And algebra was growing in importance. Although algebra had been used by the ancient Greeks, the word itself is, of course, Arabic, derived from the title of a key work by Al-Khwarizmi (790–840), a mathematical genius who lived in Cairo. He was exceptionally widely read, also deriving some of his mathematics from Hindu sources. Apart from making highly significant contributions to the fields of trigonometry, astronomy/astrology, geography and cartography, his approach to algebra established methods for solving linear and quadratic equations. His theorems were published in his book *Al-Kitab al-mukhtasar fi hisab al-jabr wa'l-muqabala*, or 'The Compendious Book on Calculation by Completion and Balancing', first translated into Latin in the twelfth century, and it was from this title that the term *al-jabr* – 'algebra' – entered the western mathematical vocabulary. His own name, rendered into

Latin, was Alghorismus – from which we get the term 'algorithm'.

Another Muslim scientist deserving our attention, albeit for not entirely positive reasons, was the Persian Jabir ibn-Hayyan, who died in 815. He is said to have been the leading chemist in the Muslim world. A Shi'ite, he became court physician to the famous Caliph Harun al-Rashid, patron of the arts and subject of *The Thousand and One Nights*. Some sources credit him with a huge number of works – around two thousand – a surprisingly large number, indeed, given that he wrote them all in longhand. My own three hundred papers, most of them pot-boilers and virtually all produced on word processor or computer, have taken a lifetime. It is fair to say that Jabir's main claim to fame was that he was the founder of alchemy, a 'science' which intrigued researchers throughout medieval times. Having mixed sulphur and mercury to form the red compound cinnabar, he was convinced that if the right proportions could be found, gold could be produced by similar means. His work was responsible, at least in part, for the phlogiston theory, thus setting European science back several hundred years.

Hunayn ibn Ishaq (808–73) is another important figure also associated with the reign of Harun al-Rashid. He was a physician; but more important than his medical skill were his talent and achievements as the leading translator in Baghdad's House of Wisdom at one of the most remarkable periods of mathematical revival, a position through which he exercised a highly significant influence on the mathematics of his time. Translation was something of a family business, his son Ishaq ibn Hunayn being famed for his rendition into Arabic of Euclid's *Elements*. Hunayn's family were Syrian Christians before the rise of Islam, and he himself was brought up as a Christian. Al-Rashid wanted access to the sources of Greek philosophy, so he organized an expedition to Greece to try to find manuscripts of the works of Aristotle, Plato, Hippocrates, Euclid, Pythagoras and others. Hunayn, being more skilled in the Greek language than any of the other scholars in Baghdad, was probably a member of this expedition. Hunayn had all the attitudes of a real academic and collector. He gives a typically obsessed account of his search for one particular medical manuscript: 'I sought for [the manuscript] earnestly and travelled in search of it in the lands of Mesopotamia, Syria, Palestine and Egypt, until I reached Alexandria, but I was not able to find anything, except about half of it at Damascus.' It puts trawling the second-hand bookshops in the Charing Cross Road in perspective. The many Greek texts, including works by Plato and Aristotle, that Hunayn translated into Arabic were disseminated throughout Mesopotamia, Syria and Egypt. It was the detective work of Muslim scholars like Hunayn that allowed many Europeans of later centuries to gain sight of Greek manuscripts that would otherwise have been lost for eternity.

One puzzle is why, given this superiority of the Muslim world in developing knowledge, modern science arose not here but in western Europe. The answer may lie in the preferred direction of study. It seems that certain Muslim leaders, like some of their counterparts in early medieval Europe, had increasingly little regard for the study of the natural world. Academic pursuits were tolerated, but learning was divided into traditional studies based on the Qur'an and 'foreign' studies – which were derivative, based on knowledge obtained from the Greeks. Although there were Arabic rationalists, many clerics saw this line of intellectual endeavour as a threat to the authority of the Qur'an. A conservative reaction in the late tenth century, together with a decline in peace and prosperity, impeded further scientific advance in the Muslim world. The orthodox view that gradually began to dominate Islam was that man was not fully rational, and that there was no room for purely systematic investigation of God's creation.

Changes in the rule of the Islamic parts of Spain were a symptom of this growing rigidity and hostility to science. For three hundred years from the early eighth century to the beginning of the eleventh, cities such as Córdoba and Toledo had enjoyed a golden age of great prosperity and intellectual vigour. In 1031 the leadership of the Moorish kingdom of Córdoba changed; and in 1085 Toledo fell to the Christian king Alfonso VI, who made it the centre of his kingdom of Castile. Over the following years Castilian rule was challenged by the fanatically religious Almoravids, and their growing dominance eclipsed the previously tolerant rulers of this region. In 1147, too, Lisbon, in Christian Portugal, fell to the puritanical Almohads. This group was initially led by Muhammad ibn Abdallah ibn Tumart, a Berber from the Atlas mountains who seems to have been the ideal fanatical reactionary. He had a strong religious background, being the son of a lamplighter from a mosque, and was noted for his piety as a teenager. So righteous was he, in fact, that he was expelled from a pilgrimage to Mecca because of his 'severe strictures on the laxity of others'. He is described by one source as 'small, ugly, and misshapen and living the life of a devotee-beggar'. His main concern was to promulgate a rigid view of the singularity of God. This denied the independent existence of the attributes of God, as being incompatible with his unity. So, unlike the Jewish God, he could not be God of both good and evil. Muhammad ibn Abdallah also believed that everything was predestined by God. It will not surprise to learn that he was a strict observer of religious law.

An anecdote from this period tells us something about the changing climate of religious belief. It concerns Ibn Rushd, known in the west by his Latinized name of

341

Averroës, the great Islamic philosopher who flourished in Córdoba in the twelfth century. He was an almost exact contemporary of Maimonides, and the two men may have met. Averroës was a physician and chief judge in Córdoba. Because he integrated many Islamic traditions with Greek philosophy, particularly Plato's *Republic*, he was highly influential in the intellectual developments of this period: Averroës' commentaries exerted considerable influence on Jews and Christians for centuries thereafter. Most of his more important writings defended the philosophical study of religion against the theologians. As personal physician to the caliphs Abu Ya'qub Yusuf and his son, he came into close contact with the court, at this date in Seville. In 1169 Averroës was presented for the first time to the Caliph, who in spite of his rigid religious views was a bit of a student of philosophy himself. His opening gambit terrified Averroës. Were the heavens created or not? Averroës, trembling, declined to answer. The Caliph then answered his own question, put Averroës at ease, and discharged him carrying princely gifts after a long, engaged conversation. But Averroës' story is a sad reflection on what was happening in his world. After he praised Aristotle as 'a man chosen by God', he was violently attacked by the Muslim clergy, his doctrines condemned and his books burned. He died in Morocco in 1198. And after his death he was condemned by Christians also, because he had stated that something (such as the Creation) might be philosophically true, but theologically untrue – and that the converse could also be the case.

With religious radicals in the ascendant across the Muslim world, the golden mantle of science began to pass gradually onto European shoulders. And more Europeans were coming into contact with the wealth of Islamic knowledge in Spain. Important Arabic and classical works from Muslim libraries were translated into Latin and began to filter into centres of learning all over Europe. They arrived at a time when influential churchmen like St Anselm (1033–1109), the Burgundian priest who became Archbishop of Canterbury, were reviving the role of reason in faith; and soon after their arrival universities started to be founded.

Over the centuries, the biblical account of Creation has led to repeated attempts at various 'scientific' and not so scientific explanations of the book of Genesis. These, to my mind, are not very enlightening. It is staggering to contemplate the hours, indeed years, of human effort that have been invested in trying to rationalize what is one of the most poetic, precious and ancient of human beliefs – a belief most remarkable because, as we have repeatedly seen, it recurs in so many different accounts of the beginnings of time.

Some years ago, the rather earnest Israeli physicist Nathan Aviezer came to give a talk at my synagogue in north London. This was a slightly surreal occasion: as it was a Sabbath in an Orthodox synagogue, we could not allow him to write or show slides (as an Orthodox Jew, he would not have wanted to, anyway); so he elected to give his lecture on the physics of creation using a huge folder containing a series of giant pop-up paper charts he had prepared the day before. Professor Aviezer started in true Maimonidean mode. 'There need be no contradiction', he said, 'between modern physics and the biblical account of creation.' He maintained that the Big Bang, the dominant scientific theory of origin, proposes that the universe started between ten and twenty billion years ago from an instantaneously expanding point. From nothingness suddenly emerged a burst of energy and matter, or a primeval light ball. 'That exactly corresponds to the Bible,' said Aviezer: 'God said, "Let there be light." ' Until the twentieth century, this line had puzzled scholars, because it advances the concept that light existed before

God created the sun and the stars. But the Big Bang offers the answer: light existed before the stars. And 'everyone worries about a "day",' said Aviezer; 'but, briefly, the word "day" in Genesis doesn't mean twenty-four hours ... days are mentioned before God creates the sun, so they have no correspondence to the earth's rotation or its relation to our notion of time'.

Professor Aviezer is by no means alone among scientists in striving hard to find total agreement between religious views and scientific knowledge. I think that perhaps, on this occasion, some of his audience wondered if this was a worthwhile exercise. But at least he was free to pursue his enquiries; in former ages, as we shall see, he might not have been so fortunate.

Giordano Bruno, burned in a flower market

If you visit the Campo de' Fiori, the 'Field of Flowers', in Rome, you may be rather disappointed. This is not one of Rome's smartest places, nor is it even a field: just a rather dilapidated square lined with bars, in which a scrappy flower and vegetable market jostles for space with street sellers mostly selling cheap toys. But it was once a flower-filled field – a field in which a particularly horrible execution took place. The square is now dominated by a dark and sinister black statue: the cowled figure of Giordano Bruno. This is where he was burned at the stake in 1600.

Born within a few years of the death of Copernicus, Bruno was a man of the Renaissance, when science started to become divorced from theology. Soon scientists would no longer be content with theorizing about the natural world and start experimenting with it. They were developing new toys – chemicals, lenses, clocks, means of measurement. Bruno himself was certainly no scientist; in fact, although he espoused Copernican theories, his knowledge of astronomy was rather flawed. But he did advocate taking a rational view about how to interpret the universe.

Bruno, a Neapolitan, was a misfit from the start – first at school, then within the church, after joining the Dominican order as a friar. As a young man his heterodox views got him into trouble in Rome, and he fled to France. Shedding his Dominican habit, he headed for Paris, where he was to give lectures. He was already a figure of quite some notoriety – people said he had 'peculiar powers' which enabled him to teach how to enhance personal memory – and his reputation intrigued King Henry III, who was fascinated to see if Bruno was some kind of sorcerer. Today, seeing his eerie statue silhouetted against the sky in Rome, dressed in his dark Dominican friar's cloak, it is easy to think of him as some kind of magician. His system to improve recall would now be familiar to modern psychologists, based as it is upon organizing knowledge more effectively. But mental techniques were not Bruno's key interest. He was much more concerned with the deficiencies he saw in the church. He wrote that Christianity was entirely irrational, that it was contrary to natural philosophy and that it disagreed with other religions. Faith through so-called revelation, he maintained, had no scientific basis.

When France did not confer upon him the kind of celebrity status he had expected, Bruno left for new pastures in Protestant England. Here he started to spread the ideas of Copernicus, opinions that most Oxford academics treated with some scorn, preferring the teachings of Aristotle. In the 1580s, in fact, hardly a single teacher in Europe openly supported Copernicus. In England, Bruno had an audience with Queen Elizabeth. He seems to have been enraptured; he thought that she was a diva, a divine, a sacred Protestant Ruler. Such views, later put in writing, were hardly likely to endear him to the Pope in Rome when he was eventually brought to trial as an atheist, an infidel and a heretic. They did not even endear him to the Virgin Queen herself, who thought him wild, radical, subversive and dangerous.

Bruno was storing up trouble for himself. His flagrantly anti-establishment views, forcefully presented, were increasingly attracting notice across Europe, particularly in Rome. He was preaching a heretical philosophy which made the mysteries of the virginity of Mary, of the crucifixion and the Mass, meaningless. Only the ignorant, he asserted, could take the Bible literally. The church, he maintained, encouraged ignorance through its own self-interest.

After a trip to Germany, he decided to return to Italy – a strange choice for, as he must have realized he now was, a wanted man. Once he was there, a rich young man called Mocenigo (presumably a relative of Alvise Mocenigo, a recent Doge) invited him to Venice, that city of duplicity. Mocenigo offered a roof over his head – and then raised charges against him. The Venetian Republic imprisoned him for a while before surrendering him to the church in Rome. Between 1593 and 1600 he lay in a papal gaol; it is unknown whether he was tortured, as seems likely, for the historical records have never been published. Eventually, in 1598, Bruno was interrogated several times by the Holy Office and convicted. He was allowed forty days to 'consider his position'; initially he promised to recant, but then appears to have 'renewed his follies'. Another forty days for deliberation produced nothing but puzzlement for the Pope and bafflement for the Inquisition. After two years in custody, he was taken on 9 February 1600 to the palace of the Grand Inquisitor to hear his sentence on bended knee. Arraigned before the expert assessors and the Governor of the City, he was condemned to a cruel fate.

Bruno answered the sentence of death with what he may perhaps have forlornly regarded as psychological warfare: 'Perhaps you, my judges, pronounce this sentence against me with greater fear than I receive it.' Execution was postponed for eight more days to see whether he would repent. But it was no use. He was tied to the stake and the fires were lit; as he was dying a crucifix was presented to him, but he pushed it away defiantly.

Galileo Galilei: the admonishing finger of science

Those who drive a car in London or any other traffic-bound city may have witnessed a peculiar gesture assertively flaunted on occasion by certain motorists – or, more usually, perhaps, brandished as a signal of protest by rather coarse fellows weaving about on motorbikes. The gesture is a raised clenched fist, held high in the air, with only the wagging middle finger standing upright. Even gentler readers may recognize that this is not the Toby Tall of their nursery years. Modesty forbids my describing what the symbolism may indicate.

Florence boasts a rather nasty object just a little reminiscent of this unofficial traffic sign. In the Museum of the History of Science, some 100 metres from the more visited Uffizi Gallery, there is a small, egg-shaped glass container. It stands 6 inches high, and its glass dome can be lifted off to afford a clearer view of the item it contains: a gruesome piece of bone and skin, mummified over 360 years. Attended solemnly by the museum's staff, who handle it reverentially, and bearing an old wax seal stamped with a coat of arms at the base of the glass container, it resembles nothing so much as a holy relic of a Catholic saint.

Actually, this object is the middle finger from Galileo Galilei's right hand – still pointing vertically in some act of eternal defiance. Galileo was one of the greatest scientists of all time – mathematician, experimenter, astronomer and the founder of modern physics. And, although he was also a Catholic, Galileo's work provided the catalyst for the divorce of scientific enquiry from religion, giving it the freedom even to supersede religion as a means of explaining the world.

Galileo was greatly taken with the ideas of Copernicus who, around 1510, had arrived at the notion that the earth rotated around the sun. Copernicus did not draw attention to his theory for a variety of reasons, not least because he was aware that his astronomical model raised many questions. If the earth moves constantly, why do we not feel the wind in our hair? Why doesn't the motion cause the oceans to overspill?

Copernicus was also hampered by his method: his science was based not on detailed observations, but merely on his use of his own reason to work out what was likely. This was the kind of 'science' practised by the ancient Greeks – performed with fingers scratching chin, rather than manipulating experimental tools in a laboratory. Some of Galileo's early work was conducted in this manner, too. As a fledgling scientist, he calculated the precise location of hell using Dante's *Inferno* as a guide. Using mathematics, he concluded that hell was shaped like an inverted ice-cream cone, whose tip passed through the centre of Jerusalem, and whose roof – 5,000 kilometres thick – was a model of the Brunelleschi Dome in Florence Cathedral. But, crucially, Galileo later realized he was wrong: the maths dictated that a dome of that size could not support its own weight (and Galileo spent part of his later years secretly trying to work out an alternative argument). But the mature Galileo's method of work represented a core change for science: he conducted experiments. For the most part, his theories were based upon direct observation and testing.

In the end, Copernicus' ideas were published in the year of his death – poignantly, the printed work was brought to him on the very day he died, 24 May 1543 – with a cautious preface, written by a Lutheran minister, saying that the model he described was not intended as a description of the universe as it really is, but just as a mathematical device to make calculations simpler. There were reasons to suspect that the church authorities might not like the Copernican model at all. Martin Luther had already objected, and in later years the Catholic Church would take a similar stance – not, it has to be said, specifically because of the model's content, but because it was advanced by Giordano Bruno. As it turns out, although Bruno had been preaching the idea of a heliocentric universe, Galileo did not rate him at all. Rather like the young Bruno, Galileo had joined a monastic order as a teenager. His horrified father promptly extricated him and had him trained as a doctor. Years later, it did not help his case that he was officially listed as an unfrocked priest.

Despite his brilliance, his reputation for questioning accepted truths and a considerable degree of intellectual arrogance, Galileo left university in Pisa without a degree. He scraped a living as a private maths tutor in Florence, conducting experiments with pendulums in his spare time. Lacking any funding, he attached himself to a local aristocrat, the Marquis Guidobaldo del Monte, himself a respected authority on mechanics – and thanks to whose influence Galileo returned to the University of Pisa four years later as a professor of mathematics. Here, a flame-haired awkward customer, he distanced himself from his university colleagues with his attitudes and behaviour, preferring to drink with students in the sleazier parts of town and objecting to the obligatory wearing of academic gowns.

In 1592 Galileo took up a new post in Padua, then ruled by Venice. This brought him into contact with the great and good of Venetian society. He became a close friend of the Venetian playboy Gianfrancesco Segafredo, famous for hosting wild parties at his stylish villa on the Brenta Canal. Galileo became a regular visitor to Segafredo's bacchanalia and adopted a libertine lifestyle. He took up with Marina Gamba and had three children by her, even though they never lived in the same house or married.

He is also credited with various inventions, including the telescope. In fact, a Dutch spectacle-maker had developed the first such device in 1608, and by 1609 copies were being sold as novelty gifts in Paris. Galileo heard rumours of this toy and discussed them with one of his well-connected coterie, the Friar Paolo Sarpi, theological adviser to the Doge of Venice. Sarpi – whose views were regarded by church leaders as bordering on Protestantism – had been excommunicated by Pope Paul V in 1608, along with the Doge himself. As Galileo inched closer towards his most famous discoveries, political circumstances were creating an atmosphere in which they would almost certainly be unwelcome. Galileo was irked to find that his friend – recovering from an assassination attempt by agents of the Pope – had known of the telescope's existence for some time, but not discussed it with him. Putting this irritation aside, he swiftly realized how vital an instrument this could be to his patrons. For Venice, which depended on maritime trade, a device that enables one to see what ships are on the horizon would be a great advantage.

To his horror, while on a trip to Venice, Galileo learned that a Dutchman had arrived in Padua with one of these new instruments. He promptly rushed back, only to learn that the Dutchman was now in Venice, attempting to hawk his wondrous invention to the Doge himself. Galileo knocked out a copy in twenty-four hours – improving it along the way so that the image in the viewfinder appeared upright. He sent a coded message to Sarpi, urging him to delay any decisions about the Dutchman's offering. Sarpi obeyed, giving Galileo time to build a further, deluxe version, with a magnifying power of ten times, which he presented to the Doge in a handsome leather case. The Doge was delighted and offered Galileo tenure at the University of Pisa, with a salary of 1,000 crowns a year. The telescope – upon which Galileo continued to improve, sending versions to various European magnates over the coming years – was to net him a considerable income. It was also to land him in hot water with the church.

Using his best instrument, which now had a magnifying power of twenty times, Galileo could observe the heavens as no man had previously been able. He discovered that the surface of the moon, thought to be perfectly smooth, had craters and mountains. He observed the motion of Jupiter's moons, and discovered that the Milky Way was made up from thousands of individual stars. He reported his findings in *Siderius Nuncius* ('The Starry Messenger'), sealing his reputation throughout the academic world – copies were even translated into Chinese within five years of its first publication in 1610.

But Galileo's observations had begun to point to one unavoidable fact above all: that the earth, along with the other planets, revolved around the sun – contradicting what was implied in the scriptures. After Giordano Bruno had been burned at the stake, Galileo was cautious about making his findings public. His view was that, if he presented his observations without drawing any stark conclusions from them, the church and everyone else would eventually see reason. But perhaps he overestimated how reasonable certain people could be. He had a foretaste of human pig-headedness when he invited a fellow professor at the university, Giulio Libri, to look through his telescope. Libri, one of Galileo's critics, replied that he did not need to, because he knew the truth already. Libri died shortly afterwards, prompting Galileo to remark that, if Libri did not care to look at the heavens while he was on earth, perhaps he might do so on his way up.

Despite Libri's refusal to face facts, Galileo had grounds to be optimistic about the Pope. Visiting Rome in 1611, as the newly appointed scientific ambassador to Tuscany, he was allowed to address Pope Paul V while standing, not kneeling – a gesture of friendship and indulgence. The Pope's chief adviser, Cardinal Bellarmine, looked through the telescope and agreed that everything stated in the *Siderius Nuncius* was true.

This initial warm reception by the Pope gave Galileo false confidence in the acceptability of his ideas about a sun-centred universe. In 1613 he first put his name to them, in an appendix to a book about sunspots. A row ensued and a couple of years later, now dogged by ill-health, Galileo visited Rome again with the intention of clearing the air. He might have hoped things would be resolved with a nod and handshake. Instead, he found the full might of the church stacked against him. Paul V, through Cardinal Bellarmine, instructed Galileo that he must not defend or teach Copernican ideas. If he refused, he would be warned formally by the Inquisition.

Even after this, Galileo was still in sufficiently good favour to have an audience with the Pope, who specifically stated that he need have no fears about his position. Bellarmine backed this up with a sworn affidavit, stating that Galileo had not been punished, but had simply been informed of the edict that was binding on all members of the Catholic faith.

Within a few years, Galileo's position had become more vulnerable. Both Pope Paul V and Bellarmine died in

1621. Tuscany's leaders, rendered less secure by political instability in the region, were less inclined to offer protection to anyone incurring the wrath of the Pope. None the less, Galileo retained some important connections in high places. He had been tutoring a young nobleman by the name of Barberini, whose family were immensely grateful for the care and patience with which he treated his protégé. When, in 1623, the young man's uncle became Pope Urban VIII, with great aplomb Galileo dedicated his new book, *The Assayer*, to the new Pope and decorated it with the Barberini family crest, featuring three bees. Urban VIII was so delighted with this offering that he had the book read aloud to him while he ate.

This unctuous gesture won Galileo favour – but it was not to last. In 1629, his hopes of acceptance high, he published his *Dialogue on the Two Chief World Systems*. The book took the form of an imaginary debate between two characters, Salviati – the name of one of Galileo's deceased playboy friends – who represented the Copernican view, and Simplicio, who presented the older, Ptolemaic view. In choosing the name of a friend as the supporter of Copernican views, and giving the views supported by the church to a man whose name means 'simple', Galileo was making a clever, but unwise dig.

The book was initially approved by the papal censors, who were happy with the bulk of the text but wanted a new preface and conclusion added to the book, making it clear that the Copernican view was presented purely as a theory, not as truth. Fatally, Galileo was given permission to alter or embellish the wording of these additions, as long as the sense was retained. Taking this provision rather too literally, he cast the preface in a different typeface, thereby signalling that the views presented were not his own; and, in the conclusion, where the Copernican view is refuted as a mere theory, he put the words into the mouth of Simplicio, who is, throughout the book, a bit of a numbskull.

Urban VIII, once delighted at Galileo's work, was now being told by his clerical colleagues that he was being labelled a fool – which seems not to have been Galileo's intention. He ordered the Jesuits to gather any damning material on Galileo. They swiftly unearthed the minutes of the meeting in March 1616 at which the errant scientist had been warned not to hold, defend or teach the Copernican model of the universe. This warning had been cleverly worded so that Galileo was, officially at least, forbidden to mention Copernican views, even if his intention was to refute them. A final denunciation was delivered by Father Melchior Inchofer of the Jesuits, who in 1631 declared: 'The opinion of the earth's motion is of all heresies the most abominable, the most pernicious, the most scandalous; the immovability of the earth is thrice sacred; argument against the immortality of the soul, the existence of God, and the incarnation, should be tolerated sooner than an argument to prove that the earth moves.' Galileo was summoned to Rome to stand trial for heresy. Outbreaks of plague delayed him, but eventually, in 1633, he travelled, with a heavy heart and now crippled by arthritis, to answer the accusations.

Of course, Galileo held one trump card – the sworn affidavit from Cardinal Bellarmine. But this would hold no weight with the Inquisition. He had been publicly summoned to stand trial for heresy. If he was not guilty, that meant the Inquisition itself was at fault – which would send out a dangerous message to anyone else in Europe contemplating a heresy. Threatened with torture, sixty-nine years old and wracked with pain, Galileo eventually swore an oath to 'abjure, curse and detest the false opinion . . .' The inflexible decree was lifetime imprisonment.

In fact, this sentence wasn't quite so harsh as it sounds.

Galileo was allowed to remain at his own, well-appointed home near Arcetri, a large, galleried villa with a fine garden in the hills overlooking Florence and within walking distance of the city. Here he continued to work in secret until his death in 1642. A work produced during these years, *Two New Sciences*, put forward the view – later developed by Isaac Newton – that the entire universe is governed by laws which can be understood by human reason, and driven by forces which can be predicted by mathematics. This work was smuggled out of Italy and published in the Netherlands by Louis Elsevier, becoming a vital influence on the development of science across Europe – as well as giving Elsevier's house a reputation for science publishing which endures to this day.

Bizarrely, it was not until 1822 that the Catholic Church's ban on Galileo's greatest work, the *Dialogue*, was lifted. This perhaps demonstrates something about the inflexible nature of Christian belief in former centuries, as fervently opposed to science as it was to the viewpoints of other religious systems. When Cardinal Bellarmine attacked the Copernican theory, he did so on the grounds that it allowed no room for heaven and hell. Hell, he argued, must be at the centre of the earth due to 'natural reason'; it must be at the furthest possible distance from the dwelling-place of the angels and departed souls. At the same time as he put forward this view, the Islamic scholar Mullah Sadra was arguing that both heaven and hell were located within the imagination of each individual; meanwhile, Jews were happy that hell did not actually exist at all, and Jewish Kabbalists were arguing that the biblical account of Creation should be interpreted symbolically, not literally. Western Christian teaching, by contrast, had a dangerous tendency to nail down its beliefs and enshrine them in written authority, rendering them rigid and thus, perversely, open to attack.

There is a curious, ironical twist in Galileo's story. If you drive for about ninety minutes from the centre of Rome past Fiumicino Airport, you come eventually to the delightful village of Albano. Above you, on the lovely hillside, is the papal summer residence, Castelgandolfo. Looking up to the Castle you see the dome of the residence, and two rather more modern domes. These mark the rooms where the telescopes of the Papal Observatory are housed. The excellent Jesuit astronomers there have made observations that are respected around the world. How extraordinary, though, that in the larger of the two domes there is a 20 centimetre refracting telescope used to study the heliocentric universe. It was built in 1933, just 300 years after Galileo's arraignment by the Inquisition. Perhaps Galileo's rebellious finger is still wagging just a little.

God and nature: a rather English compromise

After the English Reformation there was a sizeable change in how religious Protestants began to view the Creation. One great hero of mine is Ralph Cudworth, Regius Professor of Hebrew at Cambridge University, who during the Commonwealth period advised Cromwell to allow the Jews (who had been expelled 360 years earlier) to re-enter England. Although my family were not among the very first to arrive from Amsterdam, some of my mother's forebears benefited, being admitted from Holland within five or ten years of Cudworth's death. A branch of my father's family came from Prussia about five years after that.

Cudworth was a Fellow of Emmanuel College, Cambridge. In 1645 he became Master of Clare Hall. In 1654 he transferred to Christ's College, where he was Master until he died. He typifies a certain trend in the academic thinking of his time. One of the Cambridge Platonists, he was a leading opponent of the political philosopher Thomas Hobbes. Hobbes, who had met Galileo and Descartes during his sojourn in Europe, wished to explain the entire universe on materialistic principles. His brand of atheism denied the existence of anything divine in man, whose life in its basic state he famously denigrated as 'poor, solitary, nasty, brutish and short'. Perhaps not entirely surprisingly, when actually facing death at the age of ninety Hobbes affiliated himself to a church and became rather less dismissive.

Cudworth's sermons, sometimes delivered to Parliament, were part of what was to become a long Anglican tradition of toleration and charity. His great work, entitled *The True Intellectual System of the Universe*, was published in 1678. It is, by any standards, pretty impenetrable – partly because Cudworth was so erudite that he included too much indigestible detail – but it eventually made his reputation after his death, even though he left it unfinished. In it, he proposed to prove three things: (a) that God existed; (b) that moral principles were natural; and (c) that human freedom was real. These three propositions were, in his view, opposed respectively by three false principles: atheism; religious fatalism, which refers all moral distinctions to the will of God; and the fatalism of the ancient Stoics, who recognized God and yet identified Him with nature.

The work was been heavily criticized for its lack of accessibility, one modern scholar remarking that 'its argument is buried under masses of fantastic, uncritical learning, the work of a vigorous but quite unoriginal mind'. As Henry, Lord Bolingbroke, a politician and contemporary of Cudworth, said, 'he read too much to think enough, and admired too much to think freely'. Nevertheless, Cudworth is now thought of as a man of considerable scholarship and great integrity. Professor

Andrew Dickson White, a nineteenth-century scholar of Cornell and Chicago universities, called him 'one of the greatest glories of the English Church' and said: 'His work was worthy of him. He purposed to build a fortress which should protect Christianity against all dangerous theories of the universe, ancient or modern.' His contribution was to suggest that there was a plastic nature to the universe: God set things in motion, but thereafter the universe behaved according to its own laws. This view influenced Spinoza and the nineteenth-century philosophers, and in reality is not so very different from that held by many religious people today.

Not a jot of it

The Buddha told his disciples that speculation about the nature of *nirvana* – the liberated afterlife of the soul – was 'improper': what mattered was trying to achieve it. In Islam, as we saw in chapter 6, debating the nature of Allah is condemned as *zannah* – a waste of time, equivalent in offence to worshipping false gods. And Jews, perhaps by the very nature of the history of their existence, have never felt particularly unhappy about uncertainty. Jewish tradition does not hold a single worldview; the Talmud offers many conflicting interpretations of the world in which we find ourselves.

Meanwhile, in eastern Christianity, an important distinction was drawn between *kerygma* and *dogma*. *Kerygma* is the term given to the official teachings of the church, which can be written down and passed on. But *dogma* is a far more slippery concept, expressing the deeper meaning of the church's teaching, something which can be grasped only through direct, individual experience. A useful parallel might be the difference between a proverb and a truth we have learned in and through our own lives. As a boy, I practised handwriting by copying out the phrase, hundreds and hundreds of times, 'Least said, soonest mended' and, despite the slightly archaic English, I knew perfectly well what it meant. However, it took thirty or so years of letting my mouth run away with me to embed the idea as a principle to apply to my own life. *Kerygma* is like a proverb, which can be learned by rote; *dogma* is a truth that can only be accepted through personal experience. Eastern Christianity makes use of this division even in the design of its churches – an iconostasis, or screen, separates the congregation from the altar, symbolizing the hidden mysteries of the faith.

But western Christianity has always been unhappy with mystery, seeking to struggle openly with, and interpret explicitly, the numerous paradoxes and contradictions of its theology. The idea of the Trinity – God the Father, Son and Holy Ghost – provided generations of Christian head-scratchers with ammunition for debate, because it posed such obvious philosophical problems. If God is one, how can He also be three? If Jesus is the son of God, does that mean He is somehow less than God, or equal to Him – and if the latter, how can God be described as all-powerful? If Jesus is the same as God, who is beyond time and all-powerful, how could He descend into a moment of history and die, suffering, on the cross? For the eastern church, these paradoxes were meant to be left alone – a means of holding the mind in a state of wonder, serving as a reminder that no human intellect can ever truly grasp the reality of God. To try to reduce the mystery of the Trinity to a set of logical propositions was as futile as trying to describe the smell of new-mown grass.

As we have seen, western Christianity tried to weather the storm of all this speculation by returning, at every point, to the literal meaning of the Bible. Whatever the Bible meant was true. Thus, in 1576, the Spanish Inquisitor General uttered the words: 'Nothing may be said that disagrees with the Bible, be it a single period, a single little conclusion, or a single clause, a single word of expression, a single syllable or one iota.' But eventually this notion clashed in a dangerous way with other movements in the history of ideas. During the Renaissance and Reformation periods, as we have seen, there was a revived interest in the powers of human reason celebrated by the ancient Greeks and Romans. Reformers like Luther and Calvin argued that each individual had the power, and the duty, to read the scriptures. The result was that people began to apply reason to texts that were thousands of years old but not written with reason in mind.

About the time of Galileo, the Jesuits, the super-educated unit of the Catholic Church created to combat the Protestant Reformation, had a key spokesman in Leonard Lessius of Louvain (1554–1632). Lessius was a true son of his age, who believed that entrepreneurs had a divinely given right to earn more money than workers. He is better known, however, for arguing that the existence of God could be demonstrated scientifically, through observation of the universe. The design of the natural world, from the snowflake to the structure of a mighty pine tree, points to the work of some intelligent agent. There was nothing specifically Christian about the God that Lessius envisaged – indeed, his works make little mention of Jesus. He began to speak of God as if he were an observable principle of the universe, like the law of gravity or the movements of the planets. For many people – sometimes myself included – this 'cosmological' argument for the existence of God is not unpersuasive. It's easy to feel, when looking at the intricate way a cell is formed, or at the dazzling vista from a mountain top, that some supremely clever Agent has been at work.

The danger comes when we start to try to argue the case for religion on rational grounds; and it is intensified when we try to test it by experiment and observation. Scientists like Galileo based their truth on the evidence of their own senses and their reason. They were self-sceptics, ready to abandon their theories and return to the drawing board if their tests proved them wrong. So it was that man's powers of reason, championed by the ancient Greeks, and seen by Martin Luther as a divinely given tool for understanding the scriptures, became the means with which people began to dismantle those very scriptures, and the beliefs they represented. Many scientists, as we shall see, took up Lessius' ideas and embarked on a search for God – looking for Him in the laws and forces of the observable, natural world; and, when they didn't find Him, some were forced to conclude that He'd never been there at all.

Gambling on God

I need to confess. I hardly ever travel by bus. But I was recently compelled, admittedly partly out of nostalgia, to travel on one of the last Routemasters crossing London. For those not in the know, the deep red Routemasters, with their rounded roofs and open back ends, are the iconic London double-decker buses, seen in countless films from Ealing comedies to Hollywood blockbusters. Until recently, they were one of our congested capital city's more delightful sights. Sitting on the bus with my head full of *sic transit gloria mundi*, I tuned in to the conversation going on behind me – reflecting how bus travel lends itself so much better to eavesdropping than tube travel, where everyone sits facing one another in mute and mutual embarrassment. Two young men – rather improbably for Charing Cross Road at 1 p.m. on a wet Wednesday – were discussing the afterlife (maybe it was the death of the Routemasters that had sparked their conversation). One was adamant that nothing of the sort could exist. The other also felt that it defied sense, but said he was keeping an open mind. 'If there ain't nothing, right, then I don't care, 'cos I won't know nothing, will I? But if there is something, then I'll be laughing.' His sceptical companion noted archly that the Christian religion was not that tolerant. 'You gotta *believe* in it, otherwise they won't let you in.' The second young man was undaunted. 'Makes sense to believe in it, then, dunnit? Just in case, like.'

I had just been reading an account of the life of Blaise Pascal, but I refrained from giving in to a monstrous urge. I did not turn round to tell this man that his views echoed those put forward by a devout Catholic scientist over four hundred years ago. Possibly the observation would not have been welcome. But Pascal's ideas were typical of an era when men were beginning to feel that the existence of God could no longer be taken for granted.

Pascal was born into a wealthy family in Clermont-Ferrand in 1623. Losing his mother at the age of three, he was educated at home by his free-thinking father Étienne, who decreed that the boy was not to learn mathematics until he was fifteen. Blaise began to study in secret (even maths can be fun if you make it illicit), and, when he was fifteen, amazed his father by demonstrating that he had worked out the first twenty-three geometric propositions of Euclid. At sixteen the young prodigy, despite the ill-health that often dogged him, published a paper on geometry so sophisticated that many leading scientists refused to believe it was the work of one so young.

In 1639 the Pascal family moved to Rouen, where Étienne was employed as a tax collector. His young son devised the first digital calculator to help him with the necessary sums – a model that became known as the Pascaline. It was technically advanced, but never took off – largely because it was more logical than the peculiarities of the French currency system.

When Pascal was twenty-three years old, his father

injured his leg and needed round-the-clock care. Two young men from a religious order known as the Jansenists came to the house to help the family – and they had a lasting spiritual influence on Pascal and his younger sisters. The Jansenists were a Roman Catholic order condemned by the official church as heretics because they denied the authority of the Pope, believing, like Calvinists, that God alone ordained whether or not men would enter heaven. The Jesuits were particularly opposed to their teachings, but the Jansenists had support from many of the wealthier elements of northern French society. It was said that, in their stern devotion to hard work and their avoidance of all forms of fun, the Jansenists outstripped even the Calvinists.

Pascal's scientific interests were wide-ranging, covering probability and the nature of atmospheric pressure. But his most memorable work arose out of a shattering event that occurred in October 1654. While he was crossing a bridge across the Seine, one of his horses bolted and his carriage was left dangling over the river. Miraculously, Pascal was rescued without injury; but the experience shook him profoundly. How profoundly did not become apparent until after Pascal's death in 1662, when a secret piece of paper was found in his waistcoat. This detailed a mystical experience on the night of 23 November 1654, which had fuelled Pascal in his zealous embrace of Jansenism. The fragmented, awe-filled description of this encounter, subsequently published in his *Pensées*, is powerful in itself: 'Depuis environ dix heures et demie du soir jusques environ minuit et demi. FEU. Dieu d'Abraham, Dieu d'Isaac, Dieu de Jacob – non des philosophes et des savants. Certitude. Certitude. Sentiment. Joie. Paix' ('From about half past ten in the evening until about half past twelve. FIRE. God of Abraham, God of Isaac, God of Jacob and not of the philosophers and savants. Certainty. Certainty. Emotion. Joy. Peace.')

Pascal began to pay visits to the Jansenist convent at Port Royal, outside Paris, where his sister was a nun. He became a passionate defender of their cause, publishing eighteen satirical attacks on the Jesuits. It was during this time that he also began to formulate the theory now known as Pascal's wager. Like Leonard Lessius, Pascal felt that belief in God was rational. Like my fellow passenger on the bus in Charing Cross Road, he sought to prove this by arguing that it made far more sense to believe in God than not to: 'We are incapable of knowing either what [God] is or whether he is . . . Reason cannot decide this question. Infinite chaos separates us. At the far end of this distance, a coin is being spun, which will come down heads or tails. How will you wager?'[1] Pascal argued that believing in God was a win–win situation. If He did not exist, one lost nothing by believing in Him. But if He did, one risked a great deal – eternal life – by not believing in Him.

This may seem quite reasonable and unremarkable now, but it is worth remembering that at the time Pascal put forward his argument, the idea that there was no God was probably an immensely radical suggestion. Even philosophers like Hobbes were rare. From around a hundred years before, people had begun to accuse one another of 'atheism', but they did not mean this in the strict sense of not believing in the existence of God. To do so was, literally, unthinkable. The church held sway over every corner of existence – and the 'worship by work' of the Protestants meant that even the most humdrum activities were coloured by a belief in God. To suggest that He did not exist would have been as unfamiliar as saying there was no such thing as an individual personality – an idea that severely challenges modern westerners who tackle the teachings of the Buddha.

In Pascal's time and before it, people accused one another of atheism when their opponents' views differed from established church teaching. The Romans called the Jews and Christians atheists because their ideas of a single God clashed with their own populous pantheon of divinities. Giordano Bruno was called an atheist because he was committed to a God deemed unacceptable by the Catholic Church. 'Atheist' was used as an insult, adopted by the English dramatist Thomas Nashe to brand anyone whose behaviour was beyond the pale – hypocrites, the ambitious, the greedy, the gluttonous and the promiscuous.[2] For a deeply religious man like Pascal to come up with the idea that God was worth a bet suggests that some powerful changes were afoot in the landscape of the western mind. And indeed they were: ideas related not just to the development of technology and scientific enquiry, but to the whole notion of the individual as the true centre of the universe.

Advances in communications, transport, measurement and medicine had brought about a seismic shift in the way people viewed their relationship to the universe, and, correspondingly, to God. In earlier ages, civilizations had been built upon the power of certain elites in controlling agriculture – and they had been doomed to collapse once they grew beyond the limits of their resources. Famine and flood, inherent elements of the natural world, could break an empire. But the technologically advanced western world was no longer quite so strictly tied to the vagaries of nature. The ability to store up capital and reinvest it in commercial enterprises meant that western nations could survive catastrophes like the plague intact. A mind-set emerged in which people saw themselves less as resting in the hands of God, and more as masters of their own fate. Concepts such as change, development and progress began to be seen as both natural and desirable, whereas in former ages men had thought that continuity with the past, maintained through strict loyalty to church teachings, was of paramount importance.

Technological advances also brought about increasing specialization. Today, as a researcher in human reproduction, I would be viewed askance by my scientific colleagues if I started to publish papers on engineering or heart surgery. I would be similarly sniffy if someone of the brilliance of Dawkins or Hawking began to investigate the origins of life in the womb. This outlook is a direct result of changes taking place in Pascal's era. In former centuries, it was possible for a Renaissance man like Leonardo da Vinci to interest himself in engineering, physics and anatomy, as well as painting, and to excel in all these fields. But by Pascal's time it had become increasingly hard for an expert in one discipline to consider himself an expert in any other. Different branches of the scientific community, though they would often co-operate, could now also view one another with suspicion. Thus it was possible, when the philosopher and physicist Descartes met the brilliant young Pascal, for the older man to comment, witheringly and most unfairly, 'He has too much vacuum in his head.' This attitude brought about a culture where scientists saw themselves less as part of a community and more as individual pioneers, each forging his solitary path towards the truth. They began to apply this kind of thinking to their own beliefs as well, rejecting centuries of religious teaching and tradition in favour of trying to work out for themselves whether the idea of God had any use for them.

The Intelligent Mechanick

Not every scientist found God absent from the new, observable universe. A good example of one who did not was Isaac Newton, whose unconventional faith both hampered his career and sustained him through his most important discoveries.

Born in 1643, shortly after Galileo died, Isaac Newton grew into a difficult, dreamy young man. Bullied at school because of his small stature – his mother commented that at his birth he could have fitted into a quart jar – he seemed to have no aptitude for the farming life into which he was born. His early attempts at land management ended in disaster because he spent so much time reading books instead of taking care of the family's livestock. As a result, he was fined on several occasions for allowing his animals to wander onto his neighbours' lands. None the less, he displayed an aptitude for making models, devising a working flour mill and, famously, creating a paper lantern attached to a kite, which caused mayhem in his Lincolnshire village when locals saw a ghostly light floating in the sky.

In 1661 Newton won a scholarship to Trinity College, Cambridge. His no-nonsense, widowed mother refused to fund what she saw as a time-wasting exercise, and he was compelled to enter the university as a subsizar. The subsizar system, now long defunct, was rather like the custom of fagging at England's public schools: impoverished young men like Isaac were compelled to work as unpaid manservants to other, wealthier students. Understandably, then, his early years at Trinity were pretty unhappy – but things picked up when he met a young man called Nicholas Wickes, who was to remain a friend, and a room-mate, for the next twenty years. It has often been suggested that Newton, who never married, and whose closest relationships were always with men, was gay. Perhaps he was; the detail is only significant in that, throughout his life, he was a reclusive and shy figure. In a more tolerant age, perhaps he might not have been so – but then, he might not have turned into such a brilliant scientist if he had been open to other distractions.

Settling down at Cambridge, Newton rose swiftly through the ranks of academia, gaining a reputation as an obsessive scholar, who frequently forgot to eat or sleep. While carrying out his numerous enquiries into the nature of light and gravity he nearly blinded himself twice – once while staring at the sun for too long, and once while poking around his eye with a bodkin to observe the resulting images when the eyeball was distorted.

Newton began to consider that the universe was governed by a set of observable, predictable laws. He was a religious man, but his views brought him into conflict with the university authorities. He followed the teachings of the fourth-century heretic Arius, who, as we saw in chapter 5, maintained that Jesus was not truly divine. In seventeenth-century England this was no longer a burning offence, but it did create problems for a young man making his way as an academic. At that time, Fellows of Trinity were required to become priests and to swear an oath that they would 'set Theology as the object' of their studies or 'resign from the college'. Newton was lucky in that the restored king, Charles II, took a keen interest in science and, as patron of the newly founded Royal Society, was prepared to make an exception for a man whose reputation was rising thanks to his discoveries concerning the nature of light.

During really boring debates in the House of Lords, I sometimes slip out of the Chamber into the well-stocked library. One of the treasured volumes which I have dipped into there is Newton's most famous work, bound in vellum: his mammoth, three-volume *Philosophiae Naturalis Principia Mathematica*, first published in 1687. In it, he set out three laws of motion describing the behaviour of everything in the known universe. Its success was due to the fact that, for the first time, a learned scholar was demonstrating the principle towards which generations of scientists, from Copernicus onwards, had been groping: namely, that the universe did not work according to the capricious will of God or gods, but in obedience to mechanical principles understandable by man.

Newton, however, avoided incurring the wrath of the authorities – unlike so many of his predecessors – because his theory allowed room for God. For Newton, God was an 'Intelligent Mechanick', who had devised the laws and set them in motion – rather like someone winding up a clockwork device which then runs of its own accord. His discovery of the laws of the universe merely proved that some supremely clever power must have put them there in the first place. As he said in a letter to his friend Richard Bentley, the Dean of St Paul's Cathedral, 'Gravity may put ye planets in motion, but without ye divine power it could never put them into such a circulating motion as they have about ye Sun and therefore, for this, as well as other reasons, I am compelled to ascribe ye frame of this Systeme to an intelligent Agent.'[3]

As a result of his desire for legitimacy in the eyes of his fellow Trinity scholars, Newton embarked upon several forays into theology. He believed, for example, that God was not the creator of space and time, but *was* them. Matter, on the other hand, was created by God – it was parts of space (in other words, of Himself), given the qualities of shape, density and motion. It was thus possible for Newton to remain a scientist while standing by the traditional Christian idea that God had created the world from nothing.

But Newton's focus was upon the observable world, not the heavens. He therefore considered that religion, if it had any value, must be purged of anything that ran contrary to the natural laws of existence. In the 1680s, before publishing the *Principia Mathematica*, Newton had put forward his *Philosophical Origins of Gentile Theology*. This book argued that Noah had founded an original monotheistic and rational religion, based upon the worship of a God observable in nature. Later generations had corrupted this pure message with fanciful tales of miracles and magic. God, in turn, had sent various prophets to call us back to the truth – Pythagoras was one, Jesus Christ another.

Newton's ideas were well received in an age when people were beginning to accept that religions had a history just as everything else in the universe around us did. From the 1690s into the next century, radical theologians like John Toland and Matthew Tindal urged people to return to a truer, rational faith, purged of all superstition and supernatural event. In his 1696 work *Christianity Not Mysterious* Toland argued that a preoccupation with mystery caused 'tyranny and superstition'. It was heretical, Toland argued, to suggest either that God was unable to express Himself clearly to men, or that He deliberately shrouded His message in clouds of mystery and wonder. Tindal claimed that 'there's a religion of nature and reason written in the hearts of every one of us from the first creation'.

Given that Newton was a genuinely religious man, there is a curious irony about his story that is beautifully summed up in an extraordinary painting by William Blake, now in Tate Britain. In 1795, seventy or so years after the scientist's death, Blake, the great mystic poet and artist, engraved the famous copper plate of Newton which he magnificently enriched with deep-hued watercolours. The image is familiar – it provided the inspiration for Eduardo Paolozzi's massive statue of Newton outside the British Library. In Blake's original, Newton – a huge, richly muscled naked giant of a man – is seated on a rocky outcrop and bending forward, holding the twin arms of a measuring compass in his left hand. His gaze is intent and fixed on the white area he is measuring – possibly the prism which split white light into its component parts. At first glance it looks as if Newton is a great hero, strongly built, supreme and concentrating with God-given intelligence. But the picture is deeply critical. This huge man, this 'master of the universe', is sitting on a rocky outcrop, a coral reef. As the viewer looks at the multiplicitous natural forms and the rounded excrescences of this deep blue-green reef, studded with every colour imaginable, the colours start to coruscate. But Newton has turned his back on the beauties of the natural world and is intently focused on splitting white light – white light, that mysterious emanation from creation that represents the fingers of God. Newton, Blake appears to suggest, has debased this great wonder with his technology and destroyed its mystery.

Perhaps Newton feared this himself. Certainly, he grew increasingly nervous about his science when the Catholic James II succeeded his brother, ascending the throne in 1685. Now a Senior Fellow of Trinity, Newton was among those who were alarmed when James II tried to extend his influence over the university. His protest led to his being summoned before the much-loathed 'Hanging Judge' Jeffreys to defend his viewpoint. The stress of this experience seems to have led to a nervous collapse in 1693, followed by Newton's decision to quit academia altogether. He ended his working life, not as one of the most brilliant scientists ever produced by England, but as a high-ranking official at the Royal Mint. But to this day his three laws of motion are considered reasonable approximations to the truth, and we use them in a variety of applications. When NASA scientists guided the *Galileo* space probe around Venus and the Earth to give it the necessary acceleration for its journey to Jupiter, they were making use of his laws of gravity. And, in the early twentieth century, these laws provided the foundation for what must be one of history's most bizarre experiments.

Two silver dollars for a soul

In November 1901 Duncan MacDougall, a doctor working at a hospital in Massachusetts, outlined his view that the human soul must occupy space, and could therefore be weighed. A patient dying of tuberculosis was placed upon a specially constructed bed, resting upon scales of MacDougall's own design. During the three hours before the patient's death, MacDougall repeatedly weighed the man, observing that he lost weight at a rate of one ounce per hour. Dr MacDougall concluded that this weight loss was due to the evaporation of breath and sweat.

Then, at 9.10 p.m. exactly, the patient died – and, according to MacDougall's account, 'coincident with the last movement of the facial muscles, the beam end dropped to the lower limiting bar and remained there without rebound as though a weight had been lifted off the bed'. MacDougall observed that, when he reset the scales, it took a weight equivalent to two silver dollars, or 21 grammes, to restore the scales to equilibrium. He tried the experiment with a second subject, once again a man dying from consumption. In this case, the soul appeared to be heavier – the patient lost a weight of over one ounce at the moment of death. MacDougall repeated the experiment a further four times, recording losses of half an ounce, three-eighths of an ounce and, on one occasion, no observable loss at all, because the patient died before he could calibrate his scales. Perhaps this subject had entered into a Faustian pact with the devil.

Later on, MacDougall experimented with fifteen dying dogs; devotees of the canine will be saddened to hear that these subjects showed no weight loss at the moment of death, suggesting that dogs have no souls.

MacDougall wasn't sure what his results indicated. He kept his findings to himself for five years – having experienced opposition to his experiments at his hospital. He had already noted that, while he was trying to weigh the body of one of his patients, 'there was a good deal of interference from people opposed to our work'. His hand was forced when news of his experiments eventually leaked out. Rather than be tried, sentenced and hanged in the tabloid press, MacDougall published an account of his work in a professional journal, *American Medicine,* under the title: 'Hypothesis Concerning Soul Substance Together with Experimental Evidence of the Existence of Such Substance'.

To many Christians and Jews, MacDougall's experiments would seem slightly laughable. There is a long-standing view that the soul has no material substance. Hebrew has three words for the soul: *ruach*, *nefesh* and *neshamah*. None of these terms implies anything remotely measurable in what seem the rather simplistic terms considered by Dr MacDougall. Around one thousand years earlier, somewhat greater men pondered on the nature of the soul. Saadia Gaon of Sura, who can be regarded as the founder of Jewish scientific activity, took a more sophisticated approach. He, like most later scientific and philosophical Jewish sources – men such as Solomon Ibn Gabirol, Abraham Ibn Daud, Hasdai Crescas, Judah HaLevi and Judah Abrabanel (most of whom, incidentally, as noted earlier, wrote in Arabic) – regarded the soul as immaterial, often connecting it with intellect.

In MacDougall's experiments, however, and his critical attitude to them, we see one of the crucial differences between 'religious thinking' and 'scientific thinking'. Both begin as statements of faith: a religious person might say, 'I believe that suffering is rewarded'; a scientist might say, 'I believe that objects expand when they are heated.' The difference is that a religious belief continues to be held as a means of making sense of the world, whatever happens. If my suffering is not rewarded, for example, I might conclude that I haven't suffered enough, or that

I am not trying hard enough to see what 'reward' I am being given, or that the 'reward' will occur in another life. In contrast, a scientific belief – better called a hypothesis – is one which a scientist continues to hold only in so far as his observations shore it up, and is rejected if and when his observations suggest something to the contrary.

Scientists like Newton based their findings on reason. Another way of expressing that might be to say they drew their conclusions from the evidence of their senses – whatever they could see, touch, taste, smell, hear and measure. But that does not mean that science only accepts common sense. We now know, for example, thanks to electron microscopes, that the interior of a crystal is mostly composed of empty space. But yet we experience crystals as hard, solid objects, not as full of holes. How can this be? It's because our sense-organs have developed in a certain way, a way that helps make the best possible 'sense' of the world around us. Because crystals are composed mostly of space, some particles are tiny enough to pass through them and out the other side – but we are not; so our brains, helpfully, construct objects like crystals as solid, because if they didn't, we would hurt ourselves! This feature of the brain is rather like the difference between varieties of maps. If I ask a friend to draw me a map explaining how to get somewhere, he will probably sketch something based upon the direction I am coming from, on the mode of transport I am using, and outlining the sort of landmarks he and I are likely to notice – an off-licence, for example, rather than a mulberry bush. If he just put an A–Z or an Ordnance Survey map into my hand, that might well be far less useful. Our senses function in the same way, providing a sort of helpful map, rather than a picture of the world as it really is. For that reason, science sometimes draws conclusions that defy 'common sense'.

In the rarefied world of quantum physics, which investigates the properties of matter, researchers are continually confronted with findings that shake their conceptions of reality. Can you believe, for example, that it is possible for an object to exist in two places at once? That a particle can exist without having any mass itself but, at the same time, can give mass to other particles? That light can behave like a wave and also as a particle? I find these concepts hard to grapple with, but I am compelled to accept them because the maths they are based on works. It is the same maths that permits me to fly to Los Angeles, and to listen to my personal CD player. Science can involve an act of faith – but it is a faith that's qualified, a faith that rests upon certain other things that I know, rather than just believe, to be true.

I am cautious in setting out any history of scientific discovery, because it lends itself to the idea that, at a certain point in our development, we 'started to become rational', and abandoned belief in God as a consequence. To say that would be a gross over-simplification. At certain points in evolution our brains underwent a sort of seismic shift, growing in size and giving us new abilities – such as language or tool-use. But no such shift ever occurred where the use of 'reason' was concerned; or if it did, it was back in the very distant past, not in the fifteenth, sixteenth and seventeenth centuries.

Anthropologists have devoted a lot of energy to puzzling over this conundrum. It seems odd that, if we accept that all humans have the same brains, one group went on to develop a form of science that works – cures headaches, flies planes – whereas others remain tied to beliefs in the supernatural, which frequently let them down. I trust no-one will accuse me of being 'ethnocentric' if I say that blaming a child's sickness on angry ancestor-spirits might serve a social purpose, but does not cure the child. The school of thought known as 'cultural relativism' suggests that I am wrong – or rather, that I am asking the wrong question. There is no objective truth, the cultural relativist would counter; my scientific truth is simply a western notion incompatible with the magical–religious truth of distant tribes. In answer to this, I would have to side with religion's most vigorous critic, Richard Dawkins, who wrote: 'Show me a cultural relativist at 30,000 feet and I'll show you a hypocrite. If you are flying to an international congress of anthropologists ... the reason you don't plummet into a ploughed field is that a lot of Western scientifically trained engineers have got their sums right.'[4] Dawkins is not wrong – and, of course, his criticism is being directed at western academics who, in his view, should know better. But given that science works, does that mean anyone who believes in the supernatural is plain daft?

A virus or reason gone wrong?

In *The Golden Bough*, published in 1908, James George Frazer argued that magic, religion and science were points upon a line of evolution in human thought. He built upon the work of the eighteenth-century philosopher David Hume, who argued that there were three ways in which the human mind connects ideas: by resemblance, by contiguity in time and space, and by cause and effect. Magic, argued Frazer, made use of all of these, and was therefore pre-eminently rational. Its crucial mistake was to imagine a sort of cause-and-effect connection between things which are similar, or in contact with one another. For example, there is a folk belief that placing a frog upon one's tongue will cure a sore throat. This is based upon an idea that the croaking frog has a sore throat itself, and that contact of like with like will bring about a cure. In the same way, the ancient Greeks used to show a certain bird with yellow feathers – a stone curlew – to people suffering from jaundice, in the belief that the two 'yellows' would somehow cancel one another out. Homoeopathy – a branch of 'folk medicine' so popular that you can buy its remedies in high street chemists today – in fact rests on the same form of magical thinking; its very name means 'cure by same'.

For Frazer, magic was simply science that didn't work. It had an underlying logic, but it got the connections wrong. But if it didn't work, why did people continue to believe in it? Why show a yellow-feathered bird to someone with jaundice if they don't get better? Why bother invoking the spirits to explain and cure an illness? The Catholic anthropologist E. E. Evans-Pritchard felt that his work among the Azande people of the Sudan provided an answer. He noted that there were two main reasons why people failed to see that magic doesn't work. The first lay in the supposed reasons *why* the 'magic medicines' didn't work. The Azande were quite prepared to acknowledge failure in some cases; but their scepticism extended only to certain medicines, the conditions they were applied in and the magicians using them. 'It failed because So-and-so had sex with his wife the day before he gathered the plants' might be a typical explanation: one which kept the overall belief system intact. It was also possible to ascribe the failure of a magical practice to the stronger (magical) powers of an adversary. Second, Evans-Pritchard noted that magic always accompanied deliberate, goal-oriented action. When brewing beer, for example, a man invoked certain spirits to bless his enterprise and make sure the beer tasted good. But he still followed the correct recipe and went ahead and made the beer. No Azande man would try to make beer through magic alone.

For some scientists, it is not enough to say that belief in the supernatural is just a form of faulty science. For Oxford geneticist and professional God-baiter Professor Richard Dawkins, religion contradicts everything he stands for. He classifies it as a 'virus of the mind', which replicates itself with alarming frequency from one generation to the next. He lists some of its symptoms:

1   The patient typically finds himself impelled by some deep, inner conviction that something is true . . . a conviction that doesn't seem to owe anything to evidence or reason . . . Doctors refer to such a belief as 'faith'.
2   Patients typically make a positive virtue of faith's being strong and unshakeable, in spite of not being based upon evidence. Indeed, they may feel that the less evidence there is, the more virtuous the belief.
3   A related symptom . . . is the conviction that 'mystery' per se is a good thing. It is not a virtue to solve mysteries. Rather, we should not solve them, even revel in their insolubility.
4   The sufferer may find himself behaving intolerantly towards vectors of rival faiths . . . He may also feel hostile towards other modes of thought that are potentially inimical to his faith, such as the method of scientific reason, which could function like a piece of antiviral software.[5]

For Dawkins, nothing could be greater evidence of religion's viral qualities than the statement by Tertullian, 'Certum est quia impossibile est!' ('It is certain because it is impossible!') He sees this thinking not so much as lunacy, but as perverse mental acrobatics, with participants vying with one another – rather as the White Queen vies with Alice in Lewis Carroll's masterpiece – to see who can believe the least believable thing. 'Any wimp could believe that bread symbolically represents the body of Christ, but it takes a real, red-blooded Catholic to believe something as daft as the transubstantiation.'[6]

Professor Dawkins points to events like those of 11 September 2001 to demonstrate the danger of religious belief. He believes that the human psyche has two particularly dangerous 'sicknesses' – the urge to carry out vendettas across generations, and the urge to see people as groups rather than as individuals. Judaeo-Christian religion, in his view, mixes explosively with, and justifies, both of these, and should therefore be rooted out, rather than treated with respect. I think this a little harsh. The men who flew jet-planes into the Twin Towers had a political motive, as well as a religious one – they probably did not see themselves solely as 'agents of Islam' or entrants to paradise. They opposed America's global domination and possibly its involvement in the Middle East. Second, if Dawkins accepts that the human psyche has certain sicknesses within it, why does he suppose these vanish with the abolition of religion? The Nazi regime was deeply suspicious of Christianity, and sanctioned atrocities, not in the name of God, but in the name of a country and a people.

But I admire Professor Dawkins for many reasons. He is a deeply humane person, and this is sometimes forgotten by the religious people who criticize him. He is also one of the greatest living thinkers about science. And his certainty is, in a kind of way, admirable. For him, as for many, the Darwinian idea of evolution across successive generations, of the automatic selection of qualities best suited for survival and reproduction, provides an unassailable answer to the questions of existence, and negates any need for God. His friend, the late Douglas Adams, author of *The Hitch-hiker's Guide to the Galaxy*, describes his own brush with Darwinism thus:

> It fell into place. It was a concept of stunning simplicity, but it gave rise, naturally, to all the infinite and baffling complexity of life. The awe it inspired in me made the awe that people talk about in respect of religious experience seem frankly silly beside it. I'd take the awe of understanding over the awe of ignorance any day.[7]

Adams describes his apprehension of Darwinian theory in terms not dissimilar to those used by St Paul of his experience on the road to Damascus. But for some, the implication of this theory – that we are all, as Dawkins describes it, 'robot vehicles blindly programmed to preserve the selfish molecules known as genes' – has been deeply uncomfortable, even life-shattering.

Abolishing God: from Galapagos to Euston

In the winter of 1974 a policeman found the body of a middle-aged man who appeared to have killed himself by snipping his carotid artery open with a pair of nail scissors. He was shabbily dressed, and the possessions in his squalid Euston squat were limited to a mattress, a chair, a table and several ammunition boxes. When this man, George Price, was cremated, his funeral service was attended by a handful of homeless people – and two of the world's greatest living biologists.

Price had originally trained as a chemist and worked on uranium analysis for the top-secret Manhattan Project, the USA's wartime campaign to develop an atom bomb before Germany and Japan. Later he worked as a computer programmer and emigrated to London, where his initially amateur interest in biology became a professional calling. Price developed a mathematical model which explained that, in evolutionary terms, altruism could be just as beneficial to organisms as selfish behaviour.

For Price, this was a horrible discovery. Human morality might just be a matter of self-interest – it need not be God-given. A religious service in a London church prompted him to abandon control over his life and, quite literally, fling himself upon God's mercy. In a letter to his friend, the eminent biologist J. Maynard Smith, Price explained that he had given away all of his possessions to the poorest and was down to his last 15 pence; his visa was due to expire in less than a month. Horrified, Maynard Smith offered money and support. Price replied that it wasn't necessary as he still had two cans of baked beans and his Barclaycard. As Maynard Smith realized increasingly that his friend was beyond help, Price began to attract unwelcome attention for his behaviour. His habit of entertaining homeless alcoholics eventually saw him evicted from his smart Bloomsbury flat and forced to sleep on the floor of a laboratory.

Price was tormented by the idea of a world in which selfishness and altruism were equally dictated by our genes. Acts of the greatest human compassion were on a par with the savagery of the African savannahs, in that both were part of some blind, instinctual drive towards survival and reproduction. As Price saw it, all humans were deluded into thinking that they could consciously choose to be good, when in reality no-one had any choice but to be both selfish and altruistic in equal measure. This knowledge turned Price into a true fatalist and brought his life to an abrupt, lonely end.

Three centuries earlier, during Isaac Newton's lifetime, others had begun to find that the new intellectual climate brought about mental torment, rather than peace:

> I was compelled nevertheless to teach you your religion and to carry out that false duty that I had committed to as vicar of your parish ... I had the displeasure of finding myself annoyingly obliged to act and speak totally against my own feelings, to entertain you with foolish nonsense and vain superstition ... I ... never did it without great pain and extreme repugnance. This is why I hated so much the vain functions of my ministry, particularly those idolatrous and superstitious celebrations of masses, and those ... ridiculous administerings of sacraments that I had to carry out. I cursed them thousands and thousands of times in my heart ...

This tragic epistle was written by Jean Meslier, a parish priest in the Ardennes region of France. Meslier was noted for leading a life of extreme Christian rectitude, and for being exceptionally kind to the poorest of his parish. But when he died at the age of fifty-five in 1729, three copies of a 400-page manuscript were found in his home. This document, entitled 'My Confession', attacked the divisions between rich and poor, but reserved its most scathing criticism for the Christian religion. Meslier confessed that he had been forced into becoming a priest by his father and had been a committed atheist all his life. Christianity, in his view, was little more than a device used to oppress the poor, legitimize the control of the rich and perpetuate ignorance.

Copies of the manuscript were circulated with enthusiasm – one even reached the renowned philosopher Voltaire, who, though a believer in God, was seen as a spokesman for Enlightenment rationalism. Voltaire, like Newton, believed that it was possible to find God using one's powers of reason alone, and he was also keen to purge Christianity of its superstition and fable. Meslier's manuscript at first shocked him, but in later years, as Voltaire's own beliefs brought him into conflict with the Catholic Church, he began to circulate it, first removing the passages which denied the existence of God. Nevertheless, the church placed Voltaire's *L'Extrait du Testament de Jean Meslier* on the Index Expurgatorius, its list of banned books, and the Paris parliament ordered it to be burned.

The fulminations of one tortured country priest were only an acute symptom of a more chronic malaise. In the same era the Scottish philosopher David Hume, mentioned above, wrote his *Dialogues Concerning Natural Religion*, in which he argued that there were no grounds to believe anything which we did not directly experience. If the order and wonder of nature did point to the presence of an 'Intelligent Mechanick', as Newton claimed, how do we account for disorder, calamity, disease and suffering?

Meanwhile, the French philosopher Denis Diderot found himself in gaol for asking similar questions. Diderot was not an atheist, but he felt that the question of God's existence was irrelevant. The God who had built and wound up the clockwork mechanism of the universe and set it running was not interested in our affairs – he was *Deus otiosis*, a deaf God. 'Whether God exists or does not exist, He has come to rank among the most sublime and useless truths.' Diderot had, in effect, taken Newton's ideas to their logical conclusion: the laws of the universe were a kind of 'God' in themselves, and there was no need to imagine some creative force outside them. His theories, along with a veiled attack on the mistress of a prominent French politician, landed him in prison for three months in 1749, while influential friends like Voltaire agitated for his release. Unrepentant, eleven years later Diderot published a novel, *The Nun*, in which he castigated convent life for its sexual repression and corruption.

In 1831 a young Cambridge graduate spent the summer with a heavy heart, studying rocks in Wales. After dropping out of Edinburgh University, he'd obeyed his father's instructions and gone to Cambridge to prepare himself for a life as a country clergyman. The young man, who was passionately interested in natural history, couldn't think of a job he'd like less. It must have made matters far worse that his elder brother, Erasmus, had defied the paternal order to become a doctor, and was now whooping it up in London on a generous allowance.

But Fortune can seem at her most unfair just moments before she smiles kindly upon us. So, when Charles Darwin returned full of dread to his father's home in Shropshire, he found a letter awaiting him from his old tutor in Cambridge. The tutor, George Peacock, had heard that the Admiralty was planning a surveying expedition to South America. The captain of the ship, HMS *Beagle*, was looking for a gentleman to accompany him on the voyage, and take advantage of the opportunity to study the natural history of the region. At Cambridge, the young Darwin had distinguished himself by studying botany and geology while also gaining a respectable degree in Classics. Peacock suggested that he might be just the man for the job.

Charles Darwin leaped at the idea with great enthusiasm, only to find it initially quashed by his father. A successful doctor who had married into the wealthy Wedgwood family, Robert Darwin was concerned that at least one of his sons should take up a respectable career. This trip to South America seemed potentially dangerous, too, and Robert had never recovered from the shock of losing his wife. In the end, however, he was persuaded to give his consent by his brother-in-law, Josiah Wedgwood II – but no sooner had he done so than another obstacle emerged. The ship's captain, Robert FitzRoy, began to feel that this young, well-connected dilettante was being foisted upon him, and intimated that he had already found a more suitable companion. However, FitzRoy's objections vanished when he met Charles Darwin and the two men hit it off. They set sail on a five-year voyage which took Darwin around the world – and afforded him the opportunity to develop a theory which revolutionized the way we think about creation.

We now think of Charles Darwin as the man whose theories of evolution finally abolished the view of the Creation set forth in the Bible. To this day, so-called 'Creationism' is taught in some American schools as an explanation for the world around us; but it has largely been replaced by evolutionary theory, which argues that humans and all other living creatures on earth developed over successive generations, acquiring over hundreds of thousands of years the characteristics that most aided the goals of survival and reproduction in their specific environments. In evolutionary theory, not only is the Bible wrong, but there is no need for a God at the helm. The seemingly wondrous 'design' of nature and everything in it is not the work of some intelligent Creator but merely the end result of a biological process, of genes blindly copying themselves again and again.

Charles Darwin was not the inventor of this theory. He was not even, as we shall see, its best publicist. The story began with a French count of the previous century, who had made his reputation by conducting the first scientific attempts to discover the age of the earth. The Comte de Buffon (1707–88) was a workaholic who engaged a peasant to drag him bodily out of bed in the mornings. He was familiar with the work of the Swedish clergyman Linnaeus, whose examination of fossils had led him to conclude that the biblical account of Creation got the maths wrong. The view that the earth was created in 4004 BC sat awkwardly with accounts coming from China, which suggested that there had been an Emperor on a throne three thousand years before the birth of Christ.

Buffon was a showman – in one famous experiment, he demonstrated that it was possible to set fire to wood at a distance of 200 feet, by using an array of mirrors to focus the sun's rays. He is better known for his experiments concerning heat loss, conducted by heating iron balls to great temperatures and measuring the speed at which they cooled. These experiments led Buffon to conclude that, if an object the size of the earth had been formed in a molten state, it must have taken a very long time to cool down – far longer than the time-span suggested by the theologians. His own theory was that the earth must be at least 75,000 years old. The inaccuracy is not important; what matters is that scientists were beginning to use observations to prove that things said in the Bible were simply not true. A closer – though still inaccurate – estimate came along towards the end of the eighteenth century, when the mathematician Fourier put the age of the earth at around 100 million years. Science was beginning to provide its own alternative time-scale.

Buffon's work was also significant because it was among the first to suggest that humans, like all other species, had evolved, rather than been created in their present form by God. 'We might be driven to admit that the ape is of the family of man, that he is but a degenerate man and that he and man have had a common ancestor, even as the ass and horse have had.'[8]

The French divines were just as strenuous as the Protestant Reformers in holding closely to the so-called Mosaic account of Creation. As late as the middle of the eighteenth century, when Buffon stated simple geological truths, the theological faculty of the Sorbonne forced him to make and to publish a most ignominious recantation which ended with these words: 'I abandon everything in my book respecting the formation of the earth, and generally all which may be Contrary to the narrative of Moses.'

Buffon's banner was taken up by Erasmus Darwin, Charles Darwin's grandfather. Known initially as the author of a hugely popular celebration of plant-life in verse, Erasmus also put forward the view that all life on earth must have come from a common source. Rather like Newton, Erasmus Darwin still had room for God in his scheme of things; but his was another *Deus otiosus*, a God who set things up and running and remained deaf to His creation thereafter. In his view, species adapted or became extinct according to natural laws, with no intervention from God. Adaptations occurred due to pressures of environment – for example, when an animal with a tail migrates to an area where there are no flies to shoo away or predators to warn off, the tail becomes vestigial. His theory was faulty only in that we now know such changes took place in piecemeal fashion over thousands of generations, not as leaps within individual animals.

To this body of theories his grandson Charles Darwin added his own, developing the notion of the 'survival of the fittest' to explain why both adaptations and extinctions could occur. In Darwin's view, population pressure and competition among members of the same species dictated that only the best-adapted individuals could survive and reproduce. Darwin felt that his theories were so shocking that he sat on them for twenty years, sharing them only with close friends, and publishing them only when it seemed he was about to be pipped to the post by another scientist who had come to just the same conclusions, Alfred Russel Wallace.

Interested readers can find plenty of more detailed material on Darwin's theories and on his inevitable clash with the religious authorities.[9] The tale of the 1860 Oxford debate between Darwin's most zealous publicist, Thomas Huxley, and the Bishop of Oxford, Samuel Wilberforce, is well known. But there is a sad footnote to that legendary event. During the debate, at which the Bishop of Oxford demanded to know whether it was Huxley's maternal or paternal grandparents who were the apes, a lone figure was seen to stand up. With a Bible in one hand, he waved his other fist in the air, denouncing the Darwinists. This was none other than Captain FitzRoy, Darwin's former friend and companion on the great voyage to South America. FitzRoy, always a volatile man (he had been sacked from his position as Governor of New Zealand), blamed himself for making it possible for Charles Darwin to advance his godless theories on evolution. Five years after the Oxford debate, he slit his own throat in a fit of melancholy.

It would be wrong to say that Darwin's ideas turned the

world upside down, for by the time he published the *Origin of Species* people had been inching their way towards the idea he promulgated for over a hundred years. In fact, the initial reaction to Darwinism was something of an anti-climax. At the meeting of the Geological Society to which his views were first presented, the society's president lamented that the year 'had not been marked by any of those striking discoveries which at once revolutionised, so to speak, [our] development of science'.

In time, however, Darwinian views of evolution came to be accepted as the norm. This was thanks, in part, to the evangelizing work of the naturalist Thomas Huxley, who believed it was his mission to make science applicable to the concerns of the common man. Huxley was a born teacher, whose lectures at the School of Mines in Piccadilly (now part of Imperial College on the Kensington site) were enthusiastically attended by the working classes. If Huxley had a fault, it was that he tended to over-simplify his message for his audience, allowing key tenets of Darwinism to be taken up and mis-interpreted by hack writers for the popular press. Huxley himself saw no harm in this, crowing happily in one letter to Darwin, 'By next Friday evening, they will all be con-vinced that they are monkeys.' But the ease with which Darwin's ideas were received by the populace was not, of course, solely down to Huxley. From the mid-seventeenth century onwards, people had been envisaging a universe without God; and in nineteenth-century England, travelling fairs like Wombwell's Menagerie had given the public their first live contact with apes and gorillas, whose distinctly humanoid appearance and signs of intelligence suggested to many people that it was perfectly possible these were our ancestors.

Although he was raised within a strictly Christian tradition, and married to a woman of great piety (another reason why he kept his theories to himself for so long),

Darwin was unable to retain his faith. The early death of several children, and in particular his beloved daughter Annie, as well as his own consistently poor health, made it additionally hard to accept the notion of a loving God who was interested in the affairs of men. He wrote that

> disbelief crept over me at a very slow rate . . . The rate was so slow that I felt no distress, and have never since doubted for a single second that my conclusion was correct. I can indeed hardly see how anyone ought to wish Christianity to be true; for if so, the plain language of the text seems to show that the men who do not believe, and this would include my Father, Brother and almost all my best friends, will be everlastingly punished. And this is a damnable doctrine . . . [10]

Not every contemporary Christian saw Darwinism as a threat. One of Darwin's champions was the Reverend Charles Kingsley, a novelist and social reformer who thought that evolutionary theory brought religion to a state of maturity. William Buckland, Dean of Westminster and Professor of Geology at Oxford, reconciled the two disciplines by saying that each of the 'days' described in Genesis was actually an epoch of several million years. In 1864, 716 English academics and clergymen published a declaration setting out their belief that the new scientific findings and the Bible could be reconciled.

Many Christians today accept Darwinian evolution and consider that such a wondrous process was itself set in motion by God. But the problem here is that the Divine Idea raises as many questions as it answers. Even if we think of evolution as divine handiwork, many conundra still remain: How can God be all-powerful and entirely good, yet create a system that allows for suffering and evil? How can He be eternal, yet intervene in history? For many people, Darwinism provides a straightforward exit from this kind of endless speculation by suggesting that God simply does not exist. It's not surprising that Darwin's idea of the survival of the fittest was a major influence on the work of the German philosopher Friedrich Nietzsche, whose most famous words were 'God is dead.'

The Tower of Babel and the Hadron Collider

It is recounted in the tenth chapter of Genesis that, shortly after God had inundated the earth with water – yet relented from destroying it and all its inhabitants completely with His flood – humans decided to protect themselves. Never again would the environment, or God, threaten them with extinction. They gathered in the valley of Shinar and used the latest technology – fire-baked bricks and bitumen to bind them together – to build a large city and, within it, a tower to the heavens. Pieter Breughel the Elder, like many other artists of the sixteenth century, produced several different canvases depicting this vast construction. In the most famous of his three versions of this famous biblical building site, the tower is a huge ziggurat reaching up to the clouds. Building materials are piled everywhere. Scaffolds, cables, pulleys, cranes, heavy machinery of all kinds are present on every storey of the incomplete building. Men swarm like ants over pathways, walkways and ladders. At the bottom of the picture stands the King, Nimrod, supervising his architects. This is man's overweening ambition at work; and it led to his being brought down to size again when God confounded humans by making them babble in many languages.

Today, the Tower of Babel is being rebuilt. In a vast man-made cavern, deep underground, a massive, complex structure rises above the visitor for many storeys. The noise of grinding and hammering is deafening. The towering building with its vast cylinders is surrounded with scaffolds, cables, pulleys, piping, cranes, heavy machinery of all kinds. Men and women from different countries, from many parts of the globe are working on this project, and people swarm like ants over the structure. This is the Swiss headquarters of CERN, the European Organization for Nuclear Research, the world's largest particle physics centre. These scientists too have an ambition. They have assembled here to explore what the universe is made of and what forces hold it together. Like those who built Babel, they all speak a common language, but they speak the language of physics – a language which to non-physicists borders on the incomprehensible.

The cavern contains part of the Large Hadron Collider. In this huge apparatus, matter will be accelerated to the speed of light in an underground circular tunnel, 27 kilometres in length. To ensure that any particles sent off here reach the fastest theoretical speed, the largest magnets in the world will produce a massive force field, and the path of the particles will be cooled to close to absolute zero – colder than outer space. Some time in the autumn of 2007, Dr Jim Virdee, the modern King Nimrod, with his colleagues, will throw the switch. Protons will collide with each other, releasing more and more sub-atomic particles.

Until the end of the nineteenth century, the atom was thought to be the smallest indivisible piece of matter. This assumption changed when J. J. Thompson described the electron. Since then, the proton, the atomic nucleus and the neutron have been found. But, as physics developed, it was realized that each minuscule particle is itself composed of other particles. Hadrons and quarks in turn are divided into two classes: baryons and mesons. Baryons are composed of three quarks and have a 'large' mass. To complicate things, every baryon has an antiparticle composed of three antiquarks. There are also mesons, pions, leptons, muons, tauons and neutrinos. Seemingly, most of these particles also have their corresponding antiparticles.

But there is one very elusive particle that particularly interests the scientists at CERN. This particle is hypothetical. It has never been seen or detected or definitively proved to exist by experiment. The Higgs field that it creates is perceived the same from every direction and is indistinguishable from empty space. But the existence of the Higgs boson is crucial to modern physics – so much so that some wags have called it the 'God particle'. This is the particle that should, if the scientists are right, exert a force which holds all the other particles in the universe together. If Dr Virdee and his colleagues prove its existence by finding it in this billion-pound experiment, much about the nature of the universe that is at present obscure will be understood.

This experiment will mimic precisely what is thought to have happened a few milliseconds after the Big Bang. The scientists need to have very accurate, very fast detectors, because some of these tiny particles decay extraordinarily quickly, and will do so when the same environment that formed them is re-created. Cynics might just argue that if the existence of the Higgs boson is confirmed that will imply that God was not needed for creation; that the natural world fits perfectly with laws of physics that do not require a Creator to throw the first switch. But if the Higgs particle cannot be found, or indeed, does not exist, then the whole language of modern physics and much of the meaning of mathematics collapses. In that event, could God be said to have thrown a cosmic spanner into the works of scientists?

But *is* God dead? How do we account for the fact that the Prime Minister and the Queen of England both believe in a God-made-man, born of a virgin, impregnated by a ghost, who rose from the dead? How do we explain the fact that 98 per cent of Americans claim to believe in God? Or that every day, across the world, people consult oracles in newspapers, based on the principle that the movements of the planets can determine our fates? Or that, even if God is missing from the equation, people engage in a myriad forms of behaviour which seem religious – making pilgrimages, gathering to celebrate their unity, worshipping certain 'special' individuals and places? A certain type of rigid, western Christian religion, with its devotion to every jot and tittle of its scriptures, may have been undermined by science, but religion per se is far from vanquished. In the next chapter, I will be examining the forms it takes in the modern world.

9

Religion in the Modern Age

On 23 April 1995, a Japanese news crew lingered in a Tokyo street to film a murder that was about to happen. A young Korean assassin, hired by Japanese mobsters, was waiting for the victim outside his place of work. Close-up shots of the suitcase containing the murder weapon show that the camera crew must have been tipped off. As the victim appeared, he was stabbed. The cameras remained rolling as the assassin quietly stood by awaiting the arrival of the police. The Korean was later imprisoned for murder, while the *yakuza* boss who had commissioned the killing went free.

Murdering in the name of peace

The murdered man was Murai Hideo, head of the 'science and technology' division of Aum Shinrikyo, a Buddhist-inspired movement centred on the teachings of the blind acupuncturist Asahara Shoko.[1] Murai Hideo had master-minded the atrocious sarin gas attacks which killed twelve and injured four thousand people on the Tokyo subway in

March 1995. Surprisingly, the Aum Shinrikyo movement was initially a benign circle of young, idealistic yoga practitioners who were far from hostile to wider society. In August 1987, in the first edition of the group's journal *Mahayana*, Asahara wrote, 'I could not bear the fact that only I was happy and the other people were still in the world of suffering. I began to think: "I will save other people at the sacrifice of my own self." '[2] Like the Buddha and so many other mystics before him, the movement's founder believed he had obtained enlightenment, but deliberately remained in the world to bring less fortunate others to the truth.

At the time Aum started in the late 1980s, large numbers of young Japanese, sympathetic to Buddhism but also interested in the supernatural, were seeking alternative forms of spirituality. In 1985 the New Age magazine *Twilight Zone* had published pictures depicting the blind Asahara levitating during a yoga exercise. These pictures were instrumental in recruiting readers to form the 'Aum Group of Mountain Ascetics'. After a trip to India, Asahara claimed to have achieved moksha, the state of perfect happiness and liberation from the cycle of suffering, life and death. His disciples then began to quit their jobs and homes and form religious communities modelled on the Buddhist monastic community or sangha.

After they went 'public' with their teachings, the movement dropped the 'mountain ascetics' reference and renamed themselves Aum Shinrikyo, or 'Aum Teaching of Absolute Truth'. Asahara was given audiences with the Dalai Lama and the President of Sri Lanka. His organization obtained legal recognition by the Tokyo authorities. The group's leaders maintained a high public profile – for example, meeting with Russian politicians when it established branches in what was then the Soviet Union. They also began to pursue political goals. In 1990 the Aum-funded 'Truth Party' sought representation in

Japan's lower house, campaigning for abolition of the newly established consumer tax. But at the national elections they did not win a single seat. Society, as Asahara and his disciples saw it, had rejected them. So it was time to reject society.

It seems quite bizarre that, in an environment characterized by universal education and freedom of choice, people can commit themselves to religious movements that perpetrate shocking acts at the command of charismatic leaders. Mass suicide, incest and rape, self-castration, murder – all these horrors have been associated with the activities of one such group or another over the past twenty years. In most cases, the people involved have been educated middle-class people from the developed world, seemingly with a range of life-choices at their fingertips. What motivates these people to adopt obscure beliefs, to follow someone blindly even if it means losing their own lives or liberty and depriving others of theirs?

Many factors appear to contribute. There is a sense of disillusionment with society. Long-established, organized religions seem in retreat, leaving a spiritual hunger not assuaged either by science or by the 'Buy Now, Pay Later' ethos of the market economy. People may feel they live on a conveyor belt, borne resistless from school to university and into employment, where they become mere units in a capitalist machine, working long hours to afford the latest mobile phone or car. Sometimes, these material rewards just pall. And the 'boom and bust' cycle of major capitalist economies leads to its own disenchantment, as people's ambitions and hard work are at first rewarded and then rendered pointless as recession strips away their earning power and their sense of self-worth – as in Japan, where the booming economy of the eighties gave way to massive recession in the nineties.

Second, the personal magnetism of cult leaders may fill

the vacuum left by parents, teachers, bosses, governments and established churches. Charisma is not a quality that can be taught or acquired; still, some people captivate a roomful of people merely by entering it. This quality, of obvious advantage in politics, has even greater force among religious enthusiasts who see wider society as pointless or corrupt. Asahara's followers were drawn to him by his blend of charm, wisdom and drama, reinforced by how he presented his teachings. 'Be aware', he stated in 1991, 'that you cannot get dharma [truth, or teaching] without asking it of me.'[3] Some gurus render themselves indispensable to their disciples by claiming that only they hold the keys to the truth. In this way, they enforce a relationship of dependence and obedience. A former follower of Asahara, compelled to abduct her own father in order to transfer his wealth to the group, claimed in court that 'I was his puppet and slave.'[4]

Third, these groups usually derive the content of their teachings from one or more of the prevailing religious traditions. The doctrines of the Unification Church of Reverend Moon, for example, are largely derived from Christianity, and their founder was initially a Presbyterian. The Aum Shinrikyo movement was based largely on a mixture of Buddhism, Hindu-inspired yoga and astrology, and apocalyptic teachings from Judaism and Christianity. This gave them the quality of being both familiar and new, and meant that certain key ideas were given an attractive, sometimes dangerous 'spin'. For example, Asahara gave a unique twist to a concept from Tibetan Buddhism relating to the rituals performed at death. In this system, a ritual called *poa* allows the dead person's soul to pass to a higher spiritual dimension. But in Asahara's teaching, the *poa* became, in essence, murder justified on the grounds that it was freeing people. 'When your guru orders you to take someone's life, it's an indication that the person's time is already up.

You are killing that person at exactly the right time and thereby letting that person have his *poa*.'

Aum was also technologically literate, appealing to young people who had grown up in a milieu dominated by science. A number of scientists, including the Murai Hideo who was later murdered, were attracted to the movement. Their yoga practices were aided by technical devices such as an apparatus of electrical headgear known as PSI, or Perfect Salvation Initiation. They used a variety of instruments to measure the effects on the body of their religious disciplines, which included meditating while buried underground in airtight containers. Computer programs were developed to make fast predictions about the fate of the universe. And, of course, the manufacture of the deadly sarin gas depended on technical know-how.

Fourth, how certain movements are treated and perceived by society is important. The Japanese are often stereotyped as being rigidly conformist, and breakaway religious movements are indeed treated with suspicion. Followers reject their families, discard the values of their society and adopt unfamiliar dress or unconventional behaviour. This is seen as mad or dangerous; dialogue tends to break down, so that rumour and paranoia inform the way both groups think about each other. Within the group a fortress mentality develops, wherein society is eventually seen as the enemy. This was true for the members of the Aum movement, who from its earliest days were derided by the Japanese media. In 1989, just as they received legal recognition from the authorities, a popular Sunday magazine ran a series of articles lingering delightedly on the horror of the group's religious practices, and criticizing it for encouraging young people to reject family and possessions.

Then, in 1990, media ridicule was compounded by electoral failure. The prophecies of the leader became more pessimistic. Armageddon was coming, and only

Asahara's followers would survive, in specially built, gas-proof shelters. These beliefs crystallized into expectation of a Third World War, between the USA, acting for Christendom, and Japan, the defender of eastern teachings. Japan would ultimately triumph.

Feeling sanctioned by their belief that they were releasing souls, the group had already murdered one of their own. In February 1989, Taguchi Shuji declared his intention to quit after witnessing the death of another initiate during rigorous religious 'exercises'. Because he might blab to the press and prevent Aum's official recognition as an approved religious organization, Taguchi was strangled. Aum also targeted non-members. The same year, a lawyer acting for relatives of various Aum members was killed, along with his family. He had given a highly critical interview to the Tokyo Broadcasting System, and the station had allowed senior Aumists a preview. The police made little headway in solving the lawyer's murder. Later, the head of the National Police Agency admitted that, had they acted more swiftly, the nerve gas attack on the Tokyo subway might never have taken place.

The group's hostility to Japanese society was heightened by a dispute that flared when they tried to build a countryside community. Worried villagers refused to allow Aum members to register as local citizens. Subsequently, the police raided Aum facilities, claiming breaches of building regulations, and arrested several leading members. The group began to speak of 'illegal suppression by the state' and of 'persecution'. In 1993 it purchased a sheep farm in Australia, possibly to acquire uranium. It began to create businesses, enabling it to import large amounts of chemicals and sophisticated computer equipment.

On 20 March 1995, during rush hour, five lines of the Tokyo subway were attacked by poison gas. The public mood was already deeply insecure: just two months previously, the Kobe earthquake had killed over five thousand and left more than quarter of a million homeless. Two days after the attacks, Aum headquarters were raided by police. As with Hideo's murder, news crews were 'fortuitously' on the scene, broadcasting the raids live. Search warrants were issued, not for evidence linked to the sarin incident, but because a man opposing his sister's membership of the movement had been kidnapped. Although police later found that the man was killed on Aum premises by an overdose of drugs, the media focused on the stockpiles of chemicals and laboratories for making weapons.

The blind Asahara was not arrested until May, when police – again with full media accompaniment – found him in a secret chamber. He was sentenced to death for the sarin attacks but has appealed. Police records revealed that Aum members had attacked a number of individual opponents with poison gas between 1994 and 1995. Thirty-three Aum members had died by accident, suicide or murder and a further twenty-one were missing.

The tale of this Buddhist-inspired murder movement is very modern. The close involvement of the media, the technological sophistication, the use of the subways to administer a deadly poison – all would have been impossible before the end of the twentieth century. The same goes for the background from which it emerged, an age dominated by science and the markets, in which those who became followers felt alienation and spiritual longing.

The advance of scientific knowledge has accompanied a decline in belief in the supernatural, and a decline in belief in God. Ever since the nineteenth century, theorists of religion have been arguing that a process known as *secularization* is under way – that is, that modern life is characterized by an increasing absence of religion. The

German sociologist Max Weber called this *Entzauberung* – demystification of life. Science explains everything, so we no longer need to look to the supernatural for answers. There have also been changes in the structure of our societies – in the west at least, state and church are now increasingly separate, and religious involvement has become a matter of personal choice. Mass communication, through channels such as satellite TV and the internet, enables people from varying backgrounds to be exposed to different cultures and ways of life.

But the picture is complex. Many traditional religions, it is true, are experiencing a decline in numbers. Grace Davie, a sociologist at the University of Exeter, points out in her excellent book that 'believing but not belonging' is still 'a persistent theme'.[5] She cites a survey of 1,600 people carried out by Dr Mark Abrams and colleagues.[6] The survey investigated religious commitment in Great Britain between 1981 and 1990 and found that there was little change over that decade. Around 34 per cent of people polled claimed to 'often think about the meaning of life'. More than 70 per cent claimed some belief in God, or believed in 'Sin'; 60 per cent believed in 'the Soul'; and 46 per cent said that they drew some comfort from religion even though, as Grace Davie points out, less than 15 per cent of the British population – where Christianity is still the official religion, with its own representatives in Parliament – are members of a Christian church. Other studies have shown fairly similar trends in various parts of Europe, though in general people on the continent of Europe show slightly stronger religious affiliation. In Britain, Roman Catholics are much more likely to be churchgoers than Anglicans. It is true also that church membership is very much higher in Northern Ireland and, to a lesser extent, in Scotland; but the populations of these two countries represent a tiny fraction of the total population of modern Britain.

It is undeniable that the numbers of people attending churches (and, to some extent, synagogues) in Europe and the USA are generally falling. Nevertheless, it is also very clear that beliefs and religious activities are still thriving, often by assuming new forms – both dangerous and benign. These beliefs and activities are emerging and adapting in ways that are often felt to be relevant to the modern world. From peculiar cults to New Age spirituality to fundamentalism, it is very clear that religious views remain a very active force. However much our societies may change outwardly, the Divine Idea just won't die.

Worshipping the flag

On 20 January 1961 the newly elected US President, John F. Kennedy, gave an inaugural address. He told his audience:

> We observe today not a victory of a party but a celebration of freedom – symbolising an end as well as a beginning – signifying renewal as well as change. For I have sworn before you and Almighty God the same solemn oath our forebears prescribed nearly a century and three quarters ago.
>
> The world is very different now. For man holds in his mortal hands the power to abolish all forms of human poverty and to abolish all forms of human life. And yet the same revolutionary beliefs for which our forebears fought are still at issue around the globe – the belief that the rights of man come not from the generosity of the state but from the hand of God . . .
>
> Finally, whether you are citizens of America or of the world, ask of us the same high standards of strength and sacrifice that we shall ask of you. With a good conscience our only sure reward, with history the final judge of our deeds, let us go forth and lead the land we love, asking His blessing and His help, but knowing that here on earth, God's work must truly be our own.

This prominent mention of God in an American President's inaugural address is unsurprising. The current President, George W. Bush, holds daily prayer meetings in the White House. His Deputy Under-Secretary of Defense for Intelligence, Lieutenant-General William 'Jerry' Boykin, a fundamentalist Christian who conducts prayer meetings, has attracted controversy and opprobrium from Muslims for claiming that he was victorious over a Somali warlord because 'I knew that my God was bigger than his. I knew that my God was a real God and his was an idol.' America is fiercely religious, so any politician worth his salt must make a few references to God, just to get the votes in.

But what did Kennedy say in that speech, and what didn't he say? He was a Christian, a Roman Catholic, but the speech gives no clues to that. There are no references to Jesus, or to the Pope. If there were, you would probably see them as inappropriate: the separation of church and state means, surely, that a political address is not the place to put forward one's own religious views. But if that's the case, why would a new President, in one of his most historically important speeches, dare to mention God at all?

According to Robert Bellah, creator of the so-called theory of 'civil religion', Kennedy mentioned God because, even though church and state are separate, there is nevertheless a strongly religious dimension to American political life. This dimension amounts to a religion in its own right, with its sacred texts and symbols, its myths, its sacred sites and ceremonies. Far from retreating, Bellah argues, Christian religion has fused with the American state, giving it power and legitimacy along the way.

Kennedy says, in his opening paragraph, 'I have sworn before you and Almighty God the same solemn oath our forebears prescribed'. The oath referred to is the oath of office, which includes acceptance of the obligation to uphold the US Constitution. His choice of words indicates that his obligation is not merely to the American people, but to God Himself. America is a democracy; yet, by invoking a Higher Power, Kennedy was almost saying that the will of the American people was not the most important factor in the national life. The same message occurs in the next paragraph. The 'rights of man' are nothing to do with mortal people striving on earth to establish fair governments, but are a gift from God. And in the final paragraph, the conclusion of his speech, Kennedy equates everything America does with fulfilling the work of God. This latter statement is particularly Protestant in tone, and reflects the strong beliefs of the European settlers who first began to colonize America in the seventeenth century.

As we have seen, Protestantism stressed the idea that God could be served through hard work and simple piety, and this idea was reiterated by generations of US Presidents in their inaugural speeches. That it should be echoed by a Roman Catholic in the 1960s shows how deeply ingrained it is in American minds.

In his address of 1789, George Washington said, 'No people can be bound to acknowledge and adore the Invisible Hand which conducts the affairs of man more than those of the United States.' There's a parallel here between the God of America and the God of the Old Testament, who 'chose' Israel as his special people, and acted both to reward and to punish them. In fact, we find many references to America as the new Israel. Early settlers often chose names for their communities from the Old Testament, such as Shiloh and Bethel. When the revolutionary leaders met in 1776 to devise an official symbol for the new nation, they came up with a range of biblical themes – such as a depiction of Moses dividing the Red Sea, with the motto 'Rebellion to tyrants is obedience to God'. In more recent times, the idea that 'God is on

America's side' has provided the justification for many acts abroad. In its campaigns against Afghanistan and Iraq, the motive cited was not just revenge on regimes that harbour terrorists, but also the wish to strike a blow for 'freedom'. This echoes President Lyndon B. Johnson, who said in his inaugural address: 'They made a covenant with this land. Conceived in justice, written in liberty, bound in union, it was meant to one day inspire the hopes of all mankind.' Just as Israel was called to be a 'light unto nations' (Isaiah 42: 5–8), there is a sense in which the American state views itself as having a God-given mission. America is identified with Israel as the Promised Land, a haven of liberty and equality, an example to the rest of the world, and an instrument of God's will.

This linking of religious concepts to the authority of the state is not confined to the west. It is the basis for the Shinto religion, which evolved in nineteenth-century Japan, bound up with the re-establishment of the rule of the Emperor and a rejection of Buddhism and Confucianism. In its earliest form, Shinto was an animistic practice, centred on worship of the ancestors and elements in nature, such as the wind, mountains, streams and moon. Over time, this was coupled to a belief that the Japanese Emperor was directly descended from the leading deity, the sun goddess. This in turn provided the religious basis for a political movement, culminating in the restoration of imperial rule in 1868. A new religious policy was introduced, stripping Shinto shrines of their Buddhist elements, creating a new official calendar of Shinto festivals and designating certain Shinto shrines for the honouring of the imperial family and the war dead. By the time of the Second World War, Shinto religion had evolved into a belief that both Emperor and land were divine, and that those who gave their lives in war (such as kamikaze pilots) gained a more-than-mortal status. Even as late as 2000 the Japanese Prime Minister Yoshiro Mori was able to claim, 'We hope the Japanese people acknowledge that Japan is a divine nation centring on the Emperor.'

Mori was criticized for his words, and Shinto is nowadays a movement associated with conservative and far-right elements within Japanese society. But 'civil religion' as a whole is by no means a fringe phenomenon. I recall being surprised when, during a recent television documentary series, an allegedly psychological experiment involved testing how groups of various nationalities would react when their flags were burned. This seemed like cheap entertainment with no purpose other than shocking people in order to film their reactions. But one of the American group asked to be given the burned remains of the flag, in order that they could be disposed of appropriately. This might seem like one individual's silliness, but it also attests to the reverence with which people approach their national symbols. Throughout the American way of life, the history, the power and the authority of the state are approached in an overtly 'religious' way. American schoolchildren recite the Pledge of Allegiance before a flag, just as the Lord's Prayer is recited in Christian schools in Britain. Other sacred texts include the Declaration of Independence and the Gettysburg Address, learned by schoolchildren, available on CD and recorded by famous actors and politicians (including our very own Margaret Thatcher). There is a sacred history, involving heroes like the Pilgrim Fathers and the pious infant George Washington, who 'could not tell a lie' after chopping down his father's cherry tree. And there is a ritual calendar, involving events like Thanksgiving and Memorial Day, when Americans commemorate those who have given their lives in 'sacrifice' to the ideals of the nation.

America is by no means the only country to 'worship itself' in an overtly religious way. When a British dignitary is given a state funeral, and when members of the armed forces are buried, a Union flag is placed on top of the coffin, linking a symbol of the state to a religious rite. Remembrance Sunday, an occasion replete with moving rituals, such as the releasing of thousands of poppies onto the heads of worshippers at Westminster Abbey, is a day when religion is invoked to remember those who have died in wars undertaken by the state. But, as the best-selling atheist writer Philip Pullman – no friend to organized religion – points out, a prime example of a state reworking 'religion' is the former Soviet Union. Although religion was formally outlawed there, the state promoted communism as the true faith. The state provided a church-like Politburo which required rigid conformity. Celebrations in Red Square were 'religious' holy days. Education reinforced the 'truth' in every branch of academic endeavour. In academic society (just as in many Catholic universities in Belgium and Germany until recently), if you were not a member of the 'church', it was much harder to get certain senior posts. Huge people's palaces celebrating the workers' revolution were built in every capital city in the Soviet bloc – essentially, cathedrals for the worship of communism. The Soviet Union was evangelical in attempts to convert nations coming under its rule. 'Popes' like Lenin and Stalin were embalmed in ornate tombs; faithful visitors trooped past them in their thousands, while armed guards enforced codes of silence and 'appropriate' dress. Much-loved writers like Tolstoy were co-opted as unofficial saints, their images revered as icons, their homes lovingly pre-served as they had left them and transformed into sites of pilgrimage. And heretics were pursued and rooted out by an Inquisition, the NKVD.

In modern Britain, too, cultural life is dominated by a new cult – not Christian, not Jewish, not conforming to any established religion but none the less looking outwardly very much like one. Football is a largely communal activity, conducted according to a fixed calendar, in which people gather together to watch and participate in a ritual, and celebrate their shared identity. The pilgrimage to the ground, the scarves, the chants – these form the framework of a Saturday afternoon religion. People who follow a team speak readily of idols, of worship, of states of ecstatic joy, of losing their individual identity within the crowd. Only God is missing from the equation; and the gap is filled, perhaps, by the players themselves, who are treated as being somehow more than mortal – and on occasions seem to believe this themselves. Relics of the saints, signed football shirts of particular players or a chunk of Highbury's hallowed turf, change hands for absurd sums of money – not viewed, as in former times, as having supernatural power, but none the less endowed with a meaning and significance beyond the objects themselves. I doubt whether football fans are any more likely to believe in God than any other group – I'll try a straw poll next time I go to see Arsenal – but many aspects of the sport are, on the surface at least, akin to religious practice. It seems that, in the modern age, while attending a church or adhering to a fixed religious doctrine may have gone by the board, we have found new focuses for our religious behaviour.

I say that God is missing from the equation; but I might recall the occasion when Dr George Carey, the Archbishop of Canterbury, joined Dr Jonathan Sacks, the Chief Rabbi, at an Arsenal versus Manchester United football match. Both men had been recently elected to their respective religious positions and, being fervent not only in their religious commitment but also in their allegiance to that great north London team, they decided that their first ecumenical gathering might be held in a box at the Highbury Stadium.

Both men were greeted with great enthusiasm, introduced to the directors, and led out onto the hallowed ground – the floodlit pitch – itself. The crowd cheered loudly when the loudspeakers announced the presence of these two religious leaders. As the Chief Rabbi points out, whatever religion Arsenal supporters took their theological wager in that evening, they certainly had friends in high places. And if the power of prayer counted for anything, Arsenal should have had a massive home win. In fact, that night Arsenal went down to their worst home defeat in sixty-three years, losing 6–2 to Manchester United. The following morning one newspaper carried the full story, offering the reasonable opinion that 'If the Archbishop of Canterbury and the Chief Rabbi between them cannot bring about a win for Arsenal, does this not finally prove that God does not exist?' The Chief Rabbi immediately sent the paper a note to reassure the journalist: 'On the contrary, it proves that God does exist. It's just He's a Manchester United supporter.'

One national upsurge of religious activity occurred in the wake of the death of Diana, Princess of Wales, in 1997. Despite the fact that the princess was a secular personality – and herself interested in New Age beliefs – the Prime Minister and the Church of England took over the mechanism by which the populace mourned her. Churches held memorial services in her name across the land, and 'books of remembrance' were set up inside Christian houses of worship. Even as I write, eight years later, some of the members of the public still seem animated by concerns over the construction of an appropriate site of pilgrimage, the Diana Memorial, in her honour. The references to her as 'Queen of Hearts' have strongly Christian overtones, reminiscent of the 'Sacred Heart' and 'Our Lady', even though she worked no miracles beyond being a kindly Royal with the common touch. The anthem to Diana sung at her funeral by Elton John, a reworking of a song he originally wrote about Marilyn Monroe, is replete with religious references, mentioning Diana as 'the grace that placed itself where lives were torn apart', praising the 'wings of your compassion' and lamenting 'a country lost without your soul'.[7] From state ceremonials to football to personality cults, it seems that the modern world may, in fact, be as vibrantly religious as it ever was, even if its religiosity is taking on new forms and focuses.

Mourning, religious fervour and the national sport of football have joined each other more than once – perhaps most strikingly in recent years after the FA Cup match played in April 1989, when Nottingham Forest met Liverpool at the Hillsborough Stadium. The game lasted a mere seven minutes. Too many supporters, most of them from Merseyside, had been let through the gate in a stampede, and in the ensuing mêlée ninety-four people were crushed to death, and many more seriously injured, as the crowd hit the perimeter fence at the start of play. Most of the dead and the injured were young – some were children – and many were carried onto the pitch for attempts at resuscitation. It was the worst sort of public nightmare for, as it was a Cup semi-final, millions watched the events unfold in real time on television.

The aftermath is still remembered and commemorated, in particular, of course, in Liverpool. Admittedly, Liverpool is in many ways not a typical English city. Apart from anything else, allegiances to church, to community and indeed to the football club are unusually strong – in part, perhaps, because of the deprivation that has existed there since long periods of recession and unemployment. As Grace Davie points out, in the days that followed the Hillsborough disaster there was an extraordinary outpouring of grief: part civic, part parliamentary, part sporting – but nearly always with strong religious overtones. Football regalia were brought into cathedrals and religious leaders were greeted on the pitch at Anfield, the Liverpool ground. The most public exhibition of grief was the Anfield pilgrimage, when an entire city united in its grief. On the Sunday following the disaster, 'pilgrims' from every walk of life and every background came from churches or the Catholic Cathedral to the football ground. There was an endless stream of mourners – by the end of that week, well over one million people (twice as many as live in Liverpool) had filed through the gate at Anfield, having queued for many hours to get into the ground, in order to pay respects, to lay wreaths, to hang shirts, to join in prayer. Counsellors were in attendance – some professional grief counsellors but also many distinguished professional footballers and their wives – and the scene was described by my wonderful colleague in the House of Lords, the late Bishop of Liverpool, David Sheppard – himself a great sportsman in his day:

Over the goalpost and the crush barriers hung red and blue scarves, with flags and banners portraying the Liver Bird emblem and the inevitable reassurance 'You'll Never Walk Alone'. On the turf below lay a field of flowers, more scarves and caps, mascots and souvenirs, and incredibly, kneeling among wreaths and rattles, a plaster Madonna straight from a Christmas crib...

Blasphemy, unhealthy superstition, tawdry sentimentality. Or a rich blend of personal mourning, prayerful respect and genuine faith?

One might say that Liverpool is an unusual case, but in a smaller way there have been quite similar responses from the British public after a number of calamities that have touched national sensitivities. A wayside shrine with numerous flowers and tributes was set up after the fire that killed thirty-one people at the London Underground station at King's Cross in November 1987. Similar public expressions of grief with strong religious overtones were seen at Southwark Bridge in 1989 after the *Marchioness* disaster, when fifty-one people drowned. As I write, a memorial service is about to be held for the tenth anniversary of the M5 motorway coach accident in which twelve children and a teacher from Hagley School were killed; immediately after that, the nearest motorway bridge was decorated with religious icons and floral tributes. Nor can we forget the national outpourings of religious feeling after the Dunblane killings in 1996, when an unemployed shopkeeper armed with a gun killed primary school children and their teacher. Much of the initial response to this horror had a strong religious flavour, and considerable media activity centred on how such a thing could happen in a divinely created world.

Spiritual Thatcherism: New Age religion

Judith Darlene Hamilton was born in New Mexico in 1946, the youngest of a large family of poor farm workers. She suffered emotional and physical abuse as a child. Denied her innocence and dreaming of one day finding a loving, caring man, she went to college, but dropped out due to poor health. In the 1970s, after a failed marriage, and giving up her two children, she moved to California, where she finally found a degree of financial stability selling cable television franchises. She remarried, taking as her second husband a dentist called Jeremy Wilder who was besotted with her physical beauty. Influenced by ideas from the culture of her time, Judith began to experiment by placing pyramids around her house. These resulted in a series of encounters with one Ramtha, a spirit being visible only to Judith who, she claimed, was a 35,000-year-old warrior from Atlantis. In late 1978 Judith began to conduct public talks and 'encounter sessions' with her ancient spirit guide. As these grew in popularity, she founded the 'Church I Am' in the mid-1980s. The Ramtha School of Enlightenment, based in Colorado, emerged as an educational arm of her operations, attracting a number of high-profile celebrities, including Shirley MacLaine and Linda Evans. Today, her humble origins behind her, JZ Knight lives in a highly guarded French-style chateau in Washington State. She became a focus of controversy in the mid-1990s when a German woman, Judith Ravell, also claimed to be receiving messages from the Atlantan warrior. JZ Knight marshalled her considerable resources into suing her competitor, in the end securing a verdict from the Austrian courts that she alone was allowed to receive revelations from Ramtha.

Is JZ Knight just a canny businesswoman? Or did she, as someone with a difficult emotional history, project 'Ramtha' as the ideal father-figure she never had – with great success? What matters is not, perhaps, the truth or fallacy of a 35,000-year-old warrior contacting a dentist's wife under a pyramid, but that her subsequent teachings have been so readily believed by others – to the extent of yielding JZ Knight a sizeable fortune. How is this possible in an age dominated by scepticism and reason?

In 2000 a telephone poll was conducted in conjunction with the television series *Soul of Britain*, which sought to examine current attitudes to religion in the UK. The results suggested that belief in a traditional 'God in heaven' was being replaced by alternative beliefs in a 'spirit' or 'life force'. In this rather limited survey, one in four people questioned maintained belief in a traditional God, whereas twice that number said they believed in some kind of spirit or life force. Polls like this suggest we are wrong to think that modern life is characterized by an indifference to spiritual concerns. In fact, if we look at religious life in closer detail, what we see is a shift away from worshipping a traditional God, and towards a view where the individual, and life itself, are worshipped as divine.

The *Concise Oxford Dictionary* defines religion as 'belief in a superhuman controlling power, especially in a personal God entitled to obedience and worship; expression of this in worship; a thing that one is devoted to'. In other words, religion is defined as something centred upon God, with no mention of any affirmation of life in the here and now. It seems fairly clear that this kind of religion – characterized by a God up in heaven, a preacher in the pulpit, prescribed rituals, fixed ways of believing and unchallengeable scriptures – is passing out of popularity. What about religious beliefs which celebrate the individual and the present life?

On a random stroll around north London the other day, I came across a newly refurbished property situated in a busy shopping area. It looked to all intents and purposes like a private clinic catering to the most affluent sections of society, with a smartly dressed receptionist sitting behind a desk, and comfortable leather seating in the waiting area. A glance at the windows of this establishment revealed, however, that they were not in the liposuction or nose-job business. They offered Cranio-Sacral Healing, Acupuncture, Reiki, Crystal and Colour Healing, Wiccan divination, Spiritual Counselling and probably a lot more besides. Had I not been thinking about writing this, I should probably not have bothered to look, so ubiquitous are these establishments in modern Britain. Indeed, its very respectable façade and presence alongside a range of other, more 'normal' businesses like newsagents and take-aways demonstrates how 'New Age' type religion and alternative therapies have become woven into the fabric of modern life.

In a sense it is risky to talk of such a thing as 'New Age' religion, just as it is inadvisable to describe every new religious movement as a 'cult'. The former term encompasses a vast range of behaviours, from eastern-influenced brands of yoga to Celtic divination and a belief in flying saucers and 'spirit guides'. But it is possible to trace the growth of a movement arising out of the counter-culture of the 1960s, operating on certain core premises. And it seems to be on the rise. In 1979, the magazine *Common Ground*, circulating in the Bay area of San Francisco – itself a breeding ground for New Age beliefs – contained 300 advertisements for 'spiritual businesses'. In the 1997–8 issue, there were 1,500. Between 1993 and 1997, there was a 75 per cent increase in the number of New Age/'spiritual' publications in the UK.[8]

New Age religion rests on the basic principle that modern life is somehow at fault, and that the individual has become lost, corrupted by materialism, enslaved to consumerism and warped by destructive patterns of belief and action, either as a result of their upbringing or from 'past lives'. Like older mystical traditions – such as Kabbalism and Gnosticism – at its core is a conviction that there is something wrong with us and the world around us, and that both are in need of healing. Alongside this, there is a belief that each person contains the ability to heal themselves, and to 'ascend' to a purer, better way of being. In fact, we could say that there is an identification of the Self with God, or at least with some Higher Power or Ultimate Reality. This, once again, is nothing new – in medieval times, devotees of Jewish and Islamic mysticism taught that God was to be found within the individual. But there is something about modern life which makes this idea particularly relevant. New Age religion stresses the notion that the individual, as representative of God, has the power to find his or her own path to Enlightenment. In his work on New Age religion,[9] Paul Heelas quotes a talk by healer Denise Linn in which she says, 'You'll be given suggestions. Always feel free to follow your own inner guidance . . . You are free to follow my suggestions or . . . journey in whatever way suits you and your soul.' Ramtha, Judith's Atlantan warlord, delivered the following utterance on the subject of personal choice: 'Everyone is right because everyone is a God who has the freedom to create his own truth.'

This, in turn, is why it is hard to speak of New Age religion as a single entity, because it contains within it the idea that we are free to pick and choose our own paths. Commentators often speak of religion in modern times as being like a supermarket of faiths, in which people assemble their own belief system from an array of 'goods' on offer, ranging from Jewish mysticism to Hopi Indian rituals and Tarot cards. On the one hand, this reflects the extent to which modern communications have rendered traditions and practices from other cultures accessible to us. On the other, it is the current peak of a tradition stretching back several centuries, during which the individual and his or her own powers of reason came to be venerated above the power of God, or revelation through the scriptures, or the authority of the church. From Martin Luther's injunction that everyone should read the scriptures for themselves to the French revolutionaries' declaration of the rights of man and beyond, the scene was set for the individual to become the active agent.

In New Age beliefs, self has replaced God as the source of truth. And there are a number of reasons for that, besides the philosophy of the Enlightenment. At the end of the twentieth century, the collapse of communism led to an additional disillusion with 'godless' ideologies that some people felt might have replaced traditional religion. Shifts in economic patterns brought an end to 'jobs for life', promoted mass migration and led to the breakdown of traditional community and family structures. As a consequence, people felt a loss of identity; New Age religion equates this with a sort of spiritual sickness, which can be cured through recognizing one's own inner power and identity with God. Technological advances have meant that the power to disseminate information no longer rests exclusively in the hands of an elite, but can also be exercised by groups who want to promote change. As a result, politicians, royal families and global corporations have often found themselves in the spotlight of scandal – with the result, among others, that there has been an understandable loss of faith in the structures of our society. New Age religion nips niftily into the space, telling us we can now rely only on ourselves.

New Age religion shows little preoccupation with the origins of the universe, or even what happens after we die – staying safely out of the areas where science can refute it. Instead, it offers a sort of life-centred spirituality, based very much on obtaining personal happiness and fulfilment in the here and now. This contrasts very strongly with traditional religions, such as Roman Catholicism, which essentially tells people how to prepare themselves for the afterlife.

There is a rather selfish quality to some New Age religion, a focus on individuals getting what they want. Some New Age beliefs, such as Transcendental Meditation, do aim to change the world for the better – in fact, I once attended a meeting where the teacher solemnly told us that, when just 1 per cent of the population uses TM™, the crime rate drops by 10 per cent. However, a cursory glance at the titles on offer in an average high street bookshop suggests that much New Age religion is focused upon individuals, not on society: *Empowering Your Life With Dreams*; *The Alchemy of Voice*; *Transform and Enrich Your Life Through The Power Of Your Voice*; *The Power of Oneness – Live The Life You Choose*.

Moreover, some New Age religions promise not just happiness, peace, fulfilment and so on, but often material wealth. The Sokka Gakkai movement, whose teachings are an offshoot of Buddhism, maintains that chanting the phrase *Nam-myoho-renge-kyo* can, by itself, create wealth. The publications of Louise L. Hay recommend making affirmations such as 'my income is constantly increasing'. And New Age religion is itself a big business, with books, CD-ROMs, workshops, spiritual retreats, business seminars and healing outlets raking in millions of dollars. New Age religion is a kind of 'spiritual Thatcherism', stressing both the power of individual choice and the ultimate desirability of worldly success.

Generally, New Age religion is also characteristically 'benign'. In former centuries, Judaism spoke of a vengeful God who demanded utter obedience to his Law, and Christianity saw Him as a dreadful judge with the power to condemn sinners to a sort of eternal barbecue. Arguably, mass communications have had a role to play in demolishing these ideas. In earlier times, atrocities were far from rare, but the experience of them was limited to those who had been there, and those who were able to read what others had written about them. But in the twentieth century the world's media played a vital role in broadcasting, albeit sadly belatedly, shocking images of the Nazi concentration camps, the destruction of Dresden by British bombers and the annihilation of Japanese cities by atom bombs. You did not have to be there to perceive, most vividly, that hell was possible on earth, and that there could be little more terrible than man's inhumanity to man. The wrath of God, accordingly, began to seem less of a threat than the possibility of a man-made Third World War or a fresh Holocaust. Again, in steps New Age religion, providing a revised format for the way we envisage our higher powers, stressing them as kindly, compassionate, forces for good.

But while New Age religion is ideally suited to the modern, capitalist world, the range of choice on offer can extend beyond benign forces. As we saw in the case of the Aum movement, what starts out as a tame circle of like-minded spiritual practitioners can sometimes, under stressful conditions, turn into a highly destructive movement demanding blind obedience from its followers.

Of anoraks and revelations

Nearly everybody remembers the events of 28 February 1993 at Waco, Texas. That was the day when the world's media filmed events unfolding as law-enforcement agents ended their siege of the headquarters of the Branch Davidians. By that evening, David Koresh, the leader of that curious messianic movement, was dead, together with eighty of his followers; twelve of the dead were Koresh's own children. It is arguable that if the FBI had been a little more tolerant and rather less heavy-handed, much of that bloodshed could have been avoided. But just four years later another religious movement came to prominence through multiple deaths at the hands of its members. The history of the Heaven's Gate movement in California is even more bizarre than that of the Branch Davidians, and repays retelling.

On 28 March 1997 – during Holy Week in the Christian calendar – the USA awoke to the news that thirty-nine people had been found dead on a ranch near San Diego, California. It must be remembered that it is not just Muslim suicide bombers who may be prepared to destroy themselves. These suicidal deaths were no act of frenzy, but had clearly been prepared for some weeks in advance: farewell videos had been recorded, statements issued on their official website, personal affairs put in order. Investigators discovered that the dead, a group calling themselves Heaven's Gate, had killed themselves in the belief that the comet Hale–Bopp was an alien spacecraft which would assist them to ascend to a higher spiritual plane.

At first glance, the Heaven's Gate movement offered its followers little that seems attractive. They lived a life of monastic seclusion, celibacy and abstinence. Some of the male followers had voluntarily undergone surgical castration. Members renounced their possessions and their names, and wore a simple uniform of a hooded anorak. Their philosophies were a curious cocktail of the biblical book of Revelation and science fiction. Most importantly, they believed that the death of the physical body was the only means available to help them attain enlightenment. Thirty-nine dead followers may seem like a drop in the ocean compared with the billions living on the planet. But anyone who has ever been in a position of authority will know how difficult it is to secure obedient performance of the most mundane and simple tasks. How is it that one person, in the world's most technologically advanced society at the close of the twentieth century, could motivate others to end their lives, and persuade them to believe the impossible? Can we learn anything useful about the nature of religious self-destruction from the history of these people? Perhaps the answer has less to do with the individuals concerned or the nature of their beliefs, and more to do with the modern environment in which they emerged.

At the centre of the movement were two personalities, who called themselves variously 'The Two', 'Bo' and 'Peep', and 'Nincom' and 'Poop'. Marshall Herf Applewhite was born in 1931 and trained as a Presbyterian minister. He later abandoned the religious life and pursued a career in music, subsequently holding university posts in Texas and Alabama. His sexuality was troubled and his marriage ended in 1968 after a number of gay affairs. In 1972 he met Bonnie Lu Nettles, a nurse working at a hospital where Applewhite came seeking a 'cure' for his homosexual urges. Little is known about Bonnie Lu, except that she was the daughter of a Baptist minister and belonged to a meditation group which claimed to be able to 'channel' messages from the spirit world. The two established an immediate rapport that seems to have been entirely non-sexual. They believed that their meeting had been foreordained.

After spending some time on a Texas ranch, and failing to join a Hindu-inspired yoga movement, the two became convinced that they were 'the two witnesses' mentioned in the book of Revelation. Armed with a car and a stolen credit card, they then set off to travel throughout Canada, leaving messages in churches to let people know that the events foreseen in Revelation were happening. At one point they arrived at a New Age centre to discover, to their chagrin, that another couple were there who also claimed to be the 'two witnesses'. Things took a serious downturn when the police arrested them both for credit card fraud and they were sent to prison. With time to reflect, Applewhite reformulated his beliefs, focusing on the idea of UFOs and coming to believe that aliens were a more highly evolved form of humans.

After emerging from gaol, the two organized meetings at which the public were invited to learn 'the truth' – namely, that 'The Two' were going to be taken to the 'next evolutionary level' by means of a spacecraft and then return to earth. They gave various names to their organization – the most short-lived, understandably, was the Anonymous Sexaholics Celibate Church. Applewhite and Bonnie Lu also regularly adopted comic names such as Guinea and Pig or Ti and Do, to emphasize their separateness from the human level. They were prolific publicists, conducting a total of 130 meetings in the USA and Canada and acquiring some two hundred followers.

In 1975 these followers were split into cells and settled in various locations around the country. Stringent membership rules were imposed: adherents were allowed minimal contact with the outside world; men had to shave off their beards; women were forbidden to wear jewellery; alcohol, drugs and sex were banned. Members were also required to assume new names, all of which had to end in '-ody'. By February 1976 half of the movement's members had dropped out in response to these demanding conditions. None the less the groups endured, their finances bolstered by the fact that many adherents found work in the newly emerging computer industry.

Bonnie Lu died in 1985, and the movement re-emerged in 1992, now calling itself Total Overcomers Anonymous. TOA claimed its 'crew' for the ascent to the next level was now complete, but nevertheless invited interested parties to join them, placing ads on satellite TV and in the newspaper *USA Today*. The message was now a curious cocktail of the apocalyptic and the horticultural – the earth's current civilization was about to be 'spaded under' because its inhabitants were refusing to evolve. 'Weeds' present in the global garden were about to become compost. The ad campaign was directed not so much at attracting new followers as at reclaiming those who had left the movement in the seventies. Some twenty of them now rejoined.

During the five years after its revival the group's lifestyle became progressively more and more austere; it was during this period that Applewhite and seven other members underwent castration. Applewhite believed that he was suffering from renal cancer – a diagnosis later proved to be incorrect. The final impetus behind the movement's drastic communal suicide was the sighting of the comet Hale–Bopp in 1997. By then, the internet had become the primary means of mass communication, and many of the movement's members undertook work as web designers, under the company name Higher Source. Internet rumours and a controversial book were advancing the idea that there was another object behind the Hale–Bopp comet, and Applewhite's movement, now calling itself Heaven's Gate, seized on the idea that this second object was a spacecraft, coming to facilitate their 'evolution'. Then, in March 1997, the group committed suicide at their California ranch, in the expectation that leaving their physical bodies behind would enable them to progress to the 'next level'. They left behind just one survivor, Chuck Humphrey, who called himself Rkkody, and who committed suicide a year later in an attempt to join the rest of his crew.

The terminology employed by the Heaven's Gate movement demonstrates what a quintessentially modern phenomenon they were. Followers were keenly interested in sci-fi, in particular, television programmes such as *The X Files* and *Star Trek*. Much of their language comes from this source – for instance, in referring to themselves as a 'crew'. Like the London suicide bombers who blew up three tube trains and a bus in July 2005, they were also highly educated in the use of the internet, for acquiring information (however faulty), for advertising their activities and, in their case, for earning income.

They also showed themselves to be thoroughly modern in their interpretation of the book of Revelation. As the son of a minister and a former ordinand himself, Applewhite would have been familiar with the correct, 'scholarly' way in which to approach this much-disputed book of the Bible. Opinion differs considerably over its dating and whether it was written for a Jewish or a Christian audience. But, like a truly modern man, Applewhite interpreted scripture in an entirely personal way – identifying himself and Bonnie Lu, for example, as the 'two witnesses', and interpreting the phrase 'come up here' as a summons to personal physical ascent into the sky – influenced, no doubt, by his keen interest in science fiction. Revelation makes various references to the story of Jesus – for example, in prophesying a mass resurrection of the dead after 'three and a half days'. Some scholars believe passages like this were inserted at a later date to give a Christian appeal to an essentially Jewish book; but Applewhite took his interpretation in another direction. Jesus was a human being who had been 'tagged' by alien visitors for further evolution, and reappeared on earth to demonstrate how, through the death of the physical body, other people could themselves ascend to the next level. In short, Applewhite grafted a *Star Trek*-style plot onto the text of this ancient book. In a real and very dangerous way, he 'picked and chose' beliefs, shopping in the faith supermarket, marking himself out as an entirely modern religious person.

Handsome Lake and the fundamentalist revival

At the close of the eighteenth century, a warrior called Handsome Lake of the Iroquois tribe lay dying in his cabin. He was afflicted with a wasting disease, exacerbated through long addiction to alcohol. Nursed by his married daughter, he lay sick for four years. Then, receiving a set of four revelations, he stood up from his sickbed, apparently cured, and began to proclaim his message throughout the Iroquois people, who were based in areas now occupied by New York State. Alcohol had begun to destroy their once proud people, leading to moral decay and the loss of their homelands. Handsome Lake urged a revitalization of their traditional beliefs and abstinence from intoxicants. So powerful was his campaign that President Jefferson ordered a letter to be written commending his teachings. To this day, the site of his burial is marked by a granite monument.

Handsome Lake was, in essence, a fundamentalist. My use of the term might seem peculiar, because we are used to seeing it applied almost exclusively to Christians and

Muslims in the modern age. True, the term itself is a twentieth-century coinage, deriving from a twelve-volume compendium published by a group of American Protestants between 1910 and 1915 under the title *The Fundamentals: A Testimony of the Truth*. These original fundamentalists were Christians who wanted to unify the various strands of Protestant belief in order to fend off Darwinism. Like so many other movements that we call 'fundamentalist', this one was born in conditions of struggle and rapid, unsettling change.

There is a tendency nowadays to think of fundamentalism as being solely a reaction to the conditions of modern life. For example, when Muslim girls in French schools campaign for the right to wear the veil, we see them as rejecting western dress in an attempt to protect their ethnic identity in the melting-pot of global society. Christian fundamentalists in America are known for their uncompromising stance on modern morality, and for their angry and sometimes violent protests directed at homosexuals and abortion clinics. It has also been suggested that only Judaism, Christianity and Islam can become 'fundamentalist' religions, because they base themselves on ancient texts, which are viewed as sacred and unchangeable, and which therefore continually clash with current thinking.

But this sort of reasoning is faulty. For a start, many fundamentalists are far from averse to the modern world – indeed, they use the internet, television, radio and newspapers as a means of spreading their beliefs. Terrorism, whether in the name of Judaism, Islam, Sikhism or any other cause, depends upon the media in order to make its atrocities known and felt by the maximum number of people. Second, such arguments neglect the fact that fundamentalism is as old as religion itself, and not at all restricted to the 'big three' of the Judaeo-Christian tradition.

We can understand fundamentalism a little better if we look at it in the abstract. Early humans were pretty well defenceless on their own among predators and in the difficult environment of the savannah. So, like all primates, they tended to form groups. Our evolution has left us with a strong desire to be in a group – it generally feels safer. If you take twenty modern humans, split them into Red and Blue teams and give each team a simple exercise to perform, the results would demonstrate how deeply ingrained is this 'groupishness', as the brilliant writer on genetics Matt Ridley calls it. After a very short time, members of the Red team will be better disposed to each other than they are towards the Blues. Research shows that they will soon tend to find members of their own team more intelligent, attractive and honest than those of the other team; they will also be more willing to cheat and deceive a member of the opposing team than one of their own.[10]

In every society, change takes place at an uneven rate; some groups embrace it, others feel disturbed by it. Fundamentalism comes to the fore when the latter group begin to feel as if change is being thrust upon them by the former. They feel as if they are losing power to the other group, and may adopt extreme measures to attempt to prevent this from happening. Within the losing group, there may be accusations of backsliding and betrayal. The reason 'the other lot are winning', it is said, is because certain members have become weak and decadent. This can lead to rigid physical and mental disciplines to 'strengthen' the group against the opposition – prayer, fasting, exercise or, as in the case of some Irish Republican or Islamic groups, military drill.

Fundamentalists also use myths to impel their struggle and motivate members of their groups. There is the myth of a 'Golden Age' – some indistinct past, when everyone was more pious, more faithful, more rigidly disciplined, and of course, happier; and there is the myth of a 'Glorious Future', in which the traditionalists are triumphant over their enemies and there is widespread conformity.

We generally see the same processes, the same language being employed whatever the background. The prophets of the Old Testament were opposed to the ostentatiousness and wealth of the Jerusalem-based Temple cult. Muhammad was opposed to the urban, affluent materialism of his Qurayshi tribe. Jesus Christ argued that strict adherence to the Law was not enough. The early Christian heretics felt the power and pomp of the church was obscuring Christ's central message of humility. Religions are not so much 'prey' to fundamentalism as fundamentalism is a vital part of how religious movements come into being in the first place. The same tendencies even occur in ideologies where God plays no part. When Stalin came to power in the Soviet Union, he castigated many of his former revolutionary comrades, such as Trotsky, as decadent, bourgeois intellectuals with ideals contrary to the 'true' spirit of Marxism.

Present-day Islamic fundamentalism owes much to a movement that began in the 1870s, in reaction to the dominance of colonial powers like Britain and France over the Islamic world.[11] These empires had usurped local power and wealth and imposed western systems of law and education on populations which already had their own highly developed traditions. The leader of the reaction, Jamal al-Din al-Afghani (1838–97), inspired generations of oppressed Muslims by idealizing the 'Golden Age' of the past, when Islamic empires across the world had been strong, independent and wealthy. Al-Afghani, considered by some Muslims the founding father of Islamic modernism, argued for a return to simple piety, reform of Islamic law to meet the needs of the modern age, and violent resistance to western influences. His

427

views proved to be a strong influence on the Iranian Revolution of 1978–9, when the secular Shah was deposed and the religious rule of the Ayatollah Khomeini established the first Islamic Republic. The Ayatollah's Revolutionary Guard were motivated by beliefs that they were restoring a purer, older Islamic tradition to its rightful place – creating a Glorious Future, in effect, by resurrecting the Golden Age. Similar ideas motivate armed groups like Hamas today.

Fundamentalist movements tend to be led by key charismatic individuals, whose role it is to call the faithful to action. In modern America, the phenomenon of 'televangelism' has made the likes of Billy Graham and Oral Roberts household names. But spreading the word is not confined to Christianity. In the nineteenth century, the Ghost Dance, a religious movement among the Plains Indians, sought to revitalize Native American civilization and expel the white invaders from Indian lands. The message originated in the Utah area, but was carried far and wide by its prophet Wovoka, who took his hypnotic dance across to California, Nevada, Oregon, Washington and South Dakota. In 1890 conflicts between Ghost Dancers and white troops stationed in South Dakota led to the Battle of Wounded Knee, where two hundred Native Americans were massacred.[12]

So fundamentalism – in terms of groups of people wishing to revive a glorious past – is nothing new; nor is it restricted to the major monotheistic religions. But it nevertheless does great harm to both the credibility and the achievements of religion. When, in the name of the sanctity of life, Christian activists murder doctors who carry out abortions, the paradoxical nature of their beliefs seems particularly vivid. Richard Dawkins cannot be alone in feeling that 'only the wilfully blind could fail to implicate the divisive force of religion in most, if not all, of the violent enmities of the world today'.

However, the evidence suggests that the Divine Idea – in both hateful and constructive forms – will never die.

No sensible person today can ignore the fact that religious revival – whether in New Age beliefs or the 'conventional' religions – has led, and may well continue to lead, into religious intolerance and hostility. The events in Japan that I described at the beginning of this chapter are not so dissimilar in their effects from the disputes between Muslims and Hindus in Kashmir, Shi'ite and Sunni in Iraq, the war between Catholic and Protestant in Northern Ireland, the struggle between Muslims and Jews in the Middle East. Religious intolerance is common and it pervades all societies.

Salman Rushdie's confrontation

Consider the *fatwa* for blasphemy on Salman Rushdie pronounced after the publication of *The Satanic Verses* in 1989, and the ripples that reverberated around the world. A *fatwa* in Islam is a legal pronouncement, issued by a religious law specialist, a *mufti*, on a specific issue which is unclear and which needs adjudication. The argument over Salman Rushdie's book related to an ostensibly blasphemous statement from a biography of the Prophet Muhammad, which implied that pagan goddesses were included in Islam's original monotheism.

A few months after the book appeared, Ayatollah Khomeini, Supreme Leader of Iran, using the medium of Radio Tehran, called for religious Muslims to kill Rushdie:

In the name of God Almighty. There is only one God, to whom we shall all return. I would like to inform all intrepid Muslims in the world that the author of the book entitled *The Satanic Verses*, which has been compiled, printed, and published in opposition to Islam, the Prophet, and the Qur'an, as well as those publishers who were aware of its contents, have been sentenced to death. I call on all zealous Muslims to execute them quickly, wherever they find them, so that no one will dare insult the Islamic sanctions. Whoever is killed on this path will be regarded as a martyr, God willing. In addition, anyone who has access to the author of the book, but does not possess the power to execute him, should refer him to the people so that he may be punished for his actions. May God's blessing be on you all. Ruhollah Musavi Khomeini.

Thereafter, Rushdie lived in fear. Guarded by British Special Branch, unable to leave his secret address except on rare occasions, he was never seen in public – effectively, he had to disappear. Later in 1989, at Berkeley University in California, bookstores selling the book were fire-bombed, and in the same year in Mumbai twelve people died by gunfire during a protest at the British Embassy. Similar demonstrations with fatal consequences occurred in other Muslim countries, including Egypt. Muslims throughout the world burned copies of the book at public rallies. In 1991 thirty-seven guests died when their hotel in Sivas, Turkey was burned down by locals protesting against Aziz Nesin, a political activist and supporter of Rushdie, and also his Turkish translator. In 1991 his Japanese translator, Hitoshi Igarashi, was stabbed to death in Tokyo, and the translator into Italian was beaten, then stabbed, in Milan. In 1993 his Norwegian publisher William Nygaard was shot and severely injured in Oslo.

Ayatollah Khomeini died fairly soon after he had issued his *fatwa* and in 1998 the government of Iran asserted that it was no longer enforcing it. But, as has been only too clear, that message did not in any way stop the persecution of the hounded Rushdie, nor did it prevent the attacks on his supporters and colleagues. And, to add insult to injury, the murder decree was reinstated in early

2005 by the present religious leader in Iran, Ayatollah Ali Khamenei. He stated that only the man who had originally pronounced the *fatwa* could legally annul it – a little difficult seeing that Ayatollah Khomeini had died sixteen years previously. So many people died, and Salman Rushdie lived in constant fear of his life, because of religious outrage about a book that few of the 'offended' had read, and probably far fewer actually understood.

In vitro fertilization and 'benign' fundamentalists

We should make no mistake: religious extremism and fundamentalism in many of their forms, once the terror of the Middle Ages, are still major forces in today's world. In more subtle ways than those that make the headlines, fundamentalism, with its various prejudices and narrow judgements of society, is still very much alive even in Britain. It might be pertinent at this point to recount my own experiences relating to the issue of research on the human embryo.

One of the greatest medical advances of the twentieth century was successful in vitro fertilization (IVF) in humans. Robert Edwards, the scientist who pioneered it, was a singular figure in the research and many people, myself included, believe it is a scandal that he has never been awarded the Nobel Prize. As a result of the medical development he was instrumental in bringing about, over one million healthy babies have been born on this planet – each one a person who, without this technology, would not have been born at all.

In Britain, the country which nurtured most of the early research, there was wide public acceptance that IVF was a useful technique. Most people were not hugely concerned by the treatment itself, but some were worried about the implications of the research used to establish the efficacy and safety of IVF. In 1982, in response to the recognition that there were a number of moral issues surrounding IVF treatment, the then Conservative government set up a Royal Commission to investigate the matter, chaired by Dame Mary Warnock (now Baroness Warnock), a Cambridge philosopher. The Warnock Commission report, issued in 1984, favoured the continuation of research involving human embryos up to the fourteenth day after fertilization. It advised that, as it was morally acceptable to undertake embryo research up to the fourteenth day, formal regulation to approve work within these bounds should be instituted by Parliament. It is interesting that the fourteen-day limit was a fairly arbitrary one, pragmatically derived – as religious Catholics quickly, and correctly, pointed out. But the cut-off point of fourteen days was nevertheless chosen for quite good reasons. Before fourteen days there is no development of even the most primitive nervous system – so there is no brain and no consciousness. Second, most fourteen-day embryos do not survive implantation, and before that happens a woman is not yet pregnant. Moreover, up to fourteen days the embryo can split and form identical twins. So the embryo at this stage cannot be said to be an individual – that potential comes later.

The government of the day dragged its heels. Towards the end of 1984, events took a sudden course when Mr Enoch Powell, Member of Parliament for County Down South, won the House of Commons private members' ballot, entitling him to introduce a bill of his personal choice. Although Powell was an Anglican (having been an atheist in his earlier years), I do not believe that he took up the issue of embryo research because of any deep conviction about the human soul. Indeed, he declared as much when I met him at the Cambridge Union that year. As a practising Anglican, he had certainly been lobbied, though, by very religious Christians. Possibly he introduced his private member's bill, the Unborn Children (Protection) Bill, for mostly political reasons. A most able politician, singled out by nature and intelligence as capable of holding the highest political office, he had through his intemperate speeches on race collided head-on with the official views of the Conservative Party. Resigning from the party and relinquishing his ambitions to continue to serve in government, he joined the Ulster Unionists and was elected as a Unionist MP. His 1984 bill, with its religious implications, certainly improved his standing in a very large Ulster constituency where his was a somewhat controversial presence. This constituency, one of the biggest in the British Isles, contained very many Roman Catholics who were prepared to support him as well as many religious Protestants who were impressed by his approach at the height of the 'troubles' in Northern Ireland.

At the second reading of his Bill in the House of Commons, there was a huge vote for an outright ban on embryo research, in spite of the liberal conclusions of the Warnock Commission. It rapidly became obvious that we scientists doing this research – at the time, there were only a handful of us in the world – faced a massive problem in preventing our work from being outlawed. The issue gained a very high profile in the media, and I found myself in the highly unwelcome position of being in the thick of daily national coverage. The main arguments against the research were presented for the most part – in almost every newspaper, television programme and radio phone-in – by members of the Catholic Church and a few religious Protestants. I regret that many of these people presented arguments which were untrue. They frequently and repeatedly asserted errors of fact, were often quite vicious in the personal nature of their attacks on people like myself, and carried exaggerated images of which one of the more aggressively competitive advertising companies might have been proud. For example, one Catholic group repeatedly showed large pictures of fully-grown fetuses and suggested that, effectively, we were vivisecting such humans. Another group, the Society for the Protection of the Unborn Child, posted a life-size (12-inch) plastic fetus to every Member of Parliament. No matter that the fertilized eggs which we were using were invisible to the naked eye, had perhaps four to eight cells with no organs, and were less advanced in their development than those routinely destroyed by at least four of the most popular methods of contraception. On the media, it was difficult to get the opponents of research to admit that they opposed contraception too, except the 'natural method' of temperature charting – because, clearly, they knew only too well that if they did so they would immediately alienate 90 per cent of viewers. Moreover, very few were prepared to admit that they were opposed to the generally accepted treatment of IVF, or that the Catholic Church viewed sex without procreation as against the divine command.

None of the science, patiently argued, made any difference. Essentially, the view of these religious people was simple. Human life was sacred; life began at conception; and any interference after fertilization was akin to murder. It was of no avail to point out that we too felt that human life was sacred, which was why we were engaged in this work. Nor would they accept the argument that many fertilized eggs, by reason of their chromosomal development, were quite incapable of viability. Nor that only about 82 per cent of fertilized eggs – under ideal conditions in the uterus – would be capable of growing into a full-term infant. Nor that fertilization was not an instantaneous process and there was not a single moment in time when it could reasonably be said that a human being had been formed.

Many of these religious people showed little compassion for their infertile Christian brothers and sisters, frequently stating – until they saw how hostile the public reaction was – that they should adopt rather than make forlorn attempts to have a baby of their own. And they accused the scientists, myself particularly, of lying about what we might achieve through pursuing our research to promote healthy human life; of lying when we said that we might find solutions to genetic disease, or that we might gain insights into the genesis of crippling disease in childhood, or better methods of contraception, or new insights into how cancers develop. These scientific opinions, put forward in good faith, all turned out in time to be well founded; but all were hotly denied by 'experts' wheeled in to support the religious groups.

Moreover, many of the churchmen most strongly opposed to this work refused to equip themselves with any first-hand information. Many of us offered to open our laboratories for lay inspection. We also asked priests to come to discuss the issues with our patients. Very few religious leaders who were opposed to the work came. Only more open-minded Christian leaders, including some bishops, and a number of rabbis, including the Chief Rabbi, bothered to take up these invitations – and most them left agreeing that what we were doing could hardly be classed as 'evil'. To his credit, the wonderful Duke of Norfolk, lay leader of the Catholic Church, did pay an extended visit to our unit at Hammersmith. He remained opposed to our work and voted against it in Parliament, but his visit was extremely useful because it showed his willingness to take a considered opinion. Thereafter, too, he treated the scientists with the greatest courtesy and did much to mend the hard feelings between people like myself and some of the more fundamentalist Catholics who had refused constructive dialogue.

I call these people 'fundamentalists' deliberately, partly

because of the virulent letters, sometimes unsigned, that I received. They were numerous and threatened me with personal violence or worse (and there were far more of these than I cared to admit to my own family). Some letters – clearly from religious Christians – were deeply anti-Semitic, accusing me of killing Christ, or asking why I hadn't been committed to a gas chamber, or at least suggesting repatriation to some other country (my family have been in England since around 1700). But just as appalling was the huge propaganda machine set in motion by some individual priests. All over the country, people going to church to pray on Sunday were asked to sign petitions to Parliament opposing the kind of work my colleagues and I were doing. Some of my churchgoing IVF patients, being left confused about exactly what it was they were signing and given very little information, appended their signatures together with many others. Children going to some Catholic schools were instructed about the murders people like myself were perpetrating at Hammersmith. More than one school in Acton, where I had long-standing commitments to give science tutorials, suddenly cancelled my impending visits after talks by a local priest. And my secretary at the time, also a Catholic, was put under considerable pressure (which she laudably resisted) to refuse to do my work.

The British do not like this kind of absolutist propaganda and eventually this fundamentalist approach did not win the day. The tide of public opinion turned considerably when it became clear that many of the scientists themselves had carefully thought-through religious and moral views. And there was a recognition that, in a pluralist society like the UK, it was wrong for a small minority to impose their views in so rigid a way on others who clearly had a different, equally valid, perspective. Enoch Powell was defeated and eventually Parliament enacted liberal legislation – which, incidentally, had wide support from most of the Anglican bishops in the House of Lords. But the episode was a deeply uncomfortable experience for me. I had had a huge respect for the high principles and strong family values of the Catholic Church. For a time, my outlook was badly bruised by the intolerance that I was shown, the physical blows and taunts I had received in the street, and the untruths that were peddled about the work I was doing with full ethical approval. It is not easy for anybody who believes that the work to which they have dedicated much time and thought has a moral purpose to experience being so reviled. And I have never forgotten the persistent lies that some religious people are prepared to make if they feel the means justifies the desired result.

Throughout this chapter I have referred to 'fundamentalism', and at an earlier point have explained how it emerges as a defence of religious thinking when religious thinking is threatened by the views of a secular society. But it goes further than that – it is an act of defiance against liberal views; it represents a return to the 'truth', revelation or faith unsullied by society's attempts at interpretation. God's word is absolute, and there is no uncertainty. The Christian fundamentalists I met in the Grand Canyon, who believed that this great valley had been formed by Noah's Flood, believed not merely that the Bible is true in every respect, but that it is meant to be taken literally – essentially, not to be interpreted. Even beyond this, fundamentalism, as it becomes more extreme, not only declines to incorporate modern values, but is not prepared to live alongside them. It sees modernity as a threat – as something to be increasingly resisted. Of course, in doing so it brings great comfort to adherents of a faith, because it offers certainty in an uncertain world. But it also brings grave problems when it forges links with mechanisms of the state. The Catholic fundamentalists who saw me as 'doing the work of the devil' by handling human embryos harmed only a few people – those who might, broadly, benefit from my research. But fundamentalist certainty becomes infinitely more dangerous when it espouses politics or, of course, embraces nationalism. And religious fundamentalism is at its most destructive when it ignores the basic tenets and moral principles laid out virtually universally in all religious scriptures – above all, the respect that should be accorded to human life.

It is a frightening thought that between the middle of the twentieth century and these early years of the twenty-first, we have seen many conflicts that have had their genesis in a mixture of religious fundamentalism and nationalist thinking. The Middle East, the wars in the former Yugoslavia, the religious conflicts in India, Pakistan and Sri Lanka – all have a serious religious background. Revolutionaries like Osama bin Laden, and the insurgents in Iraq, invoke a holy struggle, and all sorts of crimes – such as indiscriminate gunfire and suicide bombing – are given a holy purpose.

Equally, there are fundamentalists pursuing science and promoting its 'truth'. And they can sometimes be almost as threatening in many ways as religious fundamentalists, not least because they seem to see no limitation to their pursuits and appear to have little or no moral framework – except one that they themselves have devised. Take one example – one of the most threatening technologies devised by humankind: genetic modification of organisms. Make no mistake, I am greatly in favour of genetic modification where it can be shown to be helpful to human health, where it involves no suffering of animals and where it presents no threat to the fragile environment of this planet. But modification of the human germ-line, changing the genes in human embryos? This might in time lead to enhanced humans with greater intelligence, or strength – but it could well be intelligence

438

without wisdom, strength without moderation. The results of such tampering are truly unpredictable. And if we change the nature of the human genome too much, does that then mean that some of us are no longer human? If the central tenet of our morality is that human life is sacred, what happens to humans when they become an underclass to superhumans? The Nobel Prize winner James Watson once said of genetic modification: 'It seems obvious that germline therapy will be much more successful than somatic. The biggest ethical problem we have is not using our knowledge.' Such scientific certainty is almost as frightening and as dangerous as any religious certainty.

One of the curious features of so much religious fundamentalism is that it offers, sometimes demands, a return to a more primitive time – a time when society was less driven by science, and before scientific knowledge and enlightenment offered rationalist solutions to society's problems. Yet modern fundamentalists use the fruits of science whenever it suits them. Modern transport, weaponry, electronic communication – all are used to aid the fundamentalist in his war against liberal values. And what is perhaps most curious is that, while science and religion are essentially two different mechanisms, two different systems for looking at the natural world, both are driven by man's innate uncertainty. Science is so often portrayed as being certain, as giving absolute answers, as revealing 'the truth'. But of course it does nothing of the kind. Science is not about black and white; it cannot answer all humans' questions about their place in the universe, nor render plain the nature of that universe in all its quantum complexities. Science is pursued because we are uncertain, have doubts, want to know more. And all this is surely true of religion too. Religion is equally about human uncertainty, about our doubts, about our experiences as dwarves in a vast cosmos.

The problem is that humans are bad at dealing with

uncertainty. Ambiguity is always harder to bear than clarity; insecurity is less satisfactory than conviction. In this respect, as in many others, religion stands in relationship to God as technology does to science. Our search for God is very similar to our search for scientific knowledge. Religion can be good or bad, just as technology can be a force for good or evil. We need not be wary of science, only of the uses to which it is put. Certainty in religion is as dangerous as is certainty in science.

Not all fundamentalism is necessarily bad. The Polish Hasidic rabbi Menachem Mendel of Kotsk, who was born in 1787, was universally regarded by his contemporaries as one of the saintliest of men. He was also a true religious fundamentalist. It is said that he inscribed on his banner one word: *Emet*, 'Truth'. He believed that the way to God, to the one and only truth in life, was tortuous and complex. A man had to divest himself of all outside appearances, and all emotional attachments – he needed to be free – if he was to attain to that truth. And the proper use of freedom was not to conform, not to attempt to please oneself, not to engage in the world, but to study the Torah. But, he pointed out, while that study of the Torah may be the safest way for a Jew, it was also the most dangerous. Travelling on this ticket, a man might become self-righteous, conceited, self-important, and, in a sense, worship idols. This view harmonizes with concerns about the pursuit of science, whether by a medieval alchemist or a modern-day genetic engineer.

It is curious to think that the Kotske Rebbe, Menachem Mendel, was a contemporary of the Danish philosopher Søren Kierkegaard, and that they had many views in common. Mendel's ideas resonate strongly with those of Kierkegaard, for whom the great paradox was the relationship between the knowing mind and eternal truth. Both these men sought, as they saw it, truth. Both had a deep feeling that there was an immense gap separating the world, and its affairs, from God, who is not to be found within nature or human reason. But what different worlds they inhabited. Each day the bearded Kotske Rebbe would have donned the classical garb of his branch of Hasidim – clothing almost identical to that you can still see in parts of Stamford Hill today: a *spodik* – the tall black fur hat; a *kapote* – a long black coat (on the Sabbath, one made of silk or satin); a *gartel* – the belt tied with a knot around the middle of the waist to signify the separation of the higher holy intellect from the basic human functions, the sacred from the profane; and the trousers with bottoms tucked inside the socks, so that they most resemble the plus-fours of the traditional golfer. And Kierkegaard? The best portrait of him I have seen – a watercolour in Denmark – shows a clean-shaven, lean, handsome, ascetic face above a high wing-collar, a fashionable brown coat with wide lapels, smart drain-pipe trousers, a silver-topped cane and an elegant silk top hat. What might they have said to each other if they had ever actually met?

There is a poignant end to the tale of Menachem Mendel of Kotsk. Twenty years before his death he decided to 'leave' the world. He locked himself in a tiny room, allowing himself only to be fed through a window. He seldom emerged and was never seen in public again. His many disciples, who came from all over eastern Europe, would learn and study in a room close by. No-one knows why Mendel cut himself off, though there are various explanations – some say he was closer to God, others that he had lost his faith. Whatever the case, this religious fundamentalist had an entirely benign and positive effect on his disciples and on the society around him, and he was of great importance in the development of the Hasidic movement.

The point of life is – the search for the point of life

There is a remarkable account in Michael Ignatieff's biography of Isaiah Berlin of a minor event which the philosopher recalls towards the end of his life. Berlin made a point of attending synagogue each Jewish New Year, when Jews read the passage from Genesis which tells of the binding of Isaac and Abraham's agreement to God to offer him up as a sacrifice. Berlin, in a letter to his Oxford friend Peter Oppenheimer, says of this passage: 'the validity of a religion shd not, in my view, depend on its moral implications: it is transcendent, absolute, orders things which, in human terms, may be horrifying (as so often in the more blood-shedding exploits in the O[ld] T[estament]) but *are* the essence of a truly religious attitude'.

It is interesting that when Igor Stravinsky asked Isaiah Berlin to write a libretto for a religious cantata, Berlin immediately suggested that episode, the binding of Isaac. He was enthralled by the notion of a man blindly carrying out, in total submission, the most illogical and pointless act – an act, moreover, which, if brought to completion, is inhuman and extraordinarily cruel. God commands the unthinkable. Berlin was gripped by the notion that humans, in their search for truth, bow to things that are beyond their understanding.

These brief accounts of Berlin and Mendel represent only two recent instances of the enduring human search for the transcendental. Before the time of Abraham, nearly 5,000 years ago, probably even before my ancestors were assisting Rameses II with the construction of the Pyramids, Neolithic inhabitants of the British Isles were schlepping five-ton blocks of stone from the Prescelli Mountains in Pembrokeshire down to Salisbury Plain. That is a distance of 245 miles; rafts on the River Avon, which were probably employed for part of the journey, could have been used for only a small section of the voyage. To my mind, that arduous expedition represents a journey that humankind has taken repeatedly throughout its history. At one level it seems totally laughable that our predecessors could have carted those huge chunks of stone so far, for so long, simply to place one of them on top of two others in the circle that makes Stonehenge. Yet there is something totally marvellous about their devotion to the Divine Idea. What called for such toil and such effort, with all the injury and deprivation that must have been entailed in the enterprise, was Neolithic man's spiritual search for the transcendental.

Why is it that, today, *Macbeth* is possibly the most performed play in the English language? It is partly, I believe, because of its religious message – or rather, its lack of it. *Macbeth* is an extraordinary play partly because, like no other of William Shakespeare's dramas, it is in some way about denial of God. It is Shakespeare's blackest play, full of terror. Macbeth the man is a huge paradox. He is obsessed, consumed by magic, with a hallucinating imagination which gives him – and us – the horrors. I have seen numerous productions of *Macbeth*, but none has left so lasting an impression as Akira Kurosawa's film version, *Throne of Blood*. The horror is always there, but this cinematic masterpiece offers something more brilliantly depicted. In the film, Macbeth, a samurai chief, is surrounded by nothingness. When finally the King, Duncan, is killed and Macbeth achieves his great goal of being the next ruler, the camera slowly reveals his bare room in the palace. It contains one stool. It is black and totally devoid of any comfort: there is no proper furniture, not even trappings, hangings or a carpet. It is an empty room for an empty ambition. It symbolizes Macbeth's life, devoid of moral purpose. As Harold Bloom, that masterly commentator on Shakespeare, says, 'Macbeth allows no relevance to Christian revelation.'[13] Here he is, on hearing the news of Lady Macbeth's death:

> To-morrow, and to-morrow, and to-morrow,
> Creeps in this petty pace from day to day,
> To the last syllable of recorded time;
> And all our yesterdays have lighted fools
> The way to dusty death. Out, out brief candle!
> Life's but a walking shadow, a poor player
> That struts and frets his hour upon the stage,
> And then is heard no more: it is a tale
> Told by an idiot, full of sound and fury,
> Signifying nothing.

This is surely the ultimate atheist statement. Macbeth is a great paradox because, while he has imagination, ultimately he lacks the spiritual. For him, life has no meaning, it can be snuffed out like a candle, it is a mere actor; life is empty. He is a man completely without moral purpose, essentially because life for him has no inherent value – it has no point. Macbeth is wedded to the supernatural, he uses it as a means to an end, but he has no feeling for the transcendent. He denies everything that a Christian society holds most precious. Macbeth violates man's nature.

> What are we?
> What is our life?
> What is our piety, our righteousness, our salvation?
> What is our strength? What is our might?
> What shall we say before You?
> Are not all the mighty men as nought before You?
> The men of renown as if they never had existed,
> The wise as if devoid of knowledge,
> The intelligent as if without discernment?

This daily morning prayer, couched in very similar language, expresses exactly the opposite experience to that of Macbeth. It is the Jewish experience, not of the pointlessness of life, but of our frailty and insignificance when faced with the greater mystery of God's infinity. This is not to say that we 'signify nothing'; rather that we are small but are given a higher purpose, that we have difficulty in understanding but still have dignity. Macbeth does not try to understand; but man in this prayer is searching for what is missing in Macbeth's life, the transcendent.

This notion is reflected in the story of Abraham, the father of the world's leading monotheistic faiths. This infertile man bitterly says to God that his existence and riches are without meaning if he goes childless, giving voice to the key idea held in our spirituality and in our genes – that ultimately our greatness is in the next generations. God leads him outside (Genesis 15: 1–6) and asks him to look up at the stars and count them. So shall your children be, He says – as numberless as the stars. But what God is doing here is more significant than a mere exercise in existential mathematics. The vastness of the universe should put us in our place. God is showing Abraham the transcendent. This is again the message in the book of Job, when that relationship between God and man has been tested to its utmost limit and God answers the plea of his beleaguered disciple out of the whirlwind, almost sarcastically:

> Who is this that darkeneth counsel by words without
> knowledge? . . .
> Where wast thou when I laid the foundations of the earth?
> Declare, if thou hast understanding . . .
> Whereupon are the foundations thereof fastened? or who laid the
> corner stone thereof? . . .
> Or who shut up the sea with doors, when it brake forth, as if it had
> issued out of the womb? . . .

Hast thou entered into the springs of the sea? or hast thou walked
  in search of the depth?
Have the gates of death been opened unto thee? or hast thou seen
  the doors of the shadow of death?
Hast thou perceived the breadth of the earth? declare if
  thou knowest it all ...

Man's knowledge is incomplete. It is not that his science is unimportant – indeed, it is the most essential tool he has. But he must remember that it is limited. To forget those limits is dangerous. Science will never quite explain his personal existence, or the far-flung universe beyond his grasp. His search for the point of life must continue; but perhaps the search itself is sufficient meaning for his existence.

Notes

Prologue: Wrestling with God

1 Rabbi Solomon ben Isaac (1040–1105), who lived in Troyes, France. His extensive commentaries on the Old Testament and the Talmud are the pinnacle of exegesis. He was succinct and incisive, using meticulous examination of Hebrew grammar. No other commentator has been as influential within Judaism.

Chapter 1: Religion's Roots: What Did Prehistoric Humans Believe?

1 Malinowski wrote a monograph about his research in Melanesia entitled *The Sexual Life of Savages*, which sold rather better than one might expect works of learned scholarship to do. A friend of mine reports that his father kept a copy in a locked drawer – which leads one to wonder if a lot of purchases were inspired more by furtive curiosity than by anthropological interest.

2 Christophe Boesch and Hedwig Boesch-Achermann, *The Chimpanzees of the Taï-Forest: Behavioural Ecology and Evolution* (Oxford: Oxford University Press, 2000).

3 G. Reichel-Dolmatoff, *The Kogi: A Tribe of the Sierra Nevada de Santa Marta, Colombia,* vol. 2 (Bogota: Editorial Iqueima, 1951).

4 P. Pettit, 'When Burial Begins', *Archaeology,* 66 (August 2002).

5 It is a very unpalatable truth that the blood libel still persists. If a search is done on the internet, it is possible to find various anti-Semitic sites which accuse Jews even today of sacrificing Christians for ritual purposes.

6 E. B. Tylor, *Primitive Religion* (London, 1871).

7 Mircea Eliade, *Cosmos and History: The Myth of the Eternal Return,* trans. W. Trask. (Princeton, NJ: Princeton University Press, 2005. First publ. 1954.)

8 Ibid.

Chapter 2: A Question of Survival

1 Mishnah Parah 3: 5.

2 P. Boyer, *Religion Explained* (London: Vintage, 2001).

3 E. O. Wilson, *On Human Nature* (Cambridge, MA: Harvard University Press, 1978).

4 G. W. Allport and J. M. Ross, 'Personal Religious Orientation and Prejudice', *Journal of Personality and Social Psychology,* 5 (1967), 432–43.

5 N. G. Martin, L. J. Eaves, A. C. Heath, R. Jardine, L. M. Feingold and H. J. Eysenck, 'Transmission of Social Attitudes', *Proceedings of National Academy of Science, USA,* 83 (1986), 4364–8.

6 J. M. Beer, R. D. Arnold and J. C. Loehlin, 'Genetic and Environmental Influences on MMPI Factor Scales: Joint Model Fitting to Twin and Adoption Data', *Journal of Personality and Social Psychology,* 74 (1998), 818–27, citing R. J. Rose, 'Genetic and Environmental Variance in Content Dimensions of the MMPI', *Journal of Personality and Social Psychology,* 55 (1998), 302–11.

7 Richard Dawkins, 'What Use is Religion?', *Free Inquiry* magazine, vol. 24, no. 5, Aug. 2004.

8 E. Durkheim, *Elementary Forms of the Religious Life* (London: Allen & Unwin, 1915).

9 I'd say similar (mis)conceptions are prevalent in our society. At a recent barbecue I attended, the men involved themselves exclusively with the 'important' business of maintaining the fire and cooking meat, while the women unconsciously relegated themselves to the 'low-status' preparation of salads.

10 Turner saw the symbols employed in rituals – for instance, a certain kind of bark, a ceremonial mask – as being entirely bound up with the values of the society. He made much of the fact that the Ndembu word for a symbol is *chinjikijulu*, which roughly translates as 'blazing a trail'. In other words, it is a thing which points to something else. We could say that the symbol of the crucifix 'points to' the value of acting in an unselfish way. See Victor Turner, *The Ritual Process: Structure and Anti-Structure* (New York: Aldine de Gruyter, 1995).

11 Dean Hamer, *The God Gene: How Faith is Hardwired into Our Genes* (New York: Doubleday, 2004).

12 J. Diamond, 'The Religious Success Story', *New York Review of Books*, vol. 48, no. 17, 7 Nov. 2002.

13 E. Turiel, *The Development of Social Knowledge: Morality and Convention* (Cambridge: Cambridge University Press, 1983).

14 R. Winston, *The Human Mind* (London: Bantam, 2003).

15 N. Coleridge, *Godchildren* (London: Orion, 2002).

16 A. Bahal, *Bunker 13* (London: Faber & Faber, 2002).

17 J. Cohen, trans., *A Life of St Teresa of Avila by Herself* (London: Penguin, 1957).

18 A. Newberg, M. Pourdehnad, A. Alavi and E. G. d'Aquili, 'Cerebral Blood Flow during Meditative Prayer: Preliminary Findings and Methodological Issues,' *Perceptual and Motor Skills*, 97 (2003), 625–30.

19 Scott Atran, *In Gods We Trust: The Evolutionary Landscape of Religion* (New York and Oxford: Oxford University Press, 2002).

20 Pascal Boyer, *Religion Explained* (London: Heinemann, 2001).

21 T. B. Ward, 'Structured Imagination: The Role of Category Structure in Exemplar Generation', *Cognitive Psychology*, 27: 1 (1994), 1–40.

22 M. Kelly and F. C. Keil, 'The More Things Change ... Metamorphoses and Conceptual Structure', *Cognitive Science*, 9 (1985).

23 J. L. Barrett, 'Anthropomorphism, Intentional Agents, and Conceptualising God', unpublished PhD thesis, Ithaca, NY, Cornell University, 1996.

24 S. E. Guthrie, *Faces in the Clouds: A New Theory of Religion* (Oxford: Oxford University Press, 1993).

Chapter 3: Finding the One God

1 But Zoroastrians would say the same about Christianity, with its Trinity of God the Father, the Son and the Holy Ghost!

2 Quoted in N. Reeves, *Akhenaten: Egypt's False Prophet* (London: Thames & Hudson, 2001).

3 D. Redford, *Akhenaten: The Heretic King* (Princeton: Princeton University Press, 1994).

4 A. L. Burridge, 'Marfan Syndrome and the 18th Dynasty Royal Family of Ancient Egypt', preliminary research report (part II), *Paleopathology Newsletter*, 111 (2000), 8–13.

5 Sequencing is the technique used to analyse the composition of the DNA by spelling out the sequences of the individual base-pairs – or 'letters' of the DNA alphabet – contained in a person's cells, or in a bacterium, virus or parasite.

Chapter 4: The World's Greatest Book

1 These seven laws are derived from Genesis 2: 16 and 9: 2–7, and are detailed in the Talmud (particularly in Sanhedrin 56–60 and Avodah Zara 8): prohibitions on idolatry, blasphemy, bloodshed, sexual sin, theft and eating from a

live animal, and the positive command to establish a legal system.

2 Rabbi Judah HaLevi flourished in Córdoba and Granada at the end of the eleventh century and subsequently emigrated to Israel. He was equally familiar with Arabic and Hebrew. His greatest work was *Kuzari* – in Arabic, 'The Book of Argument and Proof in Defence of the Despised Faith'. It was first printed in Hebrew in Fano in 1506.

3 Jean Astruc (1684–1766) was probably of Jewish extraction, though his father, from whom he learned Hebrew, was a Huguenot. He was Professor of Anatomy at Toulouse and published a book on venereal disease. But he also wrote what is arguably the first 'modern' book of biblical criticism, anonymously published in Brussels and Paris: *Conjectures sur les mémoires originaux, dont il parait que Moyse s'est servi pour composer le livre de la Genèse.*

4 The letters YHWH are referred to as the Tetragrammaton. The vowels are always omitted in Hebrew and no Jew ever pronounces God's four-letter name except on the most holy occasions. 'Jahveh' is not used by Jews and is only an approximation of the Name.

5 This is indeed highly enigmatic, as the Hebrew can be read as if in the future tense: 'I will be what I will be.'

6 Moses Maimonides (1135–1204) was a renowned Jewish scholar who spent most of his life in Spain and Egypt. His work *Moreh Nevuchim* (1190) – known in English as *Guide for the Perplexed* – marked an attempt to synthesize Jewish philosophy with Islam and Aristotle and prove that all were comprehensible by reason.

7 Although they are borrowing from the tradition of their oppressors, the Christian slave-owners, in the use of that term.

8 I. M. Lewis, *Ecstatic Religion: An Anthropological Study of Spirit Possession and Shamanism* (Harmondsworth: Penguin, 1971).

9 P. Boyer, *Religion Explained* (London: Vintage, 2001).

Chapter 5: The God of Change

1 The 'Wall' referred to here is presumably the wall around Jerusalem, which has always had powerful significance in Jewish literature.

2 The worship of Mithra, the Iranian god of the sun, justice and war, was popular in second-century Rome. Its origins are obscure; possibly some of its beliefs may have had some connection with Zoroastrianism. The most important ritual was the sacrifice of a bull, which generally seems to have taken place in enclosed underground temples or in caves. No Mithraic temple seems capable of housing more than a few dozen people. Only men were admitted to this militaristic cult and initiates were taken through seven stages or grades as they rose through the ranks of the brotherhood. Frescoes at Capua show initiates blindfolded, kneeling and prostrated. The impression that one gets is that initiates in the cult, in such a confined space with so large an animal, would be covered in blood by the end of their worship – the so-called 'red baptism'. Once initiated into the cult, a member had access to esoteric wisdom. Leaving the more sanguineous aspects aside, the meetings of this closed society seem somewhat reminiscent of a Masonic gathering.

3 J. Betjeman, *Collected Poems* (London: John Murray, 2003).

4 The Talmud, succinct and enigmatic as ever, does not explain why they wanted to put a lance through the chest of the 'corpse'. Mutilation of a dead body would be frowned upon under religious law, so perhaps these men were in the pay of the besiegers, wanting to confirm the great rabbi's death.

5 In his *Guide for the Perplexed* (1190).

6 J. Telushkin, *Jewish Literacy* (New York: William Morrow, 1991).

7 Rabban Gamaliel the Elder was one of the greatest Jewish sages, a descendant of Rabbi Hillel. He filled the office of

nasi or prince and was head of the Sanhedrin, the Jewish court. He had many distinguished pupils, of whom Paul was probably one. It is said he was tolerant of the early Christians. His grandson, Rabban Gamaliel II, was even more eminent and was regarded as the leader of the Jewish community, presiding at the centre of learning at Javneh established after the fall of Jerusalem.

There are many tales of miraculous events concerning Chanina ben Dosa. One talmudic story records how he was once bitten by a scorpion when praying but refused to interrupt his prayer. His pupils found him alive and well, but the scorpion dead by its hole. 'Woe to the scorpion that bites ben Dosa,' they said.

8 Though, I suppose, it's possible that Jesus' parents might have fled to Bethlehem to *avoid* the tax! It's also worth noting that the traditional 'nativity' scene of the baby Jesus in a stable, surrounded by adoring farm animals, has no basis in the gospels.

9 These Zealots were distinctly unusual, not least because suicide has never been condoned by Judaism – even martyrdom is regarded with suspicion in most Jewish sources – and the idea of encompassing one's own death simultaneously with that of others seems totally foreign.

10 C. S. Lewis, *Mere Christianity* (London: HarperCollins, 2001), p. 52.

Chapter 6: Muhammad and Islam

1 The word *hajj* probably has the same root as the word *hag* in Hebrew, which means 'a festival' – usually, and more specifically, a festival when Jews would go on pilgrimage to the Temple in Jerusalem.

2 Rudolf Otto, who died in 1937, was one of the most influential religious philosophers of recent times. He described the 'numinous' experience that, in his view, underlies all religion. Religious experience, he argues, is 'different from any other', involves a feeling of

'overwhelming power' and has a notion of being 'merciful'. His most famous book, *Das Heilige*, was published in 1917 in Breslau, and printed in English as *The Idea of the Holy* by Oxford University Press in 1923.

3 The caliphs were 'deputies' of the Prophet who, after his death, ruled Islam on his behalf.

4 Reza Aslan, *No God But God* (London: Heinemann, 2005).

5 Quoted in L. Ridgeon, *Major World Religions* (London: Routledge, 2003). I wonder if Carlyle was deliberately repeating his words to make a point, or if his annoyance with the Qur'an was so great that he was unaware of committing the same crimes with which he charges the book.

6 Quoted in J. Freely, *Istanbul* (London: Penguin, 1998).

7 A. Shariati, *Hajj*, trans. Laleh Bakhtiar (Teheran, 1988).

8 Edward Gibbon, *The Rise and Fall of the Roman Empire* (first publ. 1776–88), vol. 3 (London, 1848).

9 Philip K. Hitti, *History of the Arabs*, 10th edn (London: Macmillan, 2002).

Chapter 7: Heresies and Schisms

1 Rav Nathan ben Moshe Hannover, *Yeven Mezulah* (Venice, 1653), pp. 31–2.

2 Rav Shabbetai ben Meir HaCohen, *Megillah Afah* (Amsterdam, 1651).

3 For Jews, the New Year is a time of uncertainty when God writes down what will happen in the ensuing twelve months – in the words of part of the New Year liturgy, 'Who shall live and who shall die. Who at the measure of man's days and who before it. Who shall be raised up and who shall be brought low. Who shall wax rich and who shall be made poor. Who shall be at rest and who shall wander . . .'

4 G. Scholem, *Encyclopaedia Judaica*, vol. xiv, p. 1235, quoted in P. Johnson, *A History of the Jews* (London: Weidenfeld & Nicolson, 1987).

5 Yair Bar-El, Rimona Durst, Gregory Katz et al., 'Jerusalem Syndrome', *British Journal of Psychiatry*, 176 (2000), pp. 86–90.

6 Judges 13–16. But Samson certainly visited Hebron and other places close by.

7 As a result of this misconceived alliance with the Catholic Church, the Frankists found themselves compelled to assist in the preparation of anti-Jewish propaganda and to support measures aimed at the destruction of Polish Jewry. There were dire and lasting consequences, and the mutual hostility which resulted left a permanent scar on the history of the Jews in Poland.

8 David Christie-Murray, *A History of Heresy* (Oxford: Oxford Paperbacks, 1989).

9 Jonathan Sumption, *The Albigensian Crusade* (London: Faber & Faber, 1999).

10 Christie-Murray, *A History of Heresy*.

11 I. M. Lewis, 'Trance, Possession, Shamanism and Sex', *Anthropology of Consciousness*, 14 (2003), 20–39.

Chapter 8: God in Retreat

1 Blaise Pascal, *Pensées*, trans. A. J. Krailsheimer (London: Penguin, 1995).

2 K. Armstrong, *A History of God* (London: Vintage, 1999).

3 A. R. Hall and L. Tilling (eds), *The Correspondence of Isaac Newton* (Cambridge: Cambridge University Press, 1959–77).

4 R. Dawkins, *River Out of Eden* (New York: Basic Books, 1995).

5 R. Dawkins, *A Devil's Chaplain* (London: Phoenix, 2004).

6 Ibid.

7 Ibid.

8 Quoted in J. Gribbin, *Science: A History* (Harmondsworth: Penguin, 2002).

9 For example, A. Brown, *The Darwin Wars* (London: Simon & Schuster, 1999); J. Gribbin and M. White, *Darwin*

(London: Simon & Schuster, 1995).

10 F. Darwin (ed.), *The Life and Letters of Charles Darwin* (London: John Murray, 1887).

Chapter 9: Religion in the Modern Age

1 I use the term 'movement' deliberately. In an article entitled 'Of Churches, Sects and Cults' by Rodney Stark and William Sims Bainbridge (*Journal for the Scientific Study of Religion*, 1979), the terminology was set out thus:

CHURCH: a conventional religious organization;

SECT: a deviant religious organization with traditional beliefs and practices;

CULT: a deviant religious organization with novel beliefs and practices.

However, I have chosen not to use the term 'cult' as it tends to be employed by the media and those with an axe to grind – for instance, the families of former and existing members of movements – in a pejorative fashion. Think of 'cult', and you instantly think of blind obedience to the point of suicide, sexual transgressions and so-called 'brainwashing' (for which there is little psychological proof). This encourages us to forget that the beliefs and practices of many so-called 'cults' are as benign as they are bizarre. I therefore prefer the term 'movement'. Many sociologists of religion use the fuller term New Religious Movement or NRM.

2 Quoted in M. Repp, 'Aum Shinrikyo and the Aum Incident: A Critical Introduction', in J. R. Lewis and J. A. Petersen (eds), *Controversial New Religions* (Oxford: Oxford University Press, 2004).

3 Ibid.

4 Ibid.

5 Grace Davie, *Religion in Britain since 1945* (Oxford: Blackwell, 1994).

6 M. Abrams, D. Gerard and N. Timms (eds), *Values and Social Change in Britain* (London: Macmillan, 1985).

7 © Elton John, Bernie Taupin 1997.

8 P. Heelas, 'The Spiritual Revolution: From "Religion" to "Spirituality"', in L. Woodhead et al. (eds), *Religions in the Modern World* (London: Routledge, 2002).

9 Ibid.

10 H. Tajfel, 'Experiments in Inter-Group Discrimination', *Scientific American*, 223 (1970), cited in P. Boyer, *Religion Explained* (London: Vintage, 2002).

11 W. O. Beeman, 'Fighting the Good Fight: Fundamentalism and Religious Revival', in J. McClancy (ed.), *Anthropology for the Real World* (Chicago: University of Chicago Press, 2001).

12 Ibid.

13 Harold Bloom, *Shakespeare: The Invention of the Human* (London: Fourth Estate, 1999).

Index

burial customs, 35, 38–40
  Abu Hureyra people,
    31–3, 38
  Atapuerca cave burials,
    40–2
  Croatia, 42–3
  Islam, 35
  Jewish, 35
  Kogi Indians, 39–40
  Mount Carmel, 38
  Neanderthal people, 42–3
  Pontnewydd, 42
  prehistoric humans, 31–3,
    38–40
  Qafzeh/Skhul, 38
Burridge, Alwyn L., 143
Bush, George W., 403

Cabarets caves, 56
Caesarea, 205
Cain and Abel, 116
Cairo, and science, 336
calendar disputes, 332–4
Calvin, and duty to read,
  360
Campbell, Joseph, 58–9
Canaan, 159
  Abraham in, 170, 171
Canaanite gods, Israelites
  and, 178–9, 193–5
Canaanite people
  Baal myths, 80
  Deuteronomy on, 182–3
cannibalism, 43–54
Carey, George, 408–9
Carlyle, Thomas, 248
Caro, Rabbi Joseph ben
  Ephraim, 283–4
Catal Huyuk, 71

Cathars, 300–5
cathenotheism, 141
Catholic Church see Roman
  Catholic Church
cave burials
  Atapuerca, 40–2
  Krapina, 42–3
  Mount Carmel, 38
  Neanderthal people, 42–3
  Pontnewydd, 42
  Qafzeh, 38
  Skhul, 38
cave paintings, Cro-Magnon
  people, 55–7, 57–9
CERN (European
  Organization for Nuclear
  Research), 391–3
certainty
  fundamentalist, 437–40
  religious, 95–101
  scientific, 439–40
Chance, Frank, 168
Chanina ben Dosa, 225
Charles II, King of England,
  368
chimpanzees, and death,
  36–7
Chmielnicki, Bogdan, 280–2,
  286
Christian fundamentalists,
  and Bible, 160–1
Christian Science movement,
  316–17
Christianity
  and cannibalism, 53
  and divinity, 236–7
  heresies, 295–9
  Paul and, 208–15
  renewal rites, 79–80

Mendel, Rabbi Menachem of
  Kotsk, 440–1
mental illness and religiosity,
  98
Merovingian dynasty, 300
Meslier, Jean, 381–2
Mesoamericans, customs,
  44–50
Mesopotamia, 169, 171
  written language
    development, 74–6
Messiah
  concept, 223
  Jesus and, 224–33
  prophecies, 201–3, 225–6
  Shabbetai Zevi, 286–92
messianic Jews, 238–9
Mexico, Aztec customs,
  44–50
Mexico City, Anthropology
  Museum, 47
mezuzah, 16
Micah, messianic prophecy,
  225–6
Michelangelo, creation
  pictures, 328
Middle East, climate change,
  31
Milvian Bridge, Battle of,
  215–16, 219
Minnesota Study of Twins
  Reared Apart (MISTRA),
  92–5
Mishnah, 163
Mithras cult, 211–12
Mocenigo of Venice, 346
Mohenjodaro, 147
monotheism
  Akhenaten and, 140–1

  development, 179–86,
    192–8
  Islam and, 255–7
  Judaism and, 145
  Shema, 137, 176, 197
  Zoroastrian, 130–1,
    136–8
Montanists, 297–9, 299–300
Moon Cult, 169, 170
Moore, Andrew, and Abu
  Hureyra people, 30–3
moral rules, 112
Mori, Yoshiro, 405–6
Moses (prophet), 158–9
  arguing with God, 175,
    183–4
  and Deuteronomy, 180–2
  linked to Akhenaten, 145,
    155
  and Torah, 166–7
mother goddesses, 67–9, 73
Mount Carmel cave burials,
  38
Mu'awiyah of Damascus,
  271, 272–5
Muhammad ibn Abdallah
  (prophet)
  and images, 312
  and Jewish faith, 254–5
  and Ka'ba, 243–4
  life, 244–6
  and materialism, 427
  misrepresented, 240–2
  and status of women,
    266–8
Muhammad ibn Abdallah
  ibn Tumart, 341
Murray, Mary, 81–2
mysticism, and sex, 113–16

HUMAN INSTINCT
How our primeval impulses shape our modern lives
By Robert Winston

From caveman to modern man . . .

What drives a happily married man to fantasize about pretty, slim, young women seen on a tube train? Why does a seriously injured mountaineer battle against impossible weather conditions to spend three days crawling down to the safety of base camp? How is it that thousands of people spend their week entirely focused on whether their team will win their next crucial match? What stimulates that urge to press the pedal as hard as possible at traffic lights to make the fastest getaway? And how is it that so many people still hold religious views when the notion of an all-powerful being is irrational? All of these impulses are driven by our human instincts – sexual drive, survival, competition, aggression and our search for knowledge.

Few people have a problem with the idea that humans are descended from the apes. But while people believe that our general shape and structure are derived from other creatures, few consider, let alone accept, the psychological implications. Man not only looks, moves and breathes like an ape, he also thinks like one. It is back in our primeval past that we find the first clues to the understanding of our human instincts.

But how well do instincts equip us for the twenty-first century? Do they help or hinder us as we deal with large anonymous cities and the fracturing of communal life, low-level stress and the battle of the sexes? In this erudite and fascinating book, Robert Winston takes us on a journey deep into the human mind in search of the answers to these questions and many more. Along the way he takes a very personal look at the relationship between science and religion and explores those instincts that make us human.

'WIDE-RANGING AND THOROUGHLY ENTERTAINING'
New Scientist

0 553 81492 3

BANTAM BOOKS

THE HUMAN MIND
And how to make the most of it
By Robert Winston

The most complex and mysterious object in the universe is unprepossessing in appearance. It is covered in a dull grey membrane, resembles a gigantic, convoluted fungus and its inscrutability has captivated scientists, philosophers and artists for centuries. It is, of course, the human brain.

New technological developments help us understand how the brain gives rise to the human mind. We can now see the extraordinary complexity of the brain's circuits, watch which regions use energy and which nerve cells generate electricity as we fall in love, tell a lie or dream of a lottery win. And inside the 100 billion cells of this rubbery network is something remarkable: you.

In this entertaining and accessible book, Robert Winston takes us deep into the workings of the human mind, revealing how our senses, emotions and personality are the result of a ballet of genes and environment that shapes the path of our lives. Here, as he explains how memories are formed and lost, and how the ever-changing brain is responsible for toddler tantrums, teenage angst and the battle of the sexes, he also reveals the truth behind extra-sensory perception, *déjà vu* and out-of-body experiences.

And as we discover how we can boost our intelligence, tap into creative powers we never knew we had, break old habits and keep our brains fit and active as we enter old age, we also face one great paradox. Because the human mind is all we have to enable us to understand it, it is perfectly possible that science may never quite explain everything about the remarkable mechanism that makes each of us unique.

'RICHLY INFORMATIVE'
Independent

0 553 81619 5

BANTAM BOOKS